Blood-Brain Barriers

Edited by
Rolf Dermietzel,
David C. Spray,
Maiken Nedergaard

Related Titles

R. A. Meyers

Encyclopedia of Molecular Cell Biology and Molecular Medicine

2nd Edition

2005. ISBN 3-527-30542-4

O. von Bohlen und Halbach, R. Dermietzel

Neurotransmitters and Neuromodulators

Handbook of Receptors and Biological Effects

2nd Edition

2006. ISBN 3-527-31307-9

G. Thiel (Ed.)

Transcription Factors in the Nervous System

Development, Brain Function, and Diseases

2006. ISBN 3-527-31285-4

M. Bähr (Ed.)

Neuroprotection

Models, Mechanisms and Therapies

2004. ISBN 3-527-30816-4

S. Frings, J. Bradley (Eds.)

Transduction Channels in Sensory Cells

2004. ISBN 3-527-30836-9

T. A. Woolsey, J. Hanaway, M. H. Gado

The Brain Atlas

A Visual Guide to the Human Central Nervous System

2002. ISBN 0-471-43058-7

S. H. Koslow, S. Subramaniam (Eds.)

Databasing the Brain: From Data to Knowledge (Neuroinformatics)

2005. ISBN 0-471-30921-4

C. U. M. Smith

Elements of Molecular Neurobiology

2002. ISBN 0-470-84353-5

R. Webster (Ed.)

Neurotransmitters, Drugs and Brain Function

2001. ISBN 0-471-97819-1

Blood-Brain Barriers

From Ontogeny to Artificial Interfaces

Volume 2

*Edited by
Rolf Dermietzel, David C. Spray,
Maiken Nedergaard*

WILEY-VCH Verlag GmbH & Co. KGaA

Editors:

Prof. Dr. Rolf Dermietzel
Department of Neuroanatomy
and Molecular Brain Research
Ruhr University Bochum
Universitätsstrasse 150
44780 Bochum
Germany

Prof. Dr. David Spray
Department of Neuroscience
Albert Einstein College of Medicine
1410 Pelham Parkway S
Bronx, NY 10464
USA

Prof. Dr. Maiken Nedergaard
School of Medicine and Dentistry
University of Rochester
601 Elmwood Avenue
Rochester, NY 14642
USA

■ All books published by Wiley-VCH are carefully produced. Nevertheless, authors, editors, and publisher do not warrant the information contained in these books, including this book, to be free of errors. Readers are advised to keep in mind that statements, data, illustrations, procedural details or other items may inadvertently be inaccurate.

Library of Congress Card No.: applied for

British Library Cataloguing-in-Publication Data:
A catalogue record for this book is available from the British Library

Bibliographic information published by Die Deutsche Bibliothek
Die Deutsche Bibliothek lists this publication in the Deutsche Nationalbibliografie; detailed bibliographic data is available in the internet at http://dnb.ddb.de

© 2006 WILEY-VCH Verlag GmbH & Co. KGaA, Weinheim, Germany

All rights reserved (including those of translation in other languages). No part of this book may be reproduced in any form – by photoprinting, microfilm, or any other means – nor transmitted or translated into a machine language without written permission from the publishers. Registered names, trademarks, etc. used in this book, even when not specifically marked as such, are not to be considered unprotected by law.

Typesetting K+V Fotosatz GmbH, Beerfelden
Printing betz-druck GmbH, Darmstadt
Binding Litges & Dopf Buchbinderei GmbH, Heppenheim

Printed in the Federal Republic of Germany
Printed on acid-free paper

ISBN-13: 978-3-527-31088-3
ISBN-10: 3-527-31088-6

This handbook is dedicated to Eva

Contents

Preface *XXIII*

List of Contributors *XXV*

VOLUME 1

Introduction
The Blood-Brain Barrier: An Integrated Concept *1*
Rolf Dermietzel, David C. Spray, and Maiken Nedergaard

Part I   Ontogeny of the Blood-Brain Barrier *9*

1   Development of the Blood-Brain Interface *11*
Britta Engelhardt

1.1   Introduction *11*
1.2   Pioneering Research on the Blood-Brain Barrier *11*
1.3   The Mature Blood-Brain Interface *13*
1.4   Development of the CNS Vasculature *17*
1.5   Differentiation of the Blood-Brain Barrier *20*
1.5.1 Permeability *20*
1.5.2 Transport Systems and Markers *21*
1.5.3 Extracellular Matrix *23*
1.5.4 Putative Inductive Mechanisms *23*
1.6   Maintenance of the Blood-Brain Barrier *27*
1.7   Outlook *28*
       References *29*

2	**Brain Angiogenesis and Barriergenesis** *41*
	Jeong Ae Park, Yoon Kyung Choi, Sae-Won Kim,
	and Kyu-Won Kim
2.1	Introduction *41*
2.2	Brain Angiogenesis *42*
2.2.1	Hypoxia-Regulated HIF-1 in the Development of the Brain *42*
2.2.2	Hypoxia-Inducible Factor *43*
2.2.3	Hypoxia-Induced VEGF *46*
2.2.4	Other Neuroglia-Derived Angiogenic Factors *47*
2.3	Oxygenation in the Brain: Brain Barriergenesis *49*
2.3.1	Cellular and Molecular Responses Following Brain Oxygenation *49*
2.3.2	Role of src-Suppressed C Kinase Substrate in the Induction of Barriergenesis *50*
2.3.3	Barriergenic Factors in Perivascular Astrocytes and Pericytes Following Brain Oxygenation *51*
2.4	Perspectives *53*
	References 55

3	**Microvascular Influences on Progenitor Cell Mobilization and Fate in the Adult Brain** *61*
	Christina Lilliehook and Steven A. Goldman
3.1	Introduction *61*
3.2	Angiogenic Foci Persist in the Adult Brain *61*
3.3	Neurotrophic Cytokines Can Be of Vascular Origin *62*
3.4	Angiogenesis and Neurogenesis are Linked in the Adult Avian Brain *63*
3.5	Angiogenesis-Neurogenesis Interactions in the Adult Mammalian Brain *65*
3.6	Puringergic Signaling to Neural Progenitors Cells: the Gliovascular Unit as a Functional Entity *66*
3.7	Nitric Oxide is a Local Modulator of Progenitor Cell Mobilization *67*
3.8	Parenchymal Neural Progenitor Cells May Reside Among Microvascular Pericytes *68*
3.9	The Role of the Vasculature in Post-Ischemic Mobilization of Progenitor Cells *69*
	References 70

Part II	The Cells of the Blood-Brain Interface 75

4	**The Endothelial Frontier** 77
	Hartwig Wolburg

4.1	Introduction 77
4.2	The Brain Capillary Endothelial Cell 78
4.3	Endothelial Structures Regulating Transendothelial Permeability 83
4.3.1	Tight Junctions 83
4.3.1.1	Morphology of Tight Junctions 83
4.3.1.2	Molecular Biology of Tight Junctions 84
4.3.2	Caveolae 89
4.3.3	Transporters in the Blood-Brain Barrier Endothelium 92
4.4	Brief Consideration of the Neuroglio-Vascular Complex 94
4.5	Conclusions 98
	References 99

5	**Pericytes and Their Contribution to the Blood-Brain Barrier** 109
	Markus Ramsauer

5.1	Introduction 109
5.2	Pericyte Structure and Positioning 110
5.3	Pericyte Markers 113
5.4	Pericytes in Culture 113
5.5	Contractility and Regulation of Blood Flow 115
5.6	Macrophage Function 116
5.7	Regulation of Homeostasis and Integrity 118
5.8	Angiogenesis and Stability 120
5.8.1	PDGF-B and Pericyte Recruitment 120
5.8.2	TGF-β1 and Differentiation 121
5.8.3	Ang-1 and Maturation 122
5.9	Conclusion 123
	References 123

6	**Brain Macrophages: Enigmas and Conundrums** 129
	Frederic Mercier, Sebastien Mambie, and Glenn I. Hatton

6.1	Introduction 129
6.2	Different Types and Locations of Brain Macrophages 130
6.2.1	Macrophage Structure and Ultrastructure 133
6.2.1.1	Perivascular Macrophages 133
6.2.1.2	Meningeal Macrophages 138
6.2.1.3	Dendritic Cells 138
6.2.1.4	Ventricular Macrophages 139
6.2.2	Immunotyping by Cell Surface Antigens 139
6.2.3	Macrophages Contact Basal Laminae 140

6.2.4	Network of Macrophages Through the Brain	*141*
6.3	Migration of Brain Macrophages	*142*
6.4	Fast Renewal of Brain Macrophages	*142*
6.5	Functions	*143*
6.5.1	Known Functions of Brain Macrophages	*143*
6.5.1.1	Phagocytosis	*144*
6.5.1.2	Immune Function	*144*
6.5.1.3	Production of Growth Factors, Cytokines, and Chemokines	*144*
6.5.1.4	Production and Degradation of the Extracellular Matrix	*145*
6.5.1.5	Repair After Injury	*145*
6.5.2	Potential Functions of Brain Macrophages	*146*
6.5.2.1	Interactions with Meningeal/Vascular Cells, Neurons, and Astrocytes	*146*
6.5.2.2	Do Macrophages Govern the Neural Stem Cell Niche in Adulthood?	*147*
6.5.2.3	Role of Macrophages in CNS Angiogenesis	*151*
6.5.2.4	Role of Macrophages in CNS Plasticity	*152*
6.6	Conclusion: Macrophages as Architects of the CNS Throughout Adulthood	*153*
	References	*154*
7	**The Microglial Component**	*167*
	Ingo Bechmann, Angelika Rappert, Josef Priller, and Robert Nitsch	
7.1	Microglia: Intrinsic Immune Sensor Cells of the CNS	*167*
7.1.1	Development	*167*
7.1.2	Microglial Activation	*168*
7.1.3	Antigen Presentation/Cytotoxicity	*169*
7.2	Terminology: Subtypes and Their Location in Regard to Brain Vessels	*170*
7.2.1	Perivascular Macrophages	*170*
7.2.2	Juxtavascular and Other Microglia	*170*
7.3	Turnover of Brain Mononuclear Cells by Precursor Recruitment Across the BBB	*173*
7.3.1	Perivascular Cells	*173*
7.3.2	Microglia	*174*
7.3.3	Turnover Used by "Trojan Horses"	*174*
7.4	Microglial Impact on BBB Function	*175*
7.4.1	Concept of the BBB	*175*
7.4.2	Chemokines – an Overview	*176*
7.4.3	GAG/Duffy	*176*
7.4.4	Chemokine Expression in the CNS	*177*
7.4.5	CCL2 and CCR2	*177*
7.4.6	CCL3 and CCL5	*178*
7.4.7	CXCR3 and CXCL10	*178*

7.4.8	Microglia-Endothelial Cell Dialogue	179
7.4.9	Microglial Effects on Tight Junctions	179
7.5	Concluding Remarks	180
	References 181	

8   The Bipolar Astrocyte: Polarized Features of Astrocytic Glia Underlying Physiology, with Particular Reference to the Blood-Brain Barrier   189
N. Joan Abbott

8.1	Introduction	189
8.2	Formation of the Neural Tube	189
8.3	Origin of Neurons and Glia	190
8.4	Morphology of Glial Polarity in Adult CNS	193
8.5	Astrocyte Spacing and Boundary Layers	195
8.6	Origin and Molecular Basis of Cell Polarity	196
8.7	Functional Polarity of Astrocytes and Other Ependymoglial Derivatives	197
8.8	Secretory Functions of Astrocytes	199
8.9	Induction of BBB Properties in Brain Endothelium	199
8.10	Astrocyte-Endothelial Signaling	202
8.11	Conclusion	202
	References 203	

9   Responsive Astrocytic Endfeet: the Role of AQP4 in BBB Development and Functioning   209
Grazia P. Nicchia, Beatrice Nico, Laura M. A. Camassa, Maria G. Mola, Domenico Ribatti, David C. Spray, Alejandra Bosco, Maria Svelto, and Antonio Frigeri

9.1	Introduction	209
9.2	Astrocyte Endfeet and BBB Maintenance	210
9.3	Astrocyte Endfeet and BBB Development	212
9.4	Astrocyte Endfeet and BBB Damage	215
9.5	The Role of Aquaporins in BBB Maintenance and Brain Edema	216
9.5.1	AQP Expression and Functional Roles	216
9.5.2	The Role of AQP4 in Brain Edema	220
9.5.2.1	DMD Animal Models	220
9.5.2.2	The α-Syntrophin Null Mice	223
9.5.2.3	The Effect of Lipopolysaccharide on AQP4 Expression	224
9.5.2.4	AQP4 in Astrocytomas	224
9.6	AQP4 Expression in Astrocyte-Endothelial Cocultures	225
	References 230	

Part III Hormonal and Enzymatic Control of Brain Vessels *237*

10   The Role of Fibroblast Growth Factor 2 in the Establishment and Maintenance of the Blood-Brain Barrier *239*
Bernhard Reuss

10.1   Introduction *239*
10.2   Role of FGF-2 in the Regulation of BBB Formation *239*
10.2.1 Expression of FGF-2 in Astrocytes and Endothelial Cells of the Rodent Brain *239*
10.2.2 Induction of BBB Properties in Endothelial Cells by Soluble Factors *240*
10.2.3 Indirect Astrocyte Mediated Effects Seem to Play a Role in FGF-2-Dependent Changes in Endothelial Cell Differentiation *241*
10.2.4 Involvement of FGF-2 in the Regulation of BBB Properties in the Pathologically Altered Brain *242*
10.3   Future Perspectives *243*
       References *244*

11   Cytokines Interact with the Blood-Brain Barrier *247*
Weihong Pan, Shulin Xiang, Hong Tu, and Abba J. Kastin

11.1   Introduction *247*
11.2   Identification of the Phenomena *248*
11.2.1 Cytokines That Cross the BBB by Specific Transport Systems *248*
11.2.2 Cytokines That Permeate the BBB by Simple Diffusion *249*
11.2.3 Cytokines That Have Known Effects on Endothelial Cells *250*
11.3   Mechanisms of Cytokine Interactions with the BBB *253*
11.3.1 Endocytosis of Cytokines by the Apical Surface of Endothelial Cells *253*
11.3.2 Intracellular Trafficking Pathways *253*
11.3.3 Signal Transduction in Endothelial Cells *254*
11.3.4 Involvement of Other Cells Comprising the BBB *254*
11.4   Regulation of the Interactions of Cytokines with the BBB *254*
11.5   Stroke and Other Vasculopathy *255*
11.6   Neurodegenerative Disorders *256*
11.7   Summary *257*
       References *258*

12	**Insulin and the Blood-Brain Barrier** 265	
	William A. Banks and Wee Shiong Lim	
12.1	Introduction 265	
12.1.1	Early Studies 265	
12.1.2	Debates Related to the Question of Permeability of the BBB to Insulin 269	
12.1.3	Does Insulin Cross the BBB? The Middle Years 270	
12.1.4	Insulin, the BBB, and Pathophysiology: The Past Decade 272	
12.2	Pathophysiology of Insulin Transport 274	
12.2.1	Insulin, Obesity, and Diabetes 274	
12.2.2	Insulin Resistance and Inflammatory States 275	
12.2.3	Insulin and Alzheimer's Disease 277	
	References 280	

13	**Glucocorticoid Hormones and Estrogens: Their Interaction with the Endothelial Cells of the Blood-Brain Barrier** 287	
	Jean-Bernard Dietrich	
13.1	Introduction 287	
13.2	Glucocorticoids and the Endothelial Cells of the BBB 288	
13.2.1	Mechanisms of Action of Glucocorticoids 288	
13.2.2	Glucocorticoids and Inflammation 289	
13.2.3	Regulation of Adhesion Molecules Expression by GC in Endothelial Cells 290	
13.2.4	Effects of GC on Leukocyte-Endothelial Cell Interactions 293	
13.2.5	Glucocorticoids, Cerebral Endothelium and Multiple Sclerosis 294	
13.3	Estrogens and the Endothelial Cells of the BBB 297	
13.3.1	Mechanisms of Action of Estrogens 297	
13.3.2	Endothelial Cells as Targets of Estrogens 298	
13.3.3	Adhesion Molecules are Regulated by Estrogens in Endothelial Cells 298	
13.3.4	Estrogens and Experimental Autoimmune Encephalomyelitis 299	
13.4	Conclusions and Perspectives 301	
	References 302	

14	**Metalloproteinases and the Brain Microvasculature** 313	
	Dorothee Krause and Christina Lohmann	
14.1	Introduction 313	
14.2	Metalloproteinases in Brain Microvessels: Types and Functions 314	
14.3	Cerebral Endothelial Cells and Metalloproteinases 318	
14.4	Perivascular Cells and Metalloproteinases 321	
14.4.1	Pericytes 321	
14.4.2	Astrocytes and Microglia 322	
14.5	Metalloproteinases and the Blood-Liquor Barrier 324	

14.6	Metalloproteinases and Brain Diseases *325*
14.6.1	Metalloproteinases and Cerebral Ischemia *326*
14.6.2	Metalloproteinases and Brain Tumors *327*
14.6.3	Metalloproteinases and Multiple Sclerosis *327*
14.6.4	MMPs and Migraine *328*
14.7	Conclusion *328*
	References 328

Part IV Culturing the Blood-Brain Barrier *335*

15   Modeling the Blood-Brain Barrier *337*
Roméo Cecchelli, Caroline Coisne, Lucie Dehouck, Florence Miller, Marie-Pierre Dehouck, Valérie Buée-Scherrer, and Bénédicte Dehouck

15.1	Introduction *337*
15.1.1	In Vitro BBB Model Interests *337*
15.1.2	The BBB: Brain Capillary Endothelial Cells and Brain Parenchyma Cells *338*
15.2	Culturing Brain Capillary Endothelial Cells *338*
15.2.1	Brain Capillary Endothelial Cell Isolation *338*
15.2.1.1	Brain Capillary Endothelial Cell Isolation *338*
15.2.1.2	Endothelial Cell Culture From Capillaries *340*
15.2.1.3	Primary Endothelial Cells and Subculture of Brain Capillary Endothelial Cells *341*
15.2.1.4	Immortalization *342*
15.2.1.5	Purity *344*
15.2.1.6	Species *346*
15.2.2	Coculture *347*
15.3	Characteristics Required for a Useful In Vitro BBB Model *347*
15.3.1	Confluent Monolayer *347*
15.3.2	Tight Junctions and Paracellular Permeability *349*
15.3.3	Transcellular Transport, Receptor Mediated Transport *350*
15.3.4	Expression of Endothelial Adhesion Molecules/Vascular Inflammatory Markers *351*
15.4	Conclusion *352*
	References 352

16   Induction of Blood-Brain Barrier Properties in Cultured Endothelial Cells *357*
Alla Zozulya, Christian Weidenfeller, and Hans-Joachim Galla

16.1	Introduction *357*
16.2	In Vitro BBB Models *359*
16.3	Hydrocortisone Reinforces the Barrier Properties of Primary Cultured Cerebral Endothelial Cells *360*

16.3.1	In Vitro Model Based on Pig Brain Capillary Endothelial Cells (PBCEC) 360
16.3.2	In Vitro Model Based on Mouse Brain Capillary Endothelial Cells (MBCEC) 361
16.4	The Involvement of Serum Effects 362
16.5	Hydrocortisone Improves the Culture Substrate by Suppressing the Expression of Matrix Metalloproteinases In Vitro 363
16.5.1	ECIS Analysis of Improved Endothelial Cell-Cell and Cell-Substrate Contacts in HC-Supplemented Medium 363
16.5.2	Low Degradation of ECM in HC-Supplemented Medium Leads to Improved Cell-Substrate Contacts of Cerebral Endothelial Cells 365
16.6	The Role of Endogenously Derived ECM for the BBB Properties of Cerebral Endothelial Cells In Vitro 368
16.7	Conclusions 370
	References 371

17	**Artificial Blood-Brain Barriers** 375
	Luca Cucullo, Emily Oby, Kerri Hallene, Barbara Aumayr, Ed Rapp, and Damir Janigro
17.1	Introduction: The Blood-Brain Barrier 375
17.2	Requirements for a Good BBB Model 378
17.3	Immobilized Artificial Membranes 379
17.4	Cell Culture-Based in vitro BBB Models 379
17.4.1	Cell Lines 380
17.4.1.1	Immortalized Rat Brain Endothelial Cells 380
17.4.1.2	Other Cells From Non-Cerebral Sources 380
17.4.1.3	Brain Capillary Endothelial Cell Cultures 381
17.4.2	Monoculture-Based in vitro BBB Models 382
17.4.3	Coculture-Based in vitro BBB Models 385
17.5	Shear Stress and Cell Differentiation 386
17.6	Flow-Based in vitro BBB Systems 387
17.6.1	Dynamic in vitro BBB: Standard Model 387
17.6.2	Dynamic in vitro BBB: New Model 390
17.7	A Look Into The Future: Automated Flow Based in vitro BBBs 392
17.8	Conclusion 392
	References 394

18	**In Silico Prediction Models for Blood-Brain Barrier Permeation** 403
	Gerhard F. Ecker and Christian R. Noe
18.1	Introduction: The In Silico World 403
18.2	The Blood-Brain Barrier 404
18.3	Data Sets Available 405
18.4	Computational Models 410
18.5	Passive Diffusion 410

18.5.1	Regression Models *410*
18.5.2	Classification Systems *419*
18.6	Field-Based Methods *421*
18.7	Active Transport *423*
18.8	Conclusions and Future Directions *426*
	References 426

VOLUME 2

Part V   Drug Delivery to the Brain *429*

**19   The Blood-Brain Barrier:
Roles of the Multidrug Resistance Transporter P-Glycoprotein** *431*
Sandra Turcotte, Michel Demeule, Anthony Régina, Chantal Fournier, Julie Jodoin, Albert Moghrabi, and Richard Béliveau

19.1	Introduction *431*
19.2	The Multidrug Transporter P-Glycoprotein *432*
19.2.1	P-gp Isoforms *432*
19.2.2	Structure *433*
19.2.3	P-gp Substrates *435*
19.3	Localization and Transport Activity of P-gp in the CNS *438*
19.3.1	Normal Brain *438*
19.3.2	Brain Diseases *440*
19.3.2.1	Malignant Brain Tumors *440*
19.3.2.2	Brain Metastases *441*
19.3.3	Expression of Other ABC Transporters at the BBB *442*
19.3.4	Subcellular Localization of P-gp *442*
19.4	Polymorphisms of P-gp *444*
19.4.1	MDR1 Polymorphisms at the BBB *444*
19.4.2	MDR1 Polymorphism and Brain Pathologies *445*
19.5	Role of P-gp at the BBB *446*
19.5.1	Protection Against Xenobiotics *448*
19.5.2	Secretion of Endogenous Brain Substrates and Endothelial Secretion *448*
19.5.3	Caveolar Trafficking *449*
19.6	Conclusions *450*
	References 451

20   Targeting of Neuropharmaceuticals by Chemical Delivery Systems *463*
Nicholas Bodor and Peter Buchwald

20.1	Introduction *463*
20.2	The Blood-Brain Barrier *464*

20.2.1	Structural Aspects	464
20.2.2	Enzymatic and Transporter-Related Aspects	466
20.3	Brain-Targeted Drug Delivery	467
20.3.1	Lipophilicity and Its Role in CNS Entry	468
20.3.2	Quantifying Brain-Targeting: Site-Targeting Index and Targeting Enhancement Factors	470
20.4	Chemical Delivery Systems	472
20.5	Brain-Targeting CDSs	473
20.5.1	Design Principles	473
20.5.2	Zidovudine-CDS	475
20.5.3	Ganciclovir-CDS	476
20.5.4	Benzylpenicillin-CDS	478
20.5.5	Estradiol-CDS	479
20.5.6	Cyclodextrin Complexes	484
20.6	Molecular Packaging	484
20.6.1	Leu-Enkephalin Analogs	485
20.6.2	TRH Analogs	486
20.6.3	Kyotorphin Analogs	487
20.6.4	Brain-Targeted Redox Analogs	489
	References 490	

21 Drug Delivery to the Brain by Internalizing Receptors at the Blood-Brain Barrier 501
Pieter J. Gaillard, Corine C. Visser, and Albertus (Bert) G. de Boer

21.1	Introduction	501
21.2	Blood-Brain Barrier Transport Opportunities	502
21.3	Drug Delivery and Targeting Strategies to the Brain	504
21.4	Receptor-Mediated Drug Delivery to the Brain	506
21.5	Transferrin Receptor	506
21.6	Insulin Receptor	508
21.7	LRP1 and LRP2 Receptors	509
21.8	Diphtheria Toxin Receptor	511
21.9	Conclusions	512
	References 514	

Part VI Vascular Perfusion 521

22 Blood-Brain Transfer and Metabolism of Oxygen 523
Albert Gjedde

22.1	Introduction	523
22.2	Blood-Brain Transfer of Oxygen	525
22.2.1	Capillary Model of Oxygen Transfer	525
22.2.2	Compartment Model of Oxygen Transfer	528
22.3	Oxygen in Brain Tissue	529

22.3.1	Cytochrome Oxidation	529
22.3.2	Mitochondrial Oxygen Tension	531
22.3.2.1	Distributed Model of Tissue and Mitochondrial Oxygen	531
22.3.2.2	Compartment Models of Tissue and Mitochondrial Oxygen	533
22.4	Flow-Metabolism Coupling of Oxygen	535
22.5	Limits to Oxygen Supply	538
22.5.1	Distributed Model of Insufficient Oxygen Delivery	538
22.5.2	Compartment Model of Insufficient Oxygen Delivery	541
22.6	Experimental Results	542
22.6.1	Brain Tissue and Mitochondrial Oxygen Tensions	542
22.6.2	Flow-Metabolism Coupling	543
22.6.3	Ischemic Limits of Oxygen Diffusibility	546
	References	547

23 Functional Brain Imaging 551
Gerald A. Dienel

23.1	Molecular Imaging of Biological Processes in Living Brain	551
23.1.1	Introduction	551
23.1.2	Molecular Imaging	551
23.1.3	Influence of Blood-Brain Interface on Functional Imaging	552
23.2	Overview of Brain Imaging Methodologies	554
23.2.1	Computed Tomography	556
23.2.2	Magnetic Resonance Imaging	556
23.2.3	Functional MRI	557
23.2.4	Radionuclide Imaging	560
23.2.5	Optical Imaging	562
23.2.6	Thermal and Optico-Acoustic Imaging	564
23.2.7	Summary	565
23.3	Imaging Biological Processes in Living Brain: Watching and Measuring Brain Work	565
23.3.1	Functional Activity, Brain Work, and Metabolic Imaging	565
23.3.2	Quantitative Measurement of Regional Blood Flow and Metabolism in Living Brain	567
23.3.2.1	Assays at the Blood-Brain Interface: Global Methods	567
23.3.2.2	Highly Diffusible Tracers to Measure CBF	568
23.3.2.3	Metabolizable Glucose Analogs to Measure Hexokinase Activity and CMR_{glc}	569
23.3.2.4	Non-Metabolizable Analogs to Assay Transport and Tissue Concentration	572
23.3.2.5	Cellular Basis of Glucose Utilization	572
23.3.2.6	Acetate is an "Astrocyte Reporter Molecule"	573
23.3.2.7	Summary	573
23.4	Molecular Probes are Used for a Broad Spectrum of Imaging Assays in Living Brain	574

23.4.1	Potassium Uptake and CMR_{glc} During Functional Activation 574	
23.4.2	Multimodal Assays in Serial Sections of Brain 574	
23.4.3	Imaging Human Brain Tumors 576	
23.4.4	Functional Imaging Studies of Sensory and Cognitive Activity Reveal Disproportionate Increases in CBF and CMR_{glc} Compared to CMR_{O_2} During Activation 576	
23.4.5	Imaging Brain Maturation and Aging, Neurotransmitter Systems, and Effects of Drugs of Abuse 580	
23.4.6	Imaging Electrolyte Transport Across the Blood-Brain Interface and Shifts in Calcium Homeostasis Under Pathophysiological Conditions 584	
23.4.7	Summary 587	
23.5	Optical Imaging of Functional Activity by Means of Extrinsic and Intrinsic Fluorescent Compounds 587	
23.6	Tracking Dynamic Movement of Cellular Processes and Cell Types 589	
23.7	Evaluation of Exogenous Genes, Cells, and Therapeutic Efficacy 590	
23.8	Summary and Perspectives 594	
	References 595	

Part VII   Disease-Related Response 601

24   Inflammatory Response of the Blood-Brain Interface 603
Pedro M. Faustmann and Claus G. Haase

24.1	Introduction 603
24.2	Diagnostic Features of Cerebrospinal Fluid 603
24.2.1	Cell Count and Cell Differention 604
24.2.2	Protein Level 605
24.2.3	Glucose Level 605
24.3	Acute Bacterial Meningitis 605
24.3.1	Bacteria and the Blood-Brain Barrier 606
24.3.2	Leukocyte Migration into the CNS 607
24.3.3	Cytokines 611
24.4	Inflammatory Response in Acute Trauma 612
24.5	Inflammatory Response in Alzheimer's Disease 613
	References 614

25   Stroke and the Blood-Brain Interface 619
Marilyn J. Cipolla 619

25.1	Introduction 619
25.2	Brain Edema Formation During Stroke 620
25.2.1	Cytotoxic Versus Vasogenic Edema 620
25.3	Role of Astrocytes in Mediating Edema During Ischemia 621
25.3.1	Aquaporins and Cerebral Edema During Ischemia 624

25.4	Cellular Regulation of Cerebrovascular Permeability	*626*
25.4.1	Role of Actin	*626*
25.4.2	Ischemia and Hypoxia Effects on EC Actin and Permeability	*627*
25.5	Reperfusion Injury	*628*
25.5.1	Nitric Oxide and Other Reactive Oxygen Species as Mediators of EC Permeability During Ischemia And Reperfusion	*628*
25.5.2	Thresholds of Injury	*629*
25.6	Transcellular Transport as a Mechanism of BBB Disruption During Ischemia	*630*
25.7	Mediators of EC Permeability During Ischemia	*631*
25.8	Hyperglycemic Stroke	*633*
25.8.1	Role of PKC Activity in Mediating Enhanced BBB Permeability During Hyperglycemic Stroke	*633*
25.9	Hemodynamic Changes During Ischemia and Reperfusion and its Role in Cerebral Edema	*634*
	References *635*	

26 **Diabetes and the Consequences for the Blood-Brain Barrier** *649*
Arshag D. Mooradian

26.1	Introduction	*649*
26.2	Histological Changes in the Cerebral Microvessels	*650*
26.3	Functional Changes in the Blood-Brain Barrier	*650*
26.3.1	Diabetes-Related Changes in the BBB Transport Function	*651*
26.4	Potential Mechanisms of Changes in the BBB	*655*
26.4.1	Hemodynamic Changes	*655*
26.4.2	Biophysical and Biochemical Changes	*656*
26.4.3	Changes in Neurotransmitter Activity in Cerebral Capillaries	*657*
26.5	Potential Clinical Consequences of Changes in the BBB	*658*
26.6	Conclusions	*661*
	References *661*	

27 **Human Parasitic Disease in the Context of the Blood-Brain Barrier – Effects, Interactions, and Transgressions** *671*
Mahalia S. Desruisseaux, Louis M. Weiss, Herbert B. Tanowitz, Adam Mott, and Danny A. Milner

27.1	Introduction *(by D. Milner)*	*671*
27.2	Malaria: The *Plasmodium berghei* Mouse Model and the Severe Falciparum Malaria in Man *(by M.S. Desruisseaux and D. Milner)*	*683*
27.2.1	Upregulation of Intercellular Adhesion Molecules	*685*
27.2.2	Secretion of TNF-α	*686*
27.2.3	Microglial Activation	*687*
27.2.4	Vascular Damage	*688*

27.3	Trypanosomiasis: African and American Parasites of Two Distinct Flavors *(by H. Tanowitz, M. S. Desruisseaux, and A. Mott)* 689	
27.3.1	CNS Pathology in African Trypanosomiasis 690	
27.3.2	Cytokines and Endothelial Cell Activation 690	
27.3.3	Astrocytosis and Microglial Cell Activation 691	
27.3.3.1	Nitric Oxide 691	
27.3.4	American Trypanosomiasis (Chagas' Disease) 692	
27.4	Toxoplasmosis: Transgression, Quiescence, and Destructive Infections *(by L. Weiss)* 694	
27.5	Conclusion 696	
	References 696	

28   The Blood Retinal Interface: Similarities and Contrasts with the Blood-Brain Interface   701
Tailoi Chan-Ling

28.1	Introduction 701
28.2	The Inner and Outer BRB 702
28.3	The Choroidal Vasculature 702
28.4	Characteristics of Intraretinal Blood Vessels 703
28.5	Ensheathment and Induction of the Inner BRB by Astrocytes and Müller Glia 703
28.6	BRB Properties of Newly Formed Vessels 706
28.7	Pericytes and the BRB 707
28.8	Membrane Proteins of Tight Junctions 710
28.9	Localization of Occludin and Claudin-1 to Tight Junctions of Retinal Vascular Endothelial Cells 710
28.10	Expression of Occludin by RPE Cells and Lack of Occludin Expression by Choroidal Vessels 711
28.11	Inherent Weakness of the BRB and Existence of Resident MHC Class II$^+$ Cells Predisposes the Optic Nerve Head to Inflammatory Attack 712
28.11.1	Compromised BBB Where CNS Meets Peripheral Vascular Bed 713
28.12	Clinical and Experimental Determination of the Blood-Retinal Barrier 713
28.13	Conclusions 716
	References 717

Subject Index   725

Preface

Writing on the blood-brain barrier (BBB) has become a multidisciplinary task. In this two-volume handbook we try to cover all the different aspects that need to be considered to picture a modern image of this fascinating physiological phenomenon. During the past decade our knowledge on the BBB has largely improved, and as mentioned in our introduction, the focus has shifted from an endothelio-centric point of view to a more generalized perspective that includes additional cellular adjuncts like astrocytes, pericytes, microglial cells, and even stem cells. Insofar we prefer the term blood-brain interfaces when describing the various constituents that comprise this more holistic architecture of the BBB. It is useless to say that such an endeavor can never be complete, in particular when the different facets involve a host of structural, molecular, and cell biological aspects ranging from ontogenesis to the diseased BBB. The different chapters provide basic knowledge and state-of-the-art information that can be used for further perusal. Especially, the reference lists collected by the authors are treasures of their own, since they set pathways for additional in-depth studies. All contributors are experts in their field with a profound background in BBB research; and the editors are thankful for their enthusiasm which helped them to concentrate on their subjects outside the daily rush of paperwork. From this perspective, we hope this handbook will serve as a rich source of information for a wide audience, including graduate students, advanced undergraduates, and professionals.

Our sincere thanks go to all contributors for contributing excellent and comprehensive chapters, to Mrs. Monika Birkelbach for secretarial assistance and to Dr. Andreas Sendtko and his colleagues at Wiley-VCH for continuous and encouraging support throughout the preparation of this handbook.

Bochum, New York, Rochester
January 2006

Rolf Dermietzel
David C. Spray
Maiken Nedergaard

Blood-Brain Interfaces: From Ontogeny to Artificial Barriers.
Edited by R. Dermietzel, D.C. Spray, M. Nedergaard
Copyright © 2006 WILEY-VCH Verlag GmbH & Co. KGaA, Weinheim
ISBN: 3-527-31088-6

List of Contributors

Laura M.A. Camassa
Department of General and
Environmental Physiology
and Centre of Excellence in
Comparative Genomics (CEGBA)
University of Bari
Via Amendola 165/A
70126 Bari
Italy

Roméo Cecchelli
Faculté des Sciences Jean Perrin
Université d'Artois
SP 18, rue Jean Souvraz
62307 Lens
France

Tailoi Chan-Ling
Department of Anatomy
Institute for Biomedical Research
University of Sydney
Sydney, NSW 2006
Australia

Yoon Kyung Choi
Research Institute of Pharmaceutical
Sciences and College of Pharmacy
Seoul National University
Seoul, 151-742
Korea

Marilyn J. Cipolla
Department of Neurology
University of Vermont
89 Beaumont Avenue, Given C454
Burlington, VT 05405
USA

Caroline Coisne
Faculté des Sciences Jean Perrin
Université d'Artois
SP 18, rue Jean Souvraz
62307 Lens
France

Luca Cucullo
Department of Neurosurgery
Cleveland Clinic Foundation
9500 Euclid Avenue
Cleveland, OH 44106
USA

Albertus (Bert) G. de Boer
Blood-Brain Barrier Research Group
Division of Pharmacology
Leiden University
Einsteinweg 55
2333 CC, Leiden
The Netherlands

Bénédicte Dehouck
Faculté des Sciences Jean Perrin
Université d'Artois
SP 18, rue Jean Souvraz
62307 Lens
France

Lucie Dehouck
Faculté des Sciences Jean Perrin
Université d'Artois
SP 18, rue Jean Souvraz
62307 Lens
France

Marie-Pierre Dehouck
Faculté des Sciences Jean Perrin
Université d'Artois
SP 18, rue Jean Souvraz
62307 Lens
France

Michel Demeule
Département de Chimie-Biochimie
Université du Québec à Montréal
C.P. 8888, Succursale Centre-ville
Montréal, H3C 3P8
Canada

Rolf Dermietzel
Department of Neuroanatomy and
Molecular Brain Research
Ruhr University Bochum
Universitätsstrasse 150
44780 Bochum
Germany

Mahalia S. Desruisseaux
Montefiore Medical Center
Albert Einstein College of Medicine
1300 Morris Park Avenue
Bronx, NY 10461
USA

Gerald A. Dienel
University of Arkansas for Medical
Sciences
Shorey Building
4301 W. Markham Street
Little Rock, AR 72205
USA

Jean-Bernard Dietrich
Inserm U 575
Université Louis Pasteur
5, rue B. Pascal
67084 Strasbourg
France

Gerhard F. Ecker
Department of Medicinal Chemistry
University of Vienna
Althanstrasse 14
1090 Vienna
Austria

Britta Engelhardt
Theodor Kocher Institute
University of Bern
Freiestrasse 1
3012 Bern
Switzerland

Pedro M. Faustmann
Neuroanatomy/Mol. Brain Research
Ruhr University Bochum
Universitätsstrasse 150
44780 Bochum
Germany

Chantal Fournier
Département de Chimie-Biochimie
Université du Québec à Montréal
C.P. 8888, Succursale Centre-ville
Montréal, H3C 3P8
Canada

Antonio Frigeri
Department of General and
Environmental Physiology
University of Bari
Via Amendola 165/A
70126 Bari
Italy

Pieter J. Gaillard
to-BBB technologies BV
Bio Science Park Leiden
Gorlaeus Laboratories,
LACDR Facilities-FCOL
Einsteinweg 55
2333 CC, Leiden
The Netherlands

Hans-Joachim Galla
Department of Biochemistry
University of Münster
Wilhelm-Klemm-Strasse 2
48149 Münster
Germany

Albert Gjedde
Pathophysiology and Experimental
Tomography Center and
Center of Functionally Integrative
Neuroscience
Aarhus University Hospital
44 Norrebrogade
8000 Aarhus
Denmark

Steven Goldman
Department of Neurology
University of Rochester Medical
Center
601 Elmwood Rd., Box 645
Rochester, NY 14642
USA

Claus G. Haase
Department of Neurology and Clinical
Neurophysiology
Knappschafts-Hospital
Dorstener Strasse 151
45657 Recklinghausen
Germany

Kerri Hallene
Department of Neurosurgery
Cleveland Clinic Foundation
9500 Euclid Avenue
Cleveland, OH 44106
USA

Glenn I. Hatton
Department of Cell Biology and
Neuroscience
University of California, Riverside
1208 Spieth Hall
Riverside, CA 92521
USA

Damir Janigro
Department of Neurosurgery
Cleveland Clinic Foundation
9500 Euclid Avenue
Cleveland, OH 44106
USA

Julie Jodoin
Département de Chimie-Biochimie
Université du Québec à Montréal
C.P. 8888, Succursale Centre-ville
Montréal, H3C 3P8
Canada

Abba J. Kastin
Pennington Biomedical Research
Center
Blood-Brain Barrier Laboratory
6400 Perkins Road
Baton Rouge, LA 70808
USA

Kyu-Won Kim
Research Institute of Pharmaceutical
Sciences and College of Pharmacy
Seoul National University
Seoul, 151-742
Korea

Sae-Won Kim
Research Institute of Pharmaceutical
Sciences and College of Pharmacy
Seoul National University
Seoul, 151-742
Korea

Dorothee Krause
Institute of Neuroanatomy and
Molecular Brain Research
Ruhr University Bochum
Universitätsstrasse 150
44780 Bochum
Germany

Christina Lilliehook
Department of Neurology
University of Rochester Medical
Center
601 Elmwood Road, Box 645
Rochester, NY 14642
USA

Wee Shiong Lim
Department of Geriatric Medicine
Tan Tock Seng Hospital
11 Jalan Tan Tock Seng
Singapore 308433
Republic of Singapore

Christina Lohmann
Discovery DMPK&BA
AstraZeneca R&D Lund
Scheelevägen 8
22187 Lund
Sweden

Sebastien Mambie
Pacific Biomedical Research Center
University of Hawaii
1993 East-West Road
Honolulu, HI 96822
USA

Frederic Mercier
Pacific Biomedical Research Center
University of Hawaii
1993 East-West Road
Honolulu, HI 96822
USA

Florence Miller
Faculté des Sciences Jean Perrin
Université d'Artois
SP 18, rue Jean Souvraz
62307 Lens
France

Danny A. Milner
Harvard Medical School
Department of Pathology
Brigham & Women's Hospital
75 Francis Street, Amory 3
Boston, MA 02115
USA

Albert Moghrabi
Centre de Cancérologie
Charles Bruneau
Université du Québec à Montréal
3175 Chemin Côte-Ste-Catherine
Montréal, H3T 1C5
Canada

Maria G. Mola
Department of General and
Environmental Physiology
and Centre of Excellence in
Comparative Genomics (CEGBA)
University of Bari
Via Amendola 165/A
70126 Bari
Italy

Arshag D. Mooradian
Division of Endocrinology
Saint Louis University
1402 South Grand Blvd
Saint Louis, MO 63104
USA

Adam Mott
Department of Immunology &
Infectious Diseases
Harvard School of Public Health
665 Huntington Avenue
Boston, MA 02115
USA

Maiken Nedergaard
School of Medicine and Dentistry
University of Rochester
601 Elmwood Avenue
Rochester, NY 14642
USA

Grazia P. Nicchia
Department of General and
Environmental Physiology
and Centre of Excellence in
Comparative Genomics (CEGBA)
University of Bari
Via Amendola 165/A
70126 Bari
Italy

Beatrice Nico
Department of Human Anatomy
and Histology
University of Bari
Via Amendola 165/A
70126 Bari
Italy

Robert Nitsch
Institute of Cell Biology and
Neurobiology
Charité University Hospital Berlin
Schumannstrasse 20/21
10098 Berlin
Germany

Christian R. Noe
Department of Medicinal Chemistry
University of Vienna
Althanstrasse 14
1090 Vienna
Austria

Emily Oby
Department of Neurosurgery
Cleveland Clinic Foundation
9500 Euclid Avenue
Cleveland, OH 44106
USA

Weihong Pan
Pennington Biomedical Research
Center
Blood-Brain Barrier Laboratory
6400 Perkins Road
Baton Rouge, LA 70808
USA

Jeong Ae Park
Research Institute of Pharmaceutical
Sciences and College of Pharmacy
Seoul National University
Seoul, 151-742
Korea

Josef Priller
Department of Psychiatry
Charité University Hospital Berlin
Schumannstrasse 20/21
10098 Berlin
Germany

Markus Ramsauer
Buchenrain 5
4106 Therwil
Switzerland

Ed Rapp
Cleveland Medical Devices Inc.
4415 Euclid Avenue
Cleveland, OH 44103
USA

Angelika Rappert
Institute of Cell Biology and
Neurobiology
Charité University Hospital Berlin
Schumannstrasse 20/21
10098 Berlin
Germany

Anthony Régina
Département de Chimie-Biochimie
Université du Québec à Montréal
C.P. 8888, Succursale Centre-ville
Montréal, H3C 3P8
Canada

Bernhard Reuss
Center for Anatomy
University of Göttingen
Kreuzbergring 36
37075 Göttingen
Germany

Domenico Ribatti
Department of Human Anatomy
and Histology
University of Bari
Via Amendola 165/A
70126 Bari
Italy

David C. Spray
Department of Neuroscience
Albert Einstein College of Medicine
1410 Pelham Parkway S
Bronx, NY 10464
USA

Maria Svelto
Department of General and
Environmental Physiology
and Centre of Excellence in
Comparative Genomics (CEGBA)
University of Bari
Via Amendola 165/A
70126 Bari
Italy

Herbert B. Tanowitz
Division of Infectious Diseases
Albert Einstein College of Medicine
1300 Morris Park Avenue
Bronx, NY 10461
USA

Hong Tu
Pennington Biomedical Research
Center
Blood-Brain Barrier Laboratory
6400 Perkins Road
Baton Rouge, LA 70808
USA

Sandra Turcotte
Département de Chimie-Biochimie
Université du Québec à Montréal
C.P. 8888, Succursale Centre-ville
Montréal, H3C 3P8
Canada

Corine C. Visser
Blood-Brain Barrier Research Group
Division of Pharmacology
Leiden University
Einsteinweg 55
2333 CC, Leiden
The Netherlands

Christian Weidenfeller
Department of Chemical and
Biological Engineering
University of Wisconsin-Madison
1415 Engineering Drive
Madison, WI 53706
USA

Louis M. Weiss
Division of Infectious Diseases
Albert Einstein College of Medicine
1300 Morris Park Avenue
Bronx, NY 10461
USA

Hartwig Wolburg
Institute of Pathology
University of Tübingen
Liebermeisterstrasse 8
72076 Tübingen
Germany

Shulin Xiang
Pennington Biomedical Research
Center
Blood-Brain Barrier Laboratory
6400 Perkins Road
Baton Rouge, LA 70808
USA

Alla Zozulya
Department of Pathology and
Laboratory Medicine
University of Wisconsin-Madison
1300 University Avenue
Madison, WI 53706
USA

Part V
Drug Delivery to the Brain

19
The Blood-Brain Barrier:
Roles of the Multidrug Resistance Transporter P-Glycoprotein

Sandra Turcotte, Michel Demeule, Anthony Régina, Chantal Fournier, Julie Jodoin, Albert Moghrabi, and Richard Béliveau

19.1
Introduction

For the brain, the blood-brain barrier (BBB) formed by the brain capillary endothelial cells (BCEC) is considered to be the major route for the uptake of endogenous and exogenous ligands into the brain parenchyma [1]. The EC of brain capillaries are closely sealed by tight junctions and constitute a continuous endothelium. Moreover, brain capillaries possess few fenestrae or endocytic vesicles as compared to the capillaries of other organs [1–3]. BCEC are surrounded by astrocytes, pericytes, microglial cells and by the extracellular matrix. The close association of BCEC with the astrocyte foot processes and the basement membrane of capillaries is important for the development and maintenance of the BBB properties that permit tight control of the blood-brain exchange of molecules [1–4].

The restrictive nature of the BBB is due, in part, to the tight junctions that prevent significant passive movement of small hydrophilic molecules between blood and brain. Nutrients such as glucose and amino acids penetrate into the brain via transporters, whereas uptake of larger molecules, including insulin and transferrin, occurs via receptor-mediated endocytosis [5, 6]. Among the factors controlling the passive entry of drugs into the CNS, lipid solubility is the predominant element because of the lipidic nature of cell membranes [7]. The overall hydrophilic/lipophilic balance of a molecule appears to be a better predictor of BBB permeability than the octanol/buffer partition coefficient. Molecular size, to which the rate of solute diffusion is inversely related, also appears to be relevant for hydrophilic compounds, but does not significantly influence the BBB permeability of lipophilic compounds. Aside from passive diffusion through lipid membranes, the binding of molecules to plasma proteins, ionization at physiological pH (pK_a), affinity and capacity of transport systems and potential BBB/cerebral metabolism are also important for entry into the brain. There are also an increasing number of studies showing that the activity of the

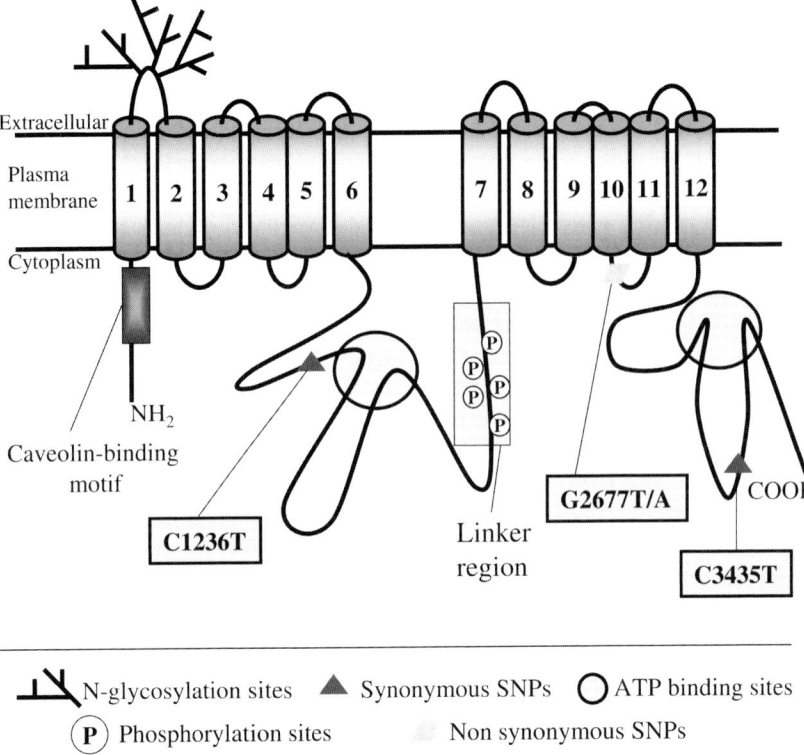

Fig. 19.1 Schematic representation of P-gp. The P-gp secondary structure, embedded in the cell membrane, is presented. P-gp possesses 12 transmembrane domains; and N-glycosylation, phosphorylation and ATP-binding sites are also indicated. The most common nonsynonymous polymorphisms, which induce encoded amino acid changes, and synonymous polymorphisms, which cause a silent mutation, are also shown.

P-gp interactions with its ligands, the direct structure-activity relationships (SAR) of P-gp remains to be clearly established. Better development of the SAR, which would increase the understanding of the pharmacological and physiological significance of P-gp, could eventually help in the prediction of drug entry into the brain through the BBB.

19.2.3
P-gp Substrates

The first molecules identified as P-gp substrates were generally from natural sources, either plants or microorganisms. Many drugs are transported by P-gp and their accumulation in the brain is limited (Table 19.1). Vinca alkaloids, epipodophyllotoxins, anthracyclines and taxanes are among the anticancer agents known to be transported by P-gp [39–41]. Increasingly, molecules other than anticancer agents have been identified as P-gp substrates. For example, P-gp trans-

Table 19.1 Substrates of P-gp.

Compound	Ref.	Compound	Ref.
Anticancer agents		*Ca^{2+} channel blockers*	
Actinomycin D	178	Diltiazem	179, 180
Anthracyclines	181	Mibefradil	182
Colchicine	183	Verapamil	184–187
Daunorubicin	178	*Fluorescent dyes*	
Dexamethasone	82, 188, 189	Rhodamine 123	191
Docetaxel	190	Hoechst 33342	35, 36
Doxorubicin	192, 193	Calcein-AM	195, 196
Etoposide	194	Tetracycline	197
Mitomycin C	178	Tetraphenylphosphonium	198
Paclitaxel (taxol)	178	Ramosetron	202
Tamoxifen	199–201		
Vinblastine	178, 203	*HIV protease inhibitors*	
Vincristine	178	Amprenavir	206
		Indinavir	209, 210
Immunosuppressive agents		Saquinavir	209
Cyclosporine A	57, 204, 205	Ritonavir	213
Rapamycin	207, 208		
Sirolimus	211	*Bioactive peptides*	
Tacrolimus	212	Adrenorphin	48
		Endomorphin 1 and 2	48
Others		Somatostatin	49, 218
Opioids (morphine)	219, 220	β-Amyloid	87
Erythromycin (antibiotic)	222, 223		
Okadaic acid	225	*Cardiac drugs*	
Steroids	44, 227	Digoxin	214, 215
Aldosterone	44, 228	Quinidine	178, 216
Cortisol	23, 44	Digitoxin	217
Corticosterone	229	Substance P	49
Glucocorticoids	44, 230		
Progesterone	44	*Toxic peptides*	
Sphingomyelin	231	Valinomycin	221
Lovastatin	226	Gramicidin D	224
(lipid-lowering agent)			
		Cytokines	
		Interferon-γ	45, 46
		Interleukin-2 and -4	47

ports cardiac drugs, Ca^{2+} channel blockers, HIV protease inhibitors, immunosuppressive agents, fluorescent dyes and cyclic and linear peptides [40–43]. In addition to xenobiotics, various endogenous substrates for P-gp have been identified in normal tissues, including several steroids such as cortisol, corticosterone, progesterone and aldosterone [23, 44]. Also, cytokines (IL-2, IL-4, IFN-γ) and bilirubin have also been shown to be transported by P-gp [45–47]. Recently, the transporter has been shown to have high affinity for endogenous bioactive peptides, such as adrenorphin, endomorphin 1 and 2, somatostatin and substance P [48, 49].

Studies have shown that molecular weight (MW), surface area, aromaticity, amphiphilicity, proton basicity and H-bond accepting are important in determining P-gp substrate specificity [50–52]. Recently, it was proposed that P-gp substrate specificity could be approximated by three rules obtained from the MW, the H-accepting capacity (given by the Abraham's β coefficient) and from the ionization which is represented by the acid and base pK_a values of compounds. Thus, compounds with an Abraham's β coefficient ≥ 8 (approximately the total number of N and O atoms), MW >400, and $pK_a>4$ are likely to be P-gp substrates, whereas compounds with $(N+O) \leq 4$, MW<400 and $pK_a<8$ are likely to be non-substrates [53]. The application of this model could be useful in absorption, distribution, metabolism and excretion (ADME) profiling of new drugs. However, since P-gp possesses multiple binding sites and complex mechanisms for substrate recognition and transport, SAR models remain difficult to develop. The prediction of P-gp substrate specificity is influenced by several factors, including the types of assays used, the confusion between P-gp substrates and inhibitors and the binding possibilities with other targets such as cytochrome P450 3A4 [43, 53].

Reversal agents are molecules that restore sensitivity to anticancer agents in drug-resistant cancer cells by inhibiting the transport activity of P-gp. Three generations of these compounds have been used so far (Table 19.2). For example, calcium channel blockers, calmodulin antagonists, quinolins, steroids, immunosuppressive agents, antibiotics and detergents are reversal agents known in the first generation [54–56]. However, most of these agents produce significant toxicities when used at concentrations sufficient to inhibit P-gp. Several of these compounds are themselves substrates for P-gp and for other transporters. Among them, cyclosporin A (CsA) and verapamil were most often employed but cannot be used safely for MDR reversal at the dosage required. This led to the development of second-generation P-gp modulators, such as SDZ PSC 833 (valspodar), a CsA analogue [57]. Most of these agents have the same pharmacological properties as the original molecules but with less toxicity. In spite of their efficiency, many characteristics limit their clinical usefulness. It has been demonstrated that these compounds can significantly inhibit the metabolism and excretion of cytotoxic agents [58]. The high toxicity associated with this side-effect requires a reduction of chemotherapeutic doses in clinical studies [58]. Also, several second-generation P-gp modulators are themselves often substrates for cytochrome P450 3A4 enzyme or other transporter proteins such as

Table 19.2 Modulators of P-gp properties.

Modulator	Ref.	Medical use/analogy/type [a]
First-generation compounds		
Cyclosporin A	57, 204, 205	Immunosuppressive
Nifedipine	232	Calcium channel blocker
Progesterone	233	Progestative
Quinidine	178, 216	Antiarrhythmic
Quinine	234, 235	Antimalarial
Tamoxifen	199, 200	Antioestrogen
Verapamil	184, 185	Calcium channel blocker
Second-generation compounds		
Valspodar (PSC833)	28, 236	Cyclosporin A
Cinchonine	234	Quinine
Dexniguldipine	237	Nifedipine
Dexverapamil	238, 239	Verapamil
Third-generation compounds		
Tariquidar (XR9576)	240	Anthranilamide
Zosuquidar (LY335979)	241, 242	Difluorocyclopropyldibenzosuberane
ONT-093	243	Substituted diarylimidazole
Tariquidar (XR9576)	244	Anthranilic acid derivative
Biricodar (VX710)	60, 245	Piperidine carboxylate
Elacridar (GF120918/GG918)	246	Acridone carboxamide
Natural compounds		
Curcumin	63	Polyphenol
Ginseng	85, 247, 248	Ginsenosides
Piperine	249, 250	Alkaloid (black pepper)
Catechins from green tea	66, 251	Polyphenols
Silymarin from milk thistle	252	Flavonoids
Garlic	253, 254	Organosulfur compounds

a) First-generation compounds: medical use. Second-generation compounds: analogy to first generation compound. Third-generation/natural compounds: type or chemical structure.

MRP1 [59, 60]. These compounds are in competition with the cytotoxic agent for transport by the pump, giving an unpredictable pharmacokinetic interaction. After disappointing results, a third generation of reversal agents was developed. These molecules aim to specifically inhibit P-gp function. These agents do not affect cytochrome P450 3A4 and were generally developed using SARs and combinatorial chemistry. Because they are noncompetitive inhibitors of the P-gp transporter, the use of this third generation of P-gp modulators permits a reduction in the dosage of chemotherapeutic agents.

The clinical efficacy of these reversal agents remains to be established, not only with regard to overcoming tumor resistance against chemotherapy, but also for other factors such as bypassing P-gp in the BBB. In recent years, several new approaches have been developed to reduce and inhibit MDR1/P-gp expression in cells. Among them, the use of monoclonal antibodies or immunotoxins against P-gp, antisense oligonucleotides and small interfering RNAs (siRNAs) has been investigated [61, 62]. Furthermore, natural products from dietary intake, such as curcumin ginsenosides and piperine, have been identified as inhibitors of P-gp and several flavonoids, such as quercetin and naringenin, are reported to modulate P-gp activity (Table 19.2) [63–65]. Moreover, we have also demonstrated that epigallocatechin gallate, the major polyphenol present in green tea, inhibits P-gp activity [66]. At the same time, much effort has gone into investigating and identifying new natural compounds that inhibit P-gp, reverse the MDR phenotype and sensitize cancer cells to conventional chemotherapy without toxicological effects. However, other studies are necessary to understand the mechanisms involved in P-gp modulation by these natural products and to explore their potential in chemoprevention.

19.3
Localization and Transport Activity of P-gp in the CNS

19.3.1
Normal Brain

P-gp is found in many normal tissues with excretory function, including liver, kidney and small intestine [67, 68], and at blood-tissue barriers such as the BBB, blood-testis barrier and placenta [69]. As a result of its anatomical localization, P-gp is one of the most important transporters for drug disposition in the organism. It limits drug entry into the body after oral drug administration (enterocyte luminal membrane), it promotes drug elimination into bile and urine (hepatocyte canalicular membrane, kidney proximal tubule luminal cell membrane) and it limits drug penetration into sensitive tissues (brain, testis, fetal circulation).

The expression of P-gp in human BBB endothelial cells has been described in many studies, performed in various species (human, rat, mouse, cow, pig) [70–72] and indicates that the major site of BBB P-gp expression is at the luminal membrane of capillary endothelial cells (Fig. 19.2A, left panel). Several reports have shown that P-gp could be also present in the brain parenchyma. For example, in vitro P-gp expression and activity have been demonstrated in primary astrocyte rat brain cultures [73, 74] and in microglia [75]. In vivo, a recent study examined P-gp distribution, using confocal microscopy on rat brain sections and indicated that this transporter was preferentially expressed in the endothelial component but was also present in astroglial cells [76]. Another team observed that the P-gp pattern of expression in human and primate brain was the same as that seen for the astrocyte marker GFAP [77, 78]. Based on these immunofluorescent studies, a model of MDR in brain was proposed where P-gp is localized on astrocyte foot processes at the antiluminal side of the human BBB. More recently, this group pub-

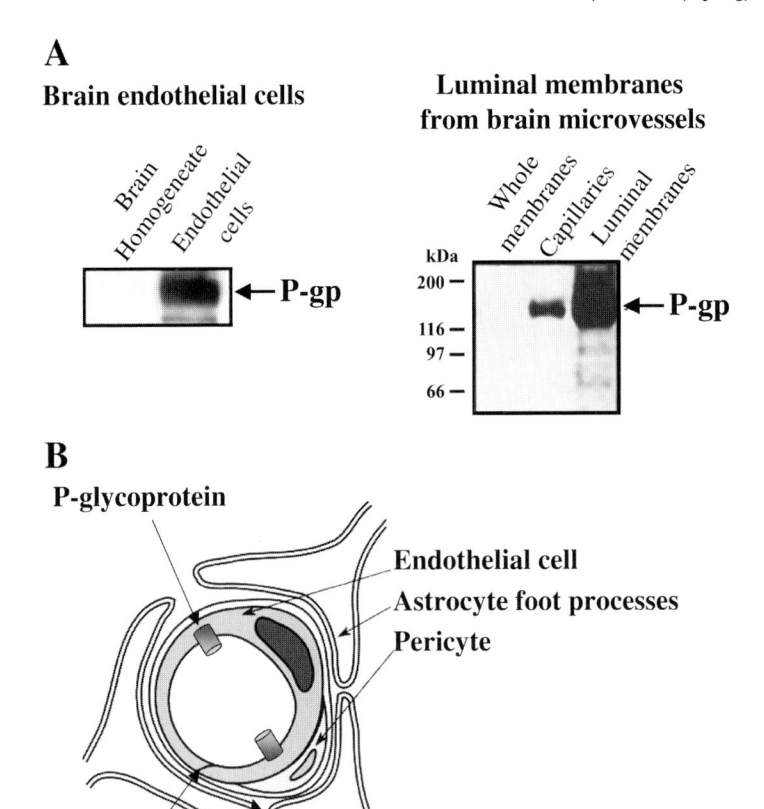

A
Brain endothelial cells　　　**Luminal membranes from brain microvessels**

← P-gp

← P-gp

B
P-glycoprotein

Endothelial cell
Astrocyte foot processes
Pericyte

Tight junction
　　　Basement membrane

Fig. 19.2 P-gp expression at the BBB. (A) Left panel: Detection of P-gp in homogenates and in endothelial cells (EC) isolated from brain using magnetic cell-sorting beads. Right panel: Proteins from whole membranes, brain capillaries and endothelial luminal membranes were separated by SDS-PAGE and immuno-detected with mAb P-gp (C219) antibody. (B) Schematic view of P-gp at the BBB. EC are sealed by continuous tight junctions and surrounded by a basal lamina. Pericytes are present at the periphery of vessels. Astrocyte foot processes are in close contact. P-gp is present in the luminal membranes of the brain vascular endothelium and impedes brain penetration by lipophilic substances.

lished results where P-gp expression was found both in astrocytes and in endothelial cells of healthy primate brain [79]. In our laboratory, a study indicated that P-gp was strongly enriched in the positive endothelial cell fraction from brain and was absent from the negative fraction in which the glial fibrillary acidic protein (GFAP), an astrocyte marker, was present [80]. It was also shown by RT-PCR analysis that the *mdr1a* gene was preferentially expressed in this enriched EC fraction from the brain. At the subcellular level, our findings demonstrated that the P-gp was localized in isolated luminal membranes from the brain vascular endothelium in rat [81] (Fig. 19.2 A, right panel).

Overall, data obtained for most studies suggest that P-gp expressed in the capillary endothelial cells of the BBB restricts the CNS accumulation of many drugs, including chemotherapeutic agents (Fig. 19.2 B). This protective action of P-gp has been demonstrated using *mdra* knockout mice [82]. In addition to the expression of P-gp at the BBB, there have been a few reports of the expression and functional activity of P-gp in the choroid plexus [76, 83, 84]. P-gp was localized at the subapical side of choroid plexus epithelia and vectorial transport experiments performed on cultured rat choroid plexus showed an apically directed efflux function for P-gp, suggesting a role in preventing the export of certain substances out of the CSF, as opposed to its action at the BBB. More studies are needed to understand and characterize the role of P-gp at the choroid plexus.

19.3.2
Brain Diseases

The expression and activity of P-gp in the CNS plays an important role in the disposition and efficacy of pharmacological agents for brain diseases, such as brain tumors, epilepsy or HIV-associated dementia [85, 86]. In addition, P-gp also seems to play a key role in the etiology and pathogenesis of certain neurological disorders, such as Alzheimer's and Parkinson's diseases [87, 88].

In brain tumors, progress in clinical treatment has been slow and one of the major problems impeding treatment of these tumors is their weak response to anticancer drugs. In fact, brain tumors are known to develop MDR quite rapidly. Furthermore, gliomas are characterized by their infiltrative pattern of growth and it is likely that the blood-brain area in the tumoral periphery, which often escapes surgical intervention, possesses a totally functional BBB. The low response to chemotherapy may also depend on tumor blood flow, the integrity of the blood-tumor barrier and an inherent or acquired MDR phenotype in cancer cells [89]. As P-gp plays a major role in the defense of the organism against xenobiotics at the BBB [9], the determination of P-gp levels in brain tumors and peritumoral tissue is crucial for evaluating the long-term efficacy of chemotherapy. P-gp has been detected in endothelial cells from newly formed microvessels of gliomas [90–92], suggesting that, despite the leaky nature of the vasculature of gliomas, angiogenic vessels have maintained some of the restrictive capacities of the BBB.

19.3.2.1 Malignant Brain Tumors
We have reported that the P-gp expression levels detected by Western blot in various human malignant brain tumors (low-, high-grade gliomas) are similar to the levels of P-gp expression found in normal brain [93]. This is in agreement with previous studies which reported the presence of P-gp in resistant

and partially chemosensitive glioblastomas by immunohistochemistry, using the monoclonal antibody C219 [94–96]. These results suggest that the poor response of brain tumors to many anticancer drugs may be related to the presence of this efflux transporter in cell populations of the primary brain tumors and that P-gp may be considered as a negative factor when predicting the outcome for patients with brain tumors. These findings also suggest that P-gp expression is maintained in both low- and high-grade gliomas. Moreover, the widespread expression of P-gp in these tumors may reflect an intrinsic resistance to anticancer drugs.

Previous immunohistochemical analyses showed that most gliomas and, more specifically, endothelial cells within the gliomas, stained positively for MDR1 P-gp [91, 92]. These studies support the concept that clinical drug resistance may be caused by P-gp expression, not only in cancer cells but also in the capillary endothelial cells of brain tumors. The role of the BBB in the low efficacy of chemotherapy is still unclear. Alterations in the brain capillary ultrastructure have been described, leading to an increase in the microvascular permeability in gliomas. In contrast, it has been reported that the neovasculature of even high-grade tumors preserves partial BBB permeability properties at the cellular level [97] and that the BBB at the tumor periphery is still intact. In addition, a study indicated that P-gp, one of the best phenotypic markers of the BBB, is expressed at the same levels in all primary tumors as in normal brain, indicating that brain tumors retain an important characteristic of the BBB which restricts the brain uptake of chemotherapeutic agents. Thus the BBB, especially at the edge of tumors, remains a formidable obstacle for drug distribution to brain regions that have been infiltrated by neoplastic cells [98].

19.3.2.2 Brain Metastases

Brain metastases occur in 20–40% of cancer patients and the estimated incidence in the United States is 170 000 new cases per year [99]. Lung cancer (9.7–64.0%), breast cancer (2–25%) and melanoma (4–20%) are the most common primary sources of metastases to the CNS [100]. Strikingly, we found that brain metastases from melanomas and lung adenocarcinomas exhibit only 5% and 40%, respectively, of the P-gp levels found in normal brain [93]. Metastatic malignant melanomas are recognized for their poor response to chemotherapy, whereas some effects of chemotherapy have been observed for lung adenocarcinomas [101]. The low expression of P-gp in these brain metastases suggests that MDR mechanisms other than P-gp could be responsible for their poor response to chemotherapy. The lack of P-gp expression in primary lung tumors and corresponding brain metastases also indicates that these brain metastases do not acquire the levels of P-gp expression found in normal brain tissue.

19.3.3
Expression of Other ABC Transporters at the BBB

It has been reported that efflux transporters other than P-gp are also expressed in brain capillaries. For instance, members of the MDR-associated protein (MRP) family have been detected at the BBB. In humans, seven MRP homologues have been identified [102]. All members of the MRP family are distributed throughout most human tissues [103]. MRP1, which was first described in 1992, was immunodetected by Western blots in human and rat choroid plexus, but the presence of MRP1 in the EC of brain capillaries remains controversial [104]. In animal models, Western blot and RT-PCR analysis suggest that MRP1 is expressed in isolated rat brain capillaries, primary cultured rat, pig and cow BCECs and immortalized rodent BCECs [105–107]. However, in isolated human brain capillaries, no expression of MRP1 was observed by immunohistochemistry [108]. The canalicular multispecific organic anion transporter (cMOAT or MRP2) was principally detected in hepatocytes, intestine and kidney but was not detected in endothelial cells of rat brain capillaries by Western blot [109]. Recently, MRP1, -4, -5 and -6 were shown to be expressed in primary BCECs by RT-PCR analysis as well as in a capillary-enriched brain extract [110]. In addition, MRP mRNA levels appeared to be closely associated with resistance to etoposide, adriamycin and vincristine in human glioma cell lines derived from patients [111]. Recently, levels of MDR1 and MRP1–MRP4 mRNA were compared between normal brain tissue and malignant gliomas [112]. The expression of both MDR1 and MRP2 were similar in normal brain and tumors, whereas MRP1 and MRP3 expression increased with tumor grade. Therefore, some of the MRPs may also confer intrinsic MDR activity in human gliomas or in metastatic brain tumors.

19.3.4
Subcellular Localization of P-gp

P-gp was also found in a specialized microdomain of plasma membranes called caveolae. This P-gp expressed in caveolae was first identified in multidrug-resistant cells [113–115], where it appears to play an important role in drug resistance development [116, 117]. Caveolae are flask-shaped plasma membrane invaginations involved in many cellular events such as transcytosis, endocytosis, cholesterol transport and signal transduction [118]. A family of proteins called caveolins comprise the structural component of caveolae. Caveolin-1 and caveolin-2 are primarily expressed in adipocytes, endothelial cells, smooth muscle cells and type I pneumocytes [119], whereas caveolin-3 is expressed in muscle and glial cells [120, 121]. Caveolin-1 possesses two isoforms (α, β) whereas three isoforms (α, β, γ) were reported for caveolin-2. MDR is a multifactorial process and recently an upregulation of caveolae and caveolar constituents, such as caveolin-1, -2 and glucosylceramide, was observed in different MDR cancer cells compared to their drug-sensitive counterparts [113, 114, 122, 123].

Localization of P-gp in caveolae has been also shown in the brain by different means. First, using a detergent-free method for caveolae isolation, our group showed the enrichment of P-gp, caveolin-1 and cholesterol in the low-density microdomains of human isolated brain capillaries and endothelial cells of an in vitro BBB model [115, 124]. Second, immunocytochemical analysis demonstrated the presence of P-gp in plasmalemmal vesicles of rat brain capillaries and in an immortalized rat brain endothelial cell line, RBE4 [125]. Then, Virgintino et al. [126] showed, by microscopy, that a large proportion of P-gp and caveolin-1 colocalize in the luminal compartment of the endothelial cells in human microvessels of the cerebral cortex. Besides endothelial cells, immunocytochemical analysis shows that, in astrocytes, a portion of P-gp is localized in caveolae [74] and colocalized with caveolin-1 [79].

In addition to the colocalization of some P-gp and caveolins in caveolae, our coimmunoprecipitation studies indicated that a population of P-gp molecules interacted with caveolin-1 in endothelial cells of the BBB [115, 124]. This coimmunoprecipitation was also reported in MDR cells [115, 127] and in astrocytes [74]. Similar to caveolin-1, caveolin-2 interacts with P-gp; and these three proteins form a high molecular mass complex at the BBB [124]. Oligomeric forms of P-gp have been observed in MDR cells and brain capillaries [28, 128] and recent data suggest that P-gp oligomerizes through indirect interactions [129]. The involvement of caveolins in P-gp oligomerization remains to be investigated, as well as the possibility of other proteins interacting with P-gp, like actin, ezrin, radixin, moesin, calnexin, Hsp70 and Hsp90 beta [130–132].

P-gp contains in its N-terminal portion a consensus caveolin-binding motif present in many proteins known to bind the scaffolding domain of caveolin-1 (Fig. 19.1). Three related caveolin-binding motifs are known ($\Phi X\Phi XXXX\Phi$, $\Phi X\Phi XXXX\Phi XX\Phi$, $\Phi X\Phi XXXX\Phi XX\Phi$, where Φ is a phenylalanine, tyrosine or tryptophan residue and X is any amino acid residue) [133]. The scaffolding domain of caveolin-1 regulates signaling molecules localized in caveolae such as eNOS, protein kinase C, insulin receptor, EGF and VEGF receptors [134, 135]. In the case of P-gp, mutation of its caveolin-binding motif decreases the interaction between P-gp and caveolin-1 and increases P-gp transport activity, indicating that caveolin-1 negatively regulates P-gp activity [124]. Moreover, overexpression of caveolin-1 in drug-resistant cells expressing P-gp causes a reduction of P-gp activity and in the cells become drug-sensitive, supporting the hypothesis that caveolin-1 inhibits drug transport by P-gp [127].

19.4
Polymorphisms of P-gp

19.4.1
MDR1 Polymorphisms at the BBB

In recent years, researchers have started to investigate the molecular mechanisms underlying inter-individual differences in the pharmacological effects of drugs. Genetic variations in drug transporters have received particular interest since they are among the factors determining the pharmacokinetic profile of drugs. Efforts have also been made to identify genetic variations of the human MDR1 (ABCB1).

Single nucleotide polymorphisms (SNPs) result in a single nucleotide substitution and possibly a change in the encoded amino acid. More than 40 SNPs and insertion/deletion polymorphisms in the ABCB1 gene have been reported. MDR1 gene SNPs are located in the coding region and in the noncoding region, including the core promoter region and the intron-exon boundaries [136–138]. The most common allelic combinations of MDR1 SNPs, which encode no amino acid changes, are synonymous polymorphisms at exon 12 (C1236T) and exon 26 (C3435T) and nonsynonymous polymorphisms (encoding amino acid changes) at exon 21 (G2677T). The localization of these major SNPs is shown in the schematic representation of P-gp (Fig. 19.1). Allele frequencies vary widely in MDR1 SNPs, particularly between populations of African descent and other ethnic groups. A large discrepancy is observed in these populations with an average of 18% and 48% frequencies, respectively, for the T allele of the three common allelic combinations of MDR1 SNPs (Table 19.3). Moreover, the segregation observed in the African American population consistently presents different, specific genetic combinations of MDR1 SNPs (haplotype) [136, 138, 139]. Consistent with wild-type allelic frequency in individuals of African origin, epidemiologic studies have observed a lower incidence of ulcerative colitis in Africans as compared with Caucasians [140]. It was suggested that the higher frequency of the wild-type allele (CC) in the African population for exon 26

Table 19.3 Most frequent allelic combinations of MDR1 genetic variations [136–138]. CA=Caucasian; AA=African American; AS=Asian American; ME=Mexican American; PA=Pacific Islander.

Exon	SNP	Allele frequency (%)				
		CA	AA	AS	ME	PA
12	C1236T	35–46	21	68	45	57
21	G2677T	42–46	10	45	40	36
21	G2677A	2–10	0.5	6–22	0	36
26	C3435T	48–56	20–23	40–49	50	50

(C3435C) SNPs may have resulted in a selective advantage against intestinal tract diseases [141]. There have also been no cases of neurotoxicity reported after treatment with ivermectin for the prevention of onchocerciasis in Africa, even though this drug causes neurotoxicity in animals with low P-gp expression [139, 142]. Thus, it has been demonstrated that some MDR1 polymorphisms have an impact on P-gp expression and function. The introduction of nucleotide changes in highly conserved regions of the MDR1 gene has a major impact on P-gp function and expression, in comparison with a nucleotide substitution introduced in less conserved regions of the gene [40, 143–145].

19.4.2
MDR1 Polymorphism and Brain Pathologies

At the BBB, impairment of P-gp function or altered P-gp expression level has been associated with severe neurotoxic side effects following administration of drugs or xenobiotics [146–148]. The SNPs in exon 26 (C3435TT or CT) genotypes are associated with low P-gp expression in the BBB in comparison to the CC genotypes references. A five-fold increased risk for developing Parkinson's disease was found in exon 26 (C3435T) heterozygous (T) and homozygous (TT) patients exposed to pesticides [88, 149]. Children with acute lymphoblastic leukemia (ALL) with the C3435TT or CT genotypes demonstrated a better response to chemotherapeutic drugs (e.g. etoposide, vincristine, doxorubicin), thus reducing the risk of CNS relapse [150]. Furthermore, the nonsynonymous G2677T SNPs in exon 21, combined with the synonymous C3435T SNPs in exon 26, increased the neurotoxicity of the immunosuppressive drug tacrolimus in liver transplant patients [151]. Patients with resistance to epileptic drugs have shown a higher frequency of the C3435CC genotype than the C3435TT SNPs. The well known C3435T polymorphism is silent (no amino acid changes) and raises the possibility that the polymorphism is not in itself causal but that different mutational groups (SNPs) forming different haplotypes in a consistent network are possibly the causal events [152]. Thus, linkage disequilibrium of the C3435T SNPs with other SNPs has underscored the importance of understanding haplotypes.

Interstudy comparison of the polymorphic effects on P-gp expression and function requires extensive haplotype analyses [136–138, 153, 154]. This will also provide a powerful tool for predicting and optimizing drug therapy, particularly for drugs with narrow therapeutic indices where induction or inhibition of transporter function can have a tremendous impact on drug efficacy and safety [136, 138, 155–157].

1. Protection against xenobiotics

2. Secretion of brain endogenous substrates

3. Endothelial secretion

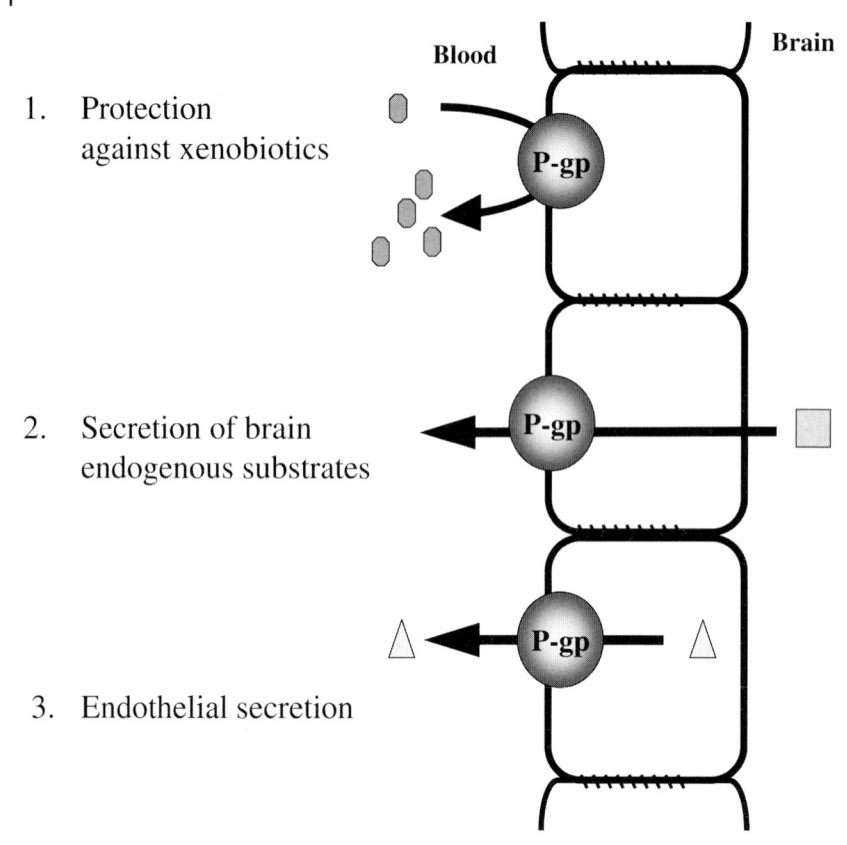

Fig. 19.3 Roles of P-gp at the BBB. P-gp at the BBB could have different physiological roles such as: (1) protection against xenobiotics, (2) secretion of brain endogenous substrates and (3) endothelial secretion.

19.5
Role of P-gp at the BBB

In brain capillaries, P-gp appears to play an important role in preventing many hydrophobic molecules from crossing the BBB and reaching the CNS. However, the exact physiological function of P-gp in the BBB is not completely understood. A growing body of evidence links P-gp to physiological roles distinct from its initially recognized function as a drug efflux system (Figs. 19.3 and 19.4).

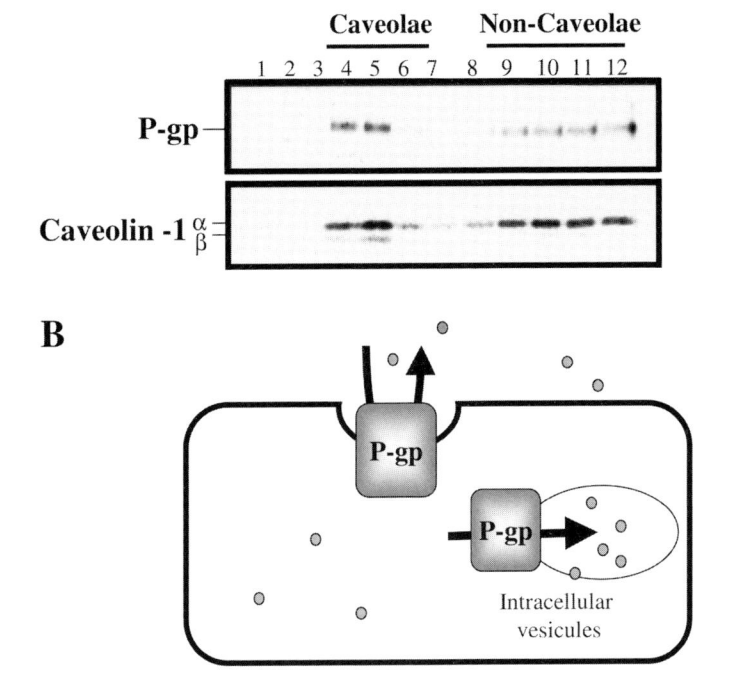

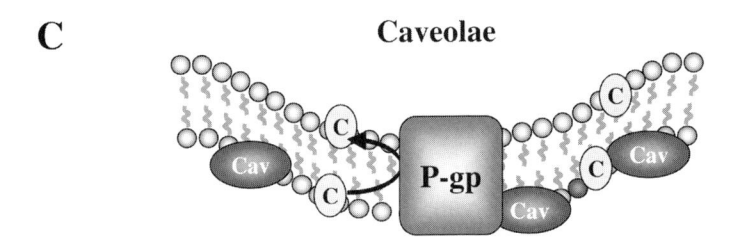

Fig. 19.4 Proposed roles of P-gp localized in caveolae.
(A) Low-density caveolae-enriched domains were isolated from bovine brain capillary endothelial cells (BBCEC) cocultured with astrocytes, using a carbonate-based fractionation method. Each fraction from a sucrose gradient was separated by SDS-PAGE and immunodetected using antibodies directed against P-gp (mAb C219) and caveolin.
(B) Since cholesterol is important for P-gp activity, the localization of P-gp in enriched cholesterol microdomains (caveolae) contributes to decreasing intracellular drug concentrations by pumping drugs inside caveolar vesicles and increasing their elimination outside the cells.
(C) P-gp mediates cholesterol redistribution from the cytosolic leaflet to the exoplasmic leaflet of the plasma membrane. C=Cholesterol; Cav=caveolin-1.

19.5.1
Protection Against Xenobiotics

Numerous reports provide functional evidence for P-gp-mediated drug efflux at the BBB. The interaction of drugs with P-gp in rat brain capillaries was demonstrated by photoaffinity labeling [158]. The generation of transgenic mice with a disruption of the *mdr1a* gene provided a pharmacological tool for the study of P-gp function in the BBB [9, 20]. These mice are viable, fertile and do not display obvious phenotypic abnormalities, indicating that this protein is not essential to their vital functions. However, P-gp substrates accumulate in the brains of these mice to a much greater extent than in wild-type animals and they are more sensitive to central neurotoxicity. For example, knockout mice are 50–100 times more sensitive to the neurotoxic effects of the pesticide ivermectin. The accumulation of this drug in brain tissue of $mdr1a^{(-/-)}$ mice was increased 80–100 times as compared to control mice. Recent application of in situ brain perfusion to wild-type and P-gp-deficient $mdr1a^{(-/-)}$ mice made it possible to assess the influence of P-gp on brain uptake of substrates without the potentially confounding differences in systemic pharmacokinetics upon P-gp distribution [159]. In summary, as indicated by Schinkel in 1999, P-gp appears to be a major efflux transporter at the BBB that acts as a guardian of the CNS by preventing the accumulation of many drugs in the brain [9].

19.5.2
Secretion of Endogenous Brain Substrates and Endothelial Secretion

In addition to its guardian role, P-gp is involved in the excretion of toxic compounds by renal proximal tubules and hepatic canalicular membranes [68, 69] and in the secretion of endogenous molecules from adrenal glands [160]. Thus, P-gp could fulfill a similar function in the BBB and be responsible for the secretion and/or excretion of brain-derived substances or metabolites into the blood (brain secretion). It could also be involved in the secretion of molecules from the endothelium itself (capillary secretion). In this respect, P-gp has been proposed to be involved in the release of neuroactive substances from the brain directly into the systemic blood following intracerebroventricular injection [161]. In addition, it has been demonstrated that β-amyloid (Aβ) is transported across the plasma membrane of P-gp-enriched vesicles in an ATP- and P-gp-dependent manner, suggesting that Aβ might be an endogenous substrate for P-gp in brain [87]. Thus, a change in MDR1 function or expression might alter the clearance of Aβ from the brain and may contribute to cerebrovascular angiopathy. Since the accumulation of Aβ in the brain is a feature of Alzheimer's disease, the mechanism of Aβ transport opens new avenues in the understanding of Alzheimer's disease.

19.5.3
Caveolar Trafficking

Why P-gp is localized in caveolar microdomains remains to be established. However, several roles can be proposed for P-gp in these membrane microdomains, which are illustrated in Fig. 19.4. As mentioned, a portion of the P-gp localized in the brain capillary endothelial cells was found in caveolae microdomains (Fig. 19.4A). It has been proposed that the P-gp localized in the caveolae of MDR cells might act to decrease intracellular drug concentrations by pumping drugs inside caveolar vesicles and increasing their elimination (Fig. 19.4B). Different observations in MDR cells support this role of P-gp in caveolae: increase of caveolae and caveolar constituents in MDR cells compared to their drug-sensitive counterparts [113, 114, 116, 117, 123, 162], localization of P-gp in caveolae in MDR cells [113, 115] and drug sequestration in P-gp-containing cytoplasmic vesicles in MDR cells [163, 164].

Studies have also reported that a portion of the P-gp expressed at the BBB is colocalized with caveolin-1 in caveolae [125, 165]. Other studies have demonstrated that the P-gp expressed at the BBB can also interact with caveolin-1 [115, 124]. Considering the protective role of P-gp at the BBB in preventing the accumulation of many hydrophobic molecules and potentially toxic substances in the brain, modification of caveolae or caveolin levels might affect brain homeostasis. In this regard, a dramatic decrease in caveolin-1 expression has been observed in brain tumor endothelial cells compared to normal brain endothelial cells [166]. Since caveolin-1 inhibits P-gp activity [124, 127], a reduction in caveolin-1 expression could affect drug transport across the BBB and decrease chemotherapy efficiency.

In addition, studies from different groups have suggested the involvement of P-gp in lipid transport (Fig. 19.4C). Studies on P-gp activity, either drug binding or drug transport, in cells where P-gp is localized in caveolae or low-density microdomains show that P-gp is functional in these cholesterol-enriched microdomains [124, 167]. Moreover, studies have shown that cholesterol is important for the activity of P-gp, suggesting that caveolae might provide a favorable environment for its activity [168–170]. Specifically, cholesterol could interact with the substrate binding site of P-gp, suggesting that cholesterol may be transported by MDR1 P-gp (Fig. 19.4C) [171]. Furthermore, one study has shown that P-gp mediates the ATP-dependent relocation of cholesterol from the cytosolic leaflet to the exoplasmic leaflet of the plasma membrane, suggesting that P-gp might contribute to stabilizing caveolae [172]. It was also reported that caveolin-1 binds cholesterol and mediates its efflux within caveolae via an identified, cytosolic caveolin-1 complex comprising heat-shock protein 56, cyclophilin A and cyclophilin 40, which carries cholesterol to the plasma membrane caveolae [173]. In addition, P-gp in caveolae might contribute towards decreasing the formation of ceramide, which is involved in apoptosis induction [174]. Elevated levels of glucosylceramide, the precursor of ceramide, were observed in MDR cells and appeared to be due to the high activity of glucosylceramide synthase activity. Also, it has been shown that sphingomyelin (SM) and the enzyme converting SM into ceramide, called sphingomyelinase, are enriched in caveolae in MDR cells [175, 176]. However, further studies are required to have a better understanding of the role of P-gp in caveolae at the BBB.

19.6
Conclusions

Overall, P-gp plays an important role in brain protection at the BBB. Its expression at the luminal side of endothelial cells in brain capillaries prevents the passage of many agents into the brain [9, 81, 177]. Moreover, many studies have proposed that the capillary EC of brain tumors participates in the resistance associated with P-gp expression, especially at the edge of tumors, where the BBB remains a formidable obstacle for the penetration of anticancer drugs into the brain regions infiltrated by cancer cells [98]. The development of P-gp inhibitors in order to reverse the MDR phenotype has been extensively investigated with generally disappointing results. The current, third-generation inhibitors present high potency and specificity for P-gp. Further studies are required to establish their contribution to potential therapeutic treatment by reversing P-gp-mediated MDR. It was recently reported, in MDR cells, that a portion of the P-gp present in the endothelial cells of the BBB is localized in caveolar microdomains [113, 115]. This particular localization could be useful for understanding the function and regulation of P-gp in drug elimination and transport across the BBB. Finally, recent observations have challenged the notion that P-gp has evolved merely to mediate the efflux of xenobiotics and raised the possibility that P-gp and related transporters might play a fundamental role in regulating cell differentiation, migration, proliferation and survival [174].

Acknowledgments

This work was supported by a grant to R.B. from the Natural Sciences and Engineering Research Council of Canada.

References

1 W. M. Pardridge **1999**, *J. Neurovirol.* 5, 556.
2 H. Kusuhara, Y. Sugiyama **2001**, *Drug Discov. Today* 6, 206.
3 H. Kusuhara, Y. Sugiyama **2001**, *Drug Discov. Today* 6, 150.
4 A. Tsuji, I. I. Tamai **1999**, *Adv. Drug Deliv. Rev.* 36, 277.
5 A. B. De Boer, E. L. De Lange, I. C. Van Der Sandt, D. D. Breimer **1998**, *Int. J. Clin. Pharmacol. Ther.* 36, 14.
6 Y. Zhang, W. M. Pardridge **2001**, *J. Neurochem.* 76, 1597.
7 M. D. Habgood, D. J. Begley, N. J. Abbott **2000**, *Cell Mol. Neurobiol.* 20, 231.
8 J. Van Asperen, U. Mayer, O. Van Tellingen, J. H. Beijnen **1997**, *J. Pharm. Sci.* 86, 881.
9 A. H. Schinkel **1999**, *Adv. Drug Deliv. Rev.* 36, 179.
10 R. L. Juliano, V. Ling **1976**, *Biochim. Biophys. Acta* 455, 152.
11 C. F. Higgins, P. D. Haag, K. Nikaido, F. Ardeshir, G. Garcia, G. F. Ames **1982**, *Nature* 298, 723.
12 S. C. Hyde, P. Emsley, M. J. Hartshorn, M. M. Mimmack, U. Gileadi, S. R. Pearce, M. P. Gallagher, D. R. Gill, R. E. Hubbard, C. F. Higgins **1990**, *Nature* 346, 362.
13 M. Dean, Y. Hamon, G. Chimini **2001**, *J. Lipid Res.* 42, 1007.
14 P. Borst, R. O. Elferink **2002**, *Annu. Rev. Biochem.* 71, 537.
15 I. Klein, B. Sarkadi, A. Varadi **1999**, *Biochim. Biophys. Acta* 1461, 237.
16 J. F. Ghersi-Egea, N. Strazielle **2002**, *J. Drug Target.* 10, 353.
17 C. Cordon-Cardo, J. P. O'Brien, D. Casals, L. Rittman-Grauer, J. L. Biedler, M. R. Melamed, J. R. Bertino **1989**, *Proc. Natl Acad. Sci. USA* 86, 695.
18 F. Thiebaut, T. Tsuruo, H. Hamada, M. M. Gottesman, I. Pastan, M. C. Willingham **1989**, *J. Histochem. Cytochem.* 37, 159.
19 J. J. Smit, A. H. Schinkel, R. P. Oude Elferink, A. K. Groen, E. Wagenaar, L. Van Deemter, C. A. Mol, R. Ottenhoff, N. M. Van Der Lugt, M. A. Van Roon, et al. **1993**, *Cell* 75, 451.
20 A. H. Schinkel, J. J. Smit, O. Van Tellingen, J. H. Beijnen, E. Wagenaar, L. Van Deemter, C. A. Mol, M. A. Van Der Valk, E. C. Robanus-Maandag, H. P. Te Riele, et al. **1994**, *Cell* 77, 491.
21 A. H. Schinkel **1997**, *Semin. Cancer Biol.* 8, 161.
22 A. H. Schinkel **1998**, *Int. J. Clin. Pharmacol. Ther.* 36, 9.
23 A. M. Karssen, O. C. Meijer, I. C. Van Der Sandt, P. J. Lucassen, E. C. De Lange, A. G. De Boer, E. R. De Kloet **2001**, *Endocrinology* 142, 2686.
24 M. M. Gottesman, C. A. Hrycyna, P. V. Schoenlein, U. A. Germann, I. Pastan **1995**, *Annu. Rev. Genet.* 29, 607.
25 A. Devault, P. Gros **1990**, *Mol. Cell Biol.* 10, 1652.
26 P. Gros, Y. B. Ben Neriah, J. M. Croop, D. E. Housman **1986**, *Nature* 323, 728.
27 D. Boscoboinik, M. T. Debanne, A. R. Stafford, C. Y. Jung, R. S. Gupta, R. M. Epand **1990**, *Biochim. Biophys. Acta* 1027, 225.
28 L. Jette, M. Potier, R. Beliveau **1997**, *Biochemistry* 36, 13929.

29 M. F. Rosenberg, R. Callaghan, R. C. Ford, C. F. Higgins **1997**, *J. Biol. Chem.* 272, 10685.
30 T. W. Loo, D. M. Clarke **1999**, *Biochim. Biophys. Acta* 1461, 315.
31 T. W. Loo, D. M. Clarke **2001**, *J. Biol. Chem.* 276, 14972.
32 T. W. Loo, D. M. Clarke **2002**, *J. Biol. Chem.* 277, 44332.
33 S. Dey, P. Hafkemeyer, I. Pastan, M. M. Gottesman **1999**, *Biochemistry* 38, 6630.
34 M. Demeule, A. Laplante, A. Sepehr-Arae, E. Beaulieu, D. Averill-Bates, R. M. Wenger, R. Beliveau **1999**, *Biochem. Cell Biol.* 77, 47.
35 A. B. Shapiro, A. B. Corder, V. Ling **1997**, *Eur. J. Biochem.* 250, 115.
36 F. Tang, H. Ouyang, J. Z. Yang, R. T. Borchardt **2004**, *J. Pharm. Sci.* 93, 1185.
37 M. F. Rosenberg, A. B. Kamis, R. Callaghan, C. F. Higgins, R. C. Ford **2003**, *J. Biol. Chem.* 278, 8294.
38 I. K. Pajeva, C. Globisch, M. Wiese **2004**, *J. Med. Chem.* 47, 2523.
39 M. M. Gottesman, I. Pastan, S. V. Ambudkar **1996**, *Curr. Opin. Genet. Dev.* 6, 610.
40 S. V. Ambudkar, S. Dey, C. A. Hrycyna, M. Ramachandra, I. Pastan, M. M. Gottesman **1999**, *Annu. Rev. Pharmacol. Toxicol.* 39, 361.
41 J. A. Silverman **1999**, *Pharm. Biotechnol.* 12, 353.
42 S. J. Thompson, K. Koszdin, C. M. Bernards **2000**, *Anesthesiology* 92, 1392.
43 M. F. Fromm **2004**, *Trends Pharmacol. Sci.* 25, 423.
44 M. Uhr, F. Holsboer, M. B. Muller **2002**, *J. Neuroendocrinol.* 14, 753.
45 H. Kawaguchi, Y. Matsui, Y. Watanabe, Y. Takakura **2004**, *J. Pharmacol. Exp. Ther.* 308, 91.
46 Y. Akazawa, H. Kawaguchi, M. Funahashi, Y. Watanabe, K. Yamaoka, M. Hashida, Y. Takakura **2002**, *J. Pharm. Sci.* 91, 2110.
47 L. Bonhomme-Faivre, A. Pelloquin, S. Tardivel, S. Urien, M. C. Mathieu, V. Castagne, B. Lacour, R. Farinotti **2002**, *Anticancer Drugs* 13, 51.
48 R. P. Oude Elferink, J. Zadina **2001**, *Peptides* 22, 2015.
49 N. Uchiyama-Kokubu, M. Naito, M. Nakajima, T. Tsuruo **2004**, *FEBS Lett.* 574, 55.
50 T. R. Stouch, O. Gudmundsson **2002**, *Adv. Drug Deliv. Rev.* 54, 315.
51 S. Ekins, C. L. Waller, P. W. Swaan, G. Cruciani, S. A. Wrighton, J. H. Wikel **2000**, *J. Pharmacol. Toxicol. Methods* 44, 251.
52 K. M. Mahar Doan, J. E. Humphreys, L. O. Webster, S. A. Wring, L. J. Shampine, C. J. Serabjit-Singh, K. K. Adkison, J. W. Polli **2002**, *J. Pharmacol. Exp. Ther.* 303, 1029.
53 R. Didziapetris, P. Japertas, A. Avdeef, A. Petrauskas **2003**, *J. Drug Target.* 11, 391.
54 J. M. Ford, W. N. Hait **1990**, *Pharmacol. Rev.* 42, 155.
55 B. L. Lum, M. P. Gosland, S. Kaubisch, B. I. Sikic **1993**, *Pharmacotherapy* 13, 88.
56 M. Raderer, W. Scheithauer **1993**, *Cancer* 72, 3553.
57 P. R. Twentyman **1992**, *Biochem. Pharmacol.* 43, 109.
58 H. Thomas, H. M. Coley **2003**, *Cancer Control* 10, 159.

59 C. Wandel, R.B. Kim, S. Kajiji, P. Guengerich, G.R. Wilkinson, A.J. Wood **1999**, *Cancer Res.* 59, 3944.
60 E.K. Rowinsky, L. Smith, Y.M. Wang, P. Chaturvedi, M. Villalona, E. Campbell, C. Aylesworth, S.G. Eckhardt, L. Hammond, M. Kraynak, R. Drengler, J. Stephenson, Jr., M.W. Harding, D.D. Von Hoff **1998**, *J. Clin. Oncol.* 16, 2964.
61 D. Xu, H. Kang, M. Fisher, R.L. Juliano **2004**, *Mol. Pharmacol.* 66, 268.
62 Z. Peng, Z. Xiao, Y. Wang, P. Liu, Y. Cai, S. Lu, W. Feng, Z.C. Han **2004**, *Cancer Gene Ther.* 11, 707.
63 N. Romiti, R. Tongiani, F. Cervelli, E. Chieli **1998**, *Life Sci.* 62, 2349.
64 S. Zhou, L.Y. Lim, B. Chowbay **2004**, *Drug Metab. Rev.* 36, 57.
65 K.A. Youdim, M.Z. Qaiser, D.J. Begley, C.A. Rice-Evans, N.J. Abbott **2004**, *Free Radic. Biol. Med.* 36, 592.
66 J. Jodoin, M. Demeule, R. Beliveau **2002**, *Biochim. Biophys. Acta* 1542, 149.
67 A.T. Fojo, K. Ueda, D.J. Slamon, D.G. Poplack, M.M. Gottesman, I. Pastan **1987**, *Proc. Natl Acad. Sci. USA* 84, 265.
68 F. Thiebaut, T. Tsuruo, H. Hamada, M.M. Gottesman, I. Pastan, M.C. Willingham **1987**, *Proc. Natl Acad. Sci. USA* 84, 7735.
69 C. Cordon-Cardo, J.P. O'Brien, J. Boccia, D. Casals, J.R. Bertino, M.R. Melamed **1990**, *J. Histochem. Cytochem.* 38, 1277.
70 L. Jette, B. Tetu, R. Beliveau **1993**, *Biochim. Biophys. Acta* 1150, 147.
71 M.A. Barrand, K.J. Robertson, S.F. Von Weikersthal **1995**, *FEBS Lett.* 374, 179.
72 A. Tsuji, T. Terasaki, Y. Takabatake, Y. Tenda, I. Tamai, T. Yamashima, S. Moritani, T. Tsuruo, J. Yamashita **1992**, *Life Sci.* 51, 1427.
73 X. Decleves, A. Regina, J.L. Laplanche, F. Roux, B. Boval, J.M. Launay, J.M. Scherrmann **2000**, *J. Neurosci. Res.* 60, 594.
74 P.T. Ronaldson, M. Bendayan, D. Gingras, M. Piquette-Miller, R. Bendayan **2004**, *J. Neurochem.* 89, 788.
75 G. Lee, L. Schlichter, M. Bendayan, R. Bendayan **2001**, *J. Pharmacol. Exp. Ther.* 299, 204.
76 C. Mercier, C. Masseguin, F. Roux, J. Gabrion, J.M. Scherrmann **2004**, *Brain Res.* 1021, 32.
77 W.M. Pardridge, P.L. Golden, Y.S. Kang, U. Bickel **1997**, *J. Neurochem.* 68, 1278.
78 P.L. Golden, W.M. Pardridge **1999**, *Brain Res.* 819, 143.
79 F. Schlachetzki, W.M. Pardridge **2003**, *Neuroreport* 14, 2041.
80 M. Demeule, M. Labelle, A. Regina, F. Berthelet, R. Beliveau **2001**, *Biochem. Biophys. Res. Commun.* 281, 827.
81 E. Beaulieu, M. Demeule, L. Ghitescu, R. Beliveau **1997**, *Biochem. J.* 326, 539.
82 A.H. Schinkel, E. Wagenaar, L. Van Deemter, C.A. Mol, P. Borst **1995**, *J. Clin. Invest.* 96, 1698.
83 V.V. Rao, J.L. Dahlheimer, M.E. Bardgett, A.Z. Snyder, R.A. Finch, A.C. Sartorelli, D. Piwnica-Worms **1999**, *Proc. Natl Acad. Sci. USA* 96, 3900.

84 K.E. Warren, M.C. Patel, C.M. Mccully, L.M. Montuenga, F.M. Balis **2000**, *Cancer Chemother. Pharmacol.* 45, 207.
85 R.B. Kim **2003**, *Top. HIV Med.* 11, 136.
86 W. Loscher, H. Potschka **2002**, *J. Pharmacol. Exp. Ther.* 301, 7.
87 F.C. Lam, R. Liu, P. Lu, A.B. Shapiro, J.M. Renoir, F.J. Sharom, P.B. Reiner **2001**, *J. Neurochem.* 76, 1121.
88 T. Furuno, M.T. Landi, M. Ceroni, N. Caporaso, I. Bernucci, G. Nappi, E. Martignoni, E. Schaeffeler, M. Eichelbaum, M. Schwab, U.M. Zanger **2002**, *Pharmacogenetics* 12, 529.
89 M. Demeule, A. Laplante, G.F. Murphy, R.M. Wenger, R. Beliveau **1998**, *Biochemistry* 37, 18110.
90 I. Sugawara, H. Hamada, T. Tsuruo, S. Mori **1990**, *Jpn J. Cancer Res.* 81, 727.
91 K. Toth, M.M. Vaughan, N.S. Peress, H.K. Slocum, Y.M. Rustum **1996**, *Am. J. Pathol.* 149, 853.
92 T. Sawada, Y. Kato, N. Sakayori, Y. Takekawa, M. Kobayashi **1999**, *Brain Tumor Pathol.* 16, 23.
93 M. Demeule, D. Shedid, E. Beaulieu, R.F. Del Maestro, A. Moghrabi, P.B. Ghosn, R. Moumdjian, F. Berthelet, R. Beliveau **2001**, *Int. J. Cancer* 93, 62.
94 I. Becker, K.F. Becker, R. Meyermann, V. Hollt **1991**, *Acta Neuropathol. (Berl.)* 82, 516.
95 J.W. Henson, C. Cordon-Cardo, J.B. Posner **1992**, *J. Neurooncol.* 14, 37.
96 F. Leweke, M.S. Damian, C. Schindler, W. Schachenmayr **1998**, *Pathol. Res. Pract.* 194, 149.
97 T. Sawada, Y. Kato, M. Kobayashi, Y. Takekekawa **2000**, *Brain Tumor Pathol.* 17, 1.
98 M. Bertossi, D. Virgintino, E. Maiorano, M. Occhiogrosso, L. Roncali **1997**, *Ultrastruct. Pathol.* 21, 41.
99 L. Gaspar, C. Scott, M. Rotman, S. Asbell, T. Phillips, T. Wasserman, W.G. Mckenna, R. Byhardt **1997**, *Int. J. Radiat. Oncol. Biol. Phys.* 37, 745.
100 A. Tosoni, M. Ermani, A.A. Brandes **2004**, *Crit. Rev. Oncol. Hematol.* 52, 199.
101 B. Savas, G. Arslan, T. Gelen, G. Karpuzoglu, C. Ozkaynak **1999**, *Anticancer Res.* 19, 4413.
102 P. Borst, R. Evers, M. Kool, J. Wijnholds **2000**, *J. Natl Cancer Inst.* 92, 1295.
103 M.J. Flens, G.J. Zaman, P. Van Der Valk, M.A. Izquierdo, A.B. Schroeijers, G.L. Scheffer, P. Van Der Groep, M. De Haas, C.J. Meijer, R.J. Scheper **1996**, *Am. J. Pathol.* 148, 1237.
104 S.P. Cole, G. Bhardwaj, J.H. Gerlach, J.E. Mackie, C.E. Grant, K.C. Almquist, A.J. Stewart, E.U. Kurz, A.M. Duncan, R.G. Deeley **1992**, *Science* 258, 1650.
105 A. Regina, A. Koman, M. Piciotti, B. El Hafny, M.S. Center, R. Bergmann, P.O. Couraud, F. Roux **1998**, *J. Neurochem.* 71, 705.
106 H. Huai-Yun, D.T. Secrest, K.S. Mark, D. Carney, C. Brandquist, W.F. Elmquist, D.W. Miller **1998**, *Biochem. Biophys. Res. Commun.* 243, 816.

107 H. Kusuhara, H. Suzuki, M. Naito, T. Tsuruo, Y. Sugiyama **1998**, *J. Pharmacol. Exp. Ther.* 285, 1260.
108 S. Seetharaman, M. A. Barrand, L. Maskell, R. J. Scheper **1998**, *J. Neurochem.* 70, 1151.
109 M. Demeule, M. Brossard, R. Beliveau **1999**, *Am. J. Physiol.* 277, F832.
110 Y. Zhang, H. Han, W. F. Elmquist, D. W. Miller **2000**, *Brain Res.* 876, 148.
111 T. Abe, S. Hasegawa, K. Taniguchi, A. Yokomizo, T. Kuwano, M. Ono, T. Mori, S. Hori, K. Kohno, M. Kuwano **1994**, *Int. J. Cancer* 58, 860.
112 S. Haga, E. Hinoshita, K. Ikezaki, M. Fukui, G. L. Scheffer, T. Uchiumi, M. Kuwano **2001**, *Jpn J. Cancer Res.* 92, 211.
113 Y. Lavie, G. Fiucci, M. Liscovitch **1998**, *J. Biol. Chem.* 273, 32380.
114 C. P. Yang, F. Galbiati, D. Volonte, S. B. Horwitz, M. P. Lisanti **1998**, *FEBS Lett.* 439, 368.
115 M. Demeule, J. Jodoin, D. Gingras, R. Beliveau **2000**, *FEBS Lett.* 466, 219.
116 Y. Lavie, G. Fiucci, M. Czarny, M. Liscovitch **1999**, *Lipids* 34[Suppl.], S57.
117 Y. Lavie, M. Liscovitch **2000**, *Glycoconj. J.* 17, 253.
118 P. W. Shaul, R. G. Anderson **1998**, *Am. J. Physiol.* 275, L843.
119 P. E. Scherer, R. Y. Lewis, D. Volonte, J. A. Engelman, F. Galbiati, J. Couet, D. S. Kohtz, E. Van Donselaar, P. Peters, M. P. Lisanti **1997**, *J. Biol. Chem.* 272, 29337.
120 Z. Tang, P. E. Scherer, T. Okamoto, K. Song, C. Chu, D. S. Kohtz, I. Nishimoto, H. F. Lodish, M. P. Lisanti **1996**, *J. Biol. Chem.* 271, 2255.
121 T. Ikezu, H. Ueda, B. D. Trapp, K. Nishiyama, J. F. Sha, D. Volonte, F. Galbiati, A. L. Byrd, G. Bassell, H. Serizawa, W. S. Lane, M. P. Lisanti, T. Okamoto **1998**, *Brain Res.* 804, 177.
122 Y. Lavie, H. Cao, S. L. Bursten, A. E. Giuliano, M. C. Cabot **1996**, *J. Biol. Chem.* 271, 19530.
123 Y. Lavie, G. Fiucci, M. Liscovitch **2001**, *Adv. Drug Deliv. Rev.* 49, 317.
124 J. Jodoin, M. Demeule, L. Fenart, R. Cecchelli, S. Farmer, K. J. Linton, C. F. Higgins, R. Beliveau **2003**, *J. Neurochem.* 87, 1010.
125 R. Bendayan, G. Lee, M. Bendayan **2002**, *Microsc. Res. Tech.* 57, 365.
126 D. Virgintino, M. Errede, D. Robertson, C. Capobianco, F. Girolamo, A. Vimercati, M. Bertossi, L. Roncali **2004**, *Histochem. Cell Biol.* 122, 51.
127 C. Cai, J. Chen **2004**, *Int. J. Cancer* 111, 522.
128 M. S. Poruchynsky, V. Ling **1994**, *Biochemistry* 33, 4163.
129 J. C. Taylor, A. R. Horvath, C. F. Higgins, G. S. Begley **2001**, *J. Biol. Chem.* 276, 36075.
130 J. Bertram, K. Palfner, W. Hiddemann, M. Kneba **1996**, *Anticancer Drugs* 7, 838.
131 T. W. Loo, D. M. Clarke **1995**, *J. Biol. Chem.* 270, 21839.
132 F. Luciani, A. Molinari, F. Lozupone, A. Calcabrini, L. Lugini, A. Stringaro, P. Puddu, G. Arancia, M. Cianfriglia, S. Fais **2002**, *Blood* 99, 641.
133 J. Couet, M. Sargiacomo, M. P. Lisanti **1997**, *J. Biol. Chem.* 272, 30429.
134 T. Okamoto, A. Schlegel, P. E. Scherer, M. P. Lisanti **1998**, *J. Biol. Chem.* 273, 5419.

184 W. S. Dalton, T. M. Grogan, P. S. Meltzer, R. J. Scheper, B. G. Durie, C. W. Taylor, T. P. Miller, S. E. Salmon **1989**, *J. Clin. Oncol.* 7, 415.

185 T. P. Miller, T. M. Grogan, W. S. Dalton, C. M. Spier, R. J. Scheper, S. E. Salmon **1991**, *J. Clin. Oncol.* 9, 17.

186 M. S. Cairo, S. Siegel, N. Anas, L. Sender **1989**, *Cancer Res.* 49, 1063.

187 M. J. Millward, B. M. Cantwell, N. C. Munro, A. Robinson, P. A. Corris, A. L. Harris **1993**, *Br. J. Cancer* 67, 1031.

188 A. Regina, I. A. Romero, J. Greenwood, P. Adamson, J. M. Bourre, P. O. Couraud, F. Roux **1999**, *J. Neurochem.* 73, 1954.

189 O. C. Meijer, E. C. De Lange, D. D. Breimer, A. G. De Boer, J. O. Workel, E. R. De Kloet **1998**, *Endocrinology* 139, 1789.

190 P. Wils, V. Phung-Ba, A. Warnery, D. Lechardeur, S. Raeissi, I. J. Hidalgo, D. Scherman **1994**, *Biochem. Pharmacol.* 48, 1528.

191 D. Kessel, W. T. Beck, D. Kukuruga, V. Schulz **1991**, *Cancer Res.* 51, 4665.

192 T. Ohnishi, I. Tamai, K. Sakanaka, A. Sakata, T. Yamashima, J. Yamashita, A. Tsuji **1995**, *Biochem. Pharmacol.* 49, 1541.

193 C. S. Hughes, S. L. Vaden, C. A. Manaugh, G. S. Price, L. C. Hudson **1998**, *J. Neurooncol.* 37, 45.

194 D. E. Burgio, M. P. Gosland, P. J. Mcnamara **1998**, *J. Pharmacol. Exp. Ther.* 287, 911.

195 A. Eneroth, E. Astrom, J. Hoogstraate, D. Schrenk, S. Conrad, H. M. Kauffmann, K. Gjellan **2001**, *Eur. J. Pharm. Sci.* 12, 205.

196 M. Essodaigui, H. J. Broxterman, A. Garnier-Suillerot **1998**, *Biochemistry* 37, 2243.

197 M. Kavallaris, J. Madafiglio, M. D. Norris, M. Haber **1993**, *Biochem. Biophys. Res. Commun.* 190, 79.

198 P. Gros, F. Talbot, D. Tang-Wai, E. Bibi, H. R. Kaback **1992**, *Biochemistry* 31, 1992.

199 J. A. Kellen, E. George, V. Ling **1991**, *Anticancer Res.* 11, 1243.

200 M. J. Millward, B. M. Cantwell, E. A. Lien, J. Carmichael, A. L. Harris **1992**, *Eur. J. Cancer* 28a, 805.

201 D. L. Trump, D. C. Smith, P. G. Ellis, M. P. Rogers, S. C. Schold, E. P. Winer, T. J. Panella, V. C. Jordan, R. L. Fine **1992**, *J. Natl Cancer Inst.* 84, 1811.

202 C. Yamamoto, H. Murakami, N. Koyabu, H. Takanaga, H. Matsuo, T. Uchiumi, M. Kuwano, M. Naito, T. Tsuruo, H. Ohtani, Y. Sawada **2002**, *J. Pharm. Pharmacol.* 54, 1055.

203 J. Van Asperen, A. H. Schinkel, J. H. Beijnen, W. J. Nooijen, P. Borst, O. Van Tellingen **1996**, *J. Natl Cancer Inst.* 88, 994.

204 A. Tsuji, I. Tamai, A. Sakata, Y. Tenda, T. Terasaki **1993**, *Biochem. Pharmacol.* 46, 1096.

205 A. H. Schinkel, E. Wagenaar, C. A. Mol, L. Van Deemter **1996**, *J. Clin. Invest.* 97, 2517.

206 J. W. Polli, J. L. Jarrett, S. D. Studenberg, J. E. Humphreys, S. W. Dennis, K. R. Brouwer, J. L. Woolley **1999**, *Pharm. Res.* 16, 1206.

207 D. S. Miller, G. Fricker, J. Drewe **1997**, *J. Pharmacol. Exp. Ther.* 282, 440.

208 R. J. Arceci, K. Stieglitz, B. E. Bierer **1992**, *Blood* 80, 1528.
209 A. E. Kim, J. M. Dintaman, D. S. Waddell, J. A. Silverman **1998**, *J. Pharmacol. Exp. Ther.* 286, 1439.
210 C. G. Lee, M. M. Gottesman, C. O. Cardarelli, M. Ramachandra, K. T. Jeang, S. V. Ambudkar, I. Pastan, S. Dey **1998**, *Biochemistry* 37, 3594.
211 V. J. Wacher, J. A. Silverman, S. Wong, P. Tran-Tau, A. O. Chan, A. Chai, X. Q. Yu, D. O'mahony, Z. Ramtoola **2002**, *J. Pharmacol. Exp. Ther.* 303, 308.
212 T. Saeki, K. Ueda, Y. Tanigawara, R. Hori, T. Komano **1993**, *J. Biol. Chem.* 268, 6077.
213 I. C. Van Der Sandt, C. M. Vos, L. Nabulsi, M. C. Blom-Roosemalen, H. H. Voorwinden, A. G. De Boer, D. D. Breimer **2001**, *Aids* 15, 483.
214 A. H. Schinkel, C. A. Mol, E. Wagenaar, L. Van Deemter, J. J. Smit, P. Borst **1995**, *Eur. J. Cancer* 31a, 1295.
215 U. Mayer, E. Wagenaar, B. Dorobek, J. H. Beijnen, P. Borst, A. H. Schinkel **1997**, *J. Clin. Invest.* 100, 2430.
216 H. Kusuhara, H. Suzuki, T. Terasaki, A. Kakee, M. Lemaire, Y. Sugiyama **1997**, *J. Pharmacol. Exp. Ther.* 283, 574.
217 C. Pauli-Magnus, T. Murdter, A. Godel, T. Mettang, M. Eichelbaum, U. Klotz, M. F. Fromm **2001**, *Naunyn Schmiedebergs Arch. Pharmacol.* 363, 337.
218 G. Fricker, S. Nobmann, D. S. Miller **2002**, *Br. J. Pharmacol.* 135, 1308.
219 R. Xie, M. Hammarlund-Udenaes, A. G. De Boer, E. C. De Lange **1999**, *Br. J. Pharmacol.* 128, 563.
220 C. Dagenais, J. Ducharme, G. M. Pollack **2001**, *Neurosci. Lett.* 301, 155.
221 F. J. Sharom, X. Yu, P. Lu, R. Liu, J. W. Chu, K. Szabo, M. Muller, C. D. Hose, A. Monks, A. Varadi, J. Seprodi, B. Sarkadi **1999**, *Biochem. Pharmacol.* 58, 571.
222 M. Takano, R. Hasegawa, T. Fukuda, R. Yumoto, J. Nagai, T. Murakami **1998**, *Eur. J. Pharmacol.* 358, 289.
223 E. G. Schuetz, K. Yasuda, K. Arimori, J. D. Schuetz **1998**, *Arch. Biochem. Biophys.* 350, 340.
224 D. W. Loe, F. J. Sharom **1994**, *Biochim. Biophys. Acta* 1190, 72.
225 H. Tohda, A. Yasui, T. Yasumoto, M. Nakayasu, H. Shima, M. Nagao, T. Sugimura **1994**, *Biochem. Biophys. Res. Commun.* 203, 1210.
226 J. Dimitroulakos, H. Yeger **1996**, *Nat. Med.* 2, 326.
227 S. Orlowski, L. M. Mir, J. Belehradek Jr., M. Garrigos **1996**, *Biochem. J.* 317, 515.
228 K. Ueda, N. Okamura, M. Hirai, Y. Tanigawara, T. Saeki, N. Kioka, T. Komano, R. Hori **1992**, *J. Biol. Chem.* 267, 24248.
229 D. C. Wolf, S. B. Horwitz **1992**, *Int. J. Cancer* 52, 141.
230 A. M. Karssen, O. C. Meijer, I. C. Van Der Sandt, A. G. De Boer, E. C. De Lange, E. R. De Kloet **2002**, *J. Endocrinol.* 175, 251.
231 A. Van Helvoort, M. L. Giudici, M. Thielemans, G. Van Meer **1997**, *J. Cell Sci.* 110, 75.
232 P. A. Philip, S. Joel, S. C. Monkman, E. Dolega-Ossowski, K. Tonkin, J. Carmichael, J. R. Idle, A. L. Harris **1992**, *Br. J. Cancer* 65, 267.

233 R. D. Christen, E. F. Mcclay, S. C. Plaxe, S. S. Yen, S. Kim, S. Kirmani, L. L. Wilgus, D. D. Heath, D. R. Shalinsky, J. L. Freddo, et al. **1993**, *J. Clin. Oncol.* 11, 2417.

234 E. Solary, B. Witz, D. Caillot, P. Moreau, B. Desablens, J. Y. Cahn, A. Sadoun, B. Pignon, C. Berthou, F. Maloisel, D. Guyotat, P. Casassus, N. Ifrah, Y. Lamy, B. Audhuy, P. Colombat, J. L. Harousseau **1996**, *Blood* 88, 1198.

235 T. P. Miller, E. M. Chase, R. Dorr, W. S. Dalton, K. S. Lam, S. E. Salmon **1998**, *Anticancer Drugs* 9, 135.

236 M. Lemaire, A. Bruelisauer, P. Guntz, H. Sato **1996**, *Cancer Chemother. Pharmacol.* 38, 481.

237 D. Ukena, C. Boewer, B. Oldenkott, F. Rathgeb, W. Wurst, K. Zech, G. W. Sybrecht **1995**, *Cancer Chemother. Pharmacol.* 36, 160.

238 W. H. Wilson, S. E. Bates, A. Fojo, G. Bryant, Z. Zhan, J. Regis, R. E. Wittes, E. S. Jaffe, S. M. Steinberg, J. Herdt, et al. **1995**, *J. Clin. Oncol.* 13, 1995.

239 R. J. Motzer, M. Mazumdar, S. C. Gulati, D. F. Bajorin, P. Lyn, V. Vlamis, G. J. Bosl **1993**, *J. Natl Cancer Inst.* 85, 1828.

240 A. Stewart, J. Steiner, G. Mellows, B. Laguda, D. Norris, P. Bevan **2000**, *Clin. Cancer Res.* 6, 4186.

241 A. H. Dantzig, R. L. Shepard, K. L. Law, L. Tabas, S. Pratt, J. S. Gillespie, S. N. Binkley, M. T. Kuhfeld, J. J. Starling, S. A. Wrighton **1999**, *J. Pharmacol. Exp. Ther.* 290, 854.

242 J. J. Starling, R. L. Shepard, J. Cao, K. L. Law, B. H. Norman, J. S. Kroin, W. J. Ehlhardt, T. M. Baughman, M. A. Winter, M. G. Bell, C. Shih, J. Gruber, W. F. Elmquist, A. H. Dantzig **1997**, *Adv. Enzyme Regul.* 37, 335.

243 M. J. Newman, J. C. Rodarte, K. D. Benbatoul, S. J. Romano, C. Zhang, S. Krane, E. J. Moran, R. T. Uyeda, R. Dixon, E. S. Guns, L. D. Mayer **2000**, *Cancer Res.* 60, 2964.

244 M. Roe, A. Folkes, P. Ashworth, J. Brumwell, L. Chima, S. Hunjan, I. Pretswell, W. Dangerfield, H. Ryder, P. Charlton **1999**, *Bioorg. Med. Chem. Lett.* 9, 595.

245 R. A. Peck, J. Hewett, M. W. Harding, Y. M. Wang, P. R. Chaturvedi, A. Bhatnagar, H. Ziessman, F. Atkins, M. J. Hawkins **2001**, *J. Clin. Oncol.* 19, 3130.

246 A. Sparreboom, A. S. Planting, R. C. Jewell, M. E. Van Der Burg, A. Van Der Gaast, P. De Bruijn, W. J. Loos, K. Nooter, L. H. Chandler, E. M. Paul, P. S. Wissel, J. Verweij **1999**, *Anticancer Drugs* 10, 719.

247 J. Molnar, D. Szabo, R. Pusztai, I. Mucsi, L. Berek, I. Ocsovszki, E. Kawata, Y. Shoyama **2000**, *Anticancer Res.* 20, 861.

248 S. W. Kim, H. Y. Kwon, D. W. Chi, J. H. Shim, J. D. Park, Y. H. Lee, S. Pyo, D. K. Rhee **2003**, *Biochem. Pharmacol.* 65, 75.

249 R. K. Bhardwaj, H. Glaeser, L. Becquemont, U. Klotz, S. K. Gupta, M. F. Fromm **2002**, *J. Pharmacol. Exp. Ther.* 302, 645.

250 C. L. Cummins, W. Jacobsen, L. Z. Benet **2002**, *J. Pharmacol. Exp. Ther.* 300, 1036.

251 E. J. Wang, M. Barecki-Roach, W. W. Johnson **2002**, *Biochem. Biophys. Res. Commun.* 297, 412.
252 S. Zhang, M. E. Morris **2003**, *J. Pharmacol. Exp. Ther.* 304, 1258.
253 B. C. Foster, M. S. Foster, S. Vandenhoek, A. Krantis, J. W. Budzinski, J. T. Arnason, K. D. Gallicano, S. Choudri **2001**, *J. Pharm. Pharm. Sci.* 4, 176.
254 M. Demeule, M. Brossard, S. Turcotte, A. Regina, J. Jodoin, R. Beliveau **2004**, *Biochem. Biophys. Res. Commun.* 324, 937.

20
Targeting of Neuropharmaceuticals by Chemical Delivery Systems

Nicholas Bodor and Peter Buchwald

20.1
Introduction

Evolution built very efficient ways to isolate and protect the brain, this delicate, but crucial organ that has many vital functions. Unfortunately, the same mechanisms that prevent intrusive blood-borne substances from accessing the brain also prevent therapeutic chemicals from doing the same. Many pharmaceuticals are ineffective in the treatment of cerebral diseases because we cannot efficiently deliver and sustain them within the brain, a considerable problem considering the prevalence of brain diseases. More than 15% of the United States adult population (>18 years) is affected by some form of mental disorder: ~5% affective disorders, ~7% anxiety disorders, ~4% substance abuse/dependence, and ~1% severe cognitive impairment (1988 data) [1]. Despite almost twice as many people suffering from central nervous system (CNS) disorders than from cardiovascular diseases, the global market for CNS drugs is only about half of that for cardiovascular drugs (in the USA) [2, 3]. In light of these numbers, it is obvious that any drug delivery method that can provide brain delivery or eventually preferential brain delivery (i.e., brain targeting) is of great interest. Here, we present a review of the field of brain-targeted chemical delivery systems (CDSs), a rational drug design approach that exploits sequential metabolism not only to deliver, but also to target drugs to their site of action. To illustrate the complexity of the problems that have to be overcome for successful brain targeting, a brief overview of the blood-brain barrier (BBB) and other issues related to brain targeting is also included.

20.2
The Blood-Brain Barrier

20.2.1
Structural Aspects

The cerebral blood compartment does not have free diffusional communication with the interstitium of the brain: unlike most other organs, the brain is tightly segregated from the circulating blood by a unique membranous barrier, the BBB [4–13]. Capillaries in the brain and spinal cord are lined with a layer of special endothelial cells that lack fenestrations and are sealed with tight junctions; and therefore, these capillaries lack the small pores that allow the rapid movement of solutes from circulation into other organs (Figs. 20.1 and 20.2). These endothelial cells, together with perivascular elements such as astrocytes and pericytes, constitute the BBB. Neuronal endings may also directly innervate the endothelium [14, 15]. A barrier seems to be present only in animals with a complex level of higher neural function, and, at least based on qualitative observations, "higher" functions of neural tissue seem to be protected by a tighter barrier [16]. The presence of a barrier only in invertebrate groups capable of complex CNS functions might also indicate that a barrier is needed when the level of integrative activity in the nervous tissue reaches a critically sophisticated level [16]. Precise synaptic signaling might indeed require a finely regulated cerebral ionic microenvironment. Nevertheless, certain areas within the brain (circumventricular organs) are "outside the barrier" in the sense that they lack a tight epithelium, but appreciable leakage across these sites is unlikely as they account for only about 0.0002 of the surface area of the blood-brain exchange [16, 17].

Intercellular clefts, pinocytosis, and fenestrae are virtually nonexistent in brain capillaries, and hence, diffusion has to take place transcellularly (Fig. 20.2). Consequently, only the lipid-soluble solutes that can freely diffuse through the capillary endothelial membrane may passively cross the BBB. In general capillaries, such an exchange is overshadowed by other nonspecific exchanges. The brain is an organ of high metabolic rate, and therefore, of high blood flow. Cerebral blood flow takes up a considerable portion of the total cardiac output, and the brain's extensive network of capillaries could bring any compound within easy reach of neurons. However, despite an estimated total length of 650 km and total surface area of 12–20 m^2 of capillaries covered by only ~1 ml of total capillary endothelial cell volume in an average human brain of 1.3 kg [15], the BBB is very efficient and makes the brain practically inaccessible for lipid-insoluble compounds, such as polar molecules and small ions. Microvessels make up an estimated 95% of the total surface area of the BBB, and they represent the principal route by which chemicals enter the brain [18]. Capillaries in the brain are separated by only about 40–70 µm, and hence, the chemicals that can pass through the BBB can rapidly reach any brain tissue by diffusion. Vessels in the brain seem to have somewhat smaller diameters and thinner walls than vessels in other organs (Fig. 20.1) [19]. Also, the

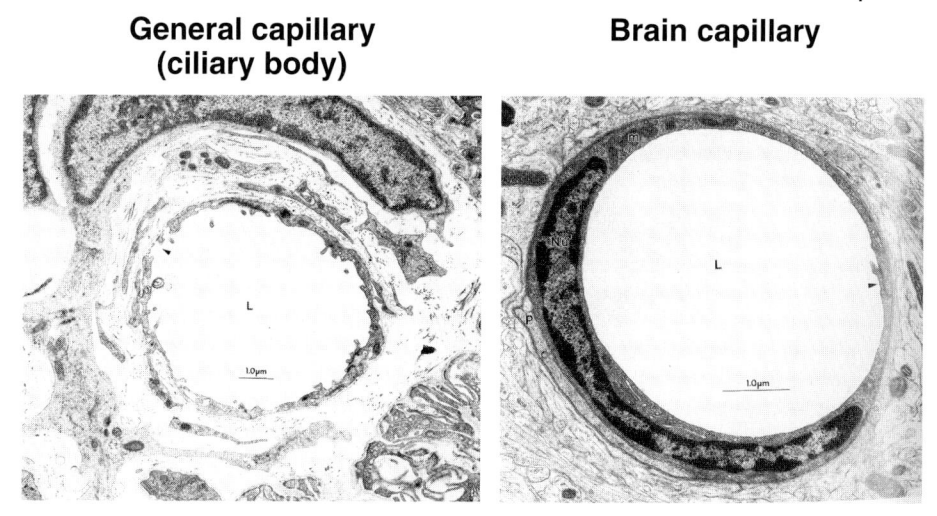

Fig. 20.1 Capillary profiles from ciliary body and brain gray matter in rat, shown at slightly different scales. L=Lumen; m=mitochondrion; Nu= endothelial nucleus; P=pericyte. (Reprinted from [19] with permission of Wiley-Liss, Wiley Publishing Inc.).

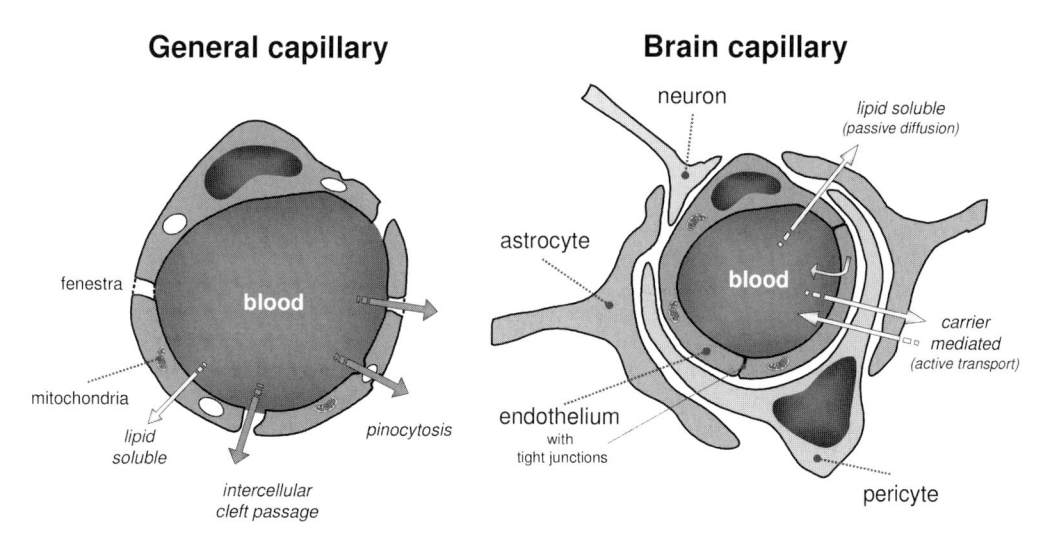

Fig. 20.2 Schematic comparison of general and brain capillaries.

mitochondrial density in brain microvessels seems to be somewhat higher than in other capillaries, not because more numerous or larger mitochondria are present, but because the small dimensions of the brain microvessels and the consequently smaller cytoplasmic area make the mitochondrial density appear higher [19]. The transendothelial electrical resistance, a measure of ionic permeability through junctions, has been estimated to be 1000–5000 Ω cm^{-2} for brain microvessels, in contrast to values of about 10 Ω cm^{-2} for most noncerebral capillaries [17].

The BBB is disrupted following acute stroke and trauma (see Chapter 25), and it is often assumed that in such cases all compounds can be delivered by an intravascular approach. However, the situation is more complex, as openings are usually localized and biphasic, and the BBB still impedes the delivery of therapeutics [17]. Brain tumors may also disrupt the BBB, but these are also local and nonhomogeneous disruptions [20]. Brain tumors probably contain capillaries with intact BBB as well as capillaries with increased permeability.

Like many other things in modern medicinal chemistry, the concept of a barrier between blood and brain can be traced back to the work of Paul Ehrlich, who noticed around 1885 that i.v. injection of dyes such as trypan blue in laboratory animals colored all tissues except the brain [21]. He originally thought that the brain has a lower affinity for these dyes than other tissues, but in 1913 Edwin Goldmann, one of his students, showed that injection of trypan blue directly into the cerebrospinal fluid of rabbits and dogs colors the brain without staining other organs [10]. This was a clear indication that the CNS is separated from the blood by some kind of a barrier; but, concerning the exact nature and localization of this barrier, a long period of uncertainty followed. For a long while, it was thought that astrocytes, members of the largest class of brain cells – the glial cells, must constitute the BBB because brain capillaries are almost completely surrounded by astrocyte foot processes. This later proved false, but the exact role of astrocytes is still not completely clarified, and most likely, they are involved in stimulating the development of barrier properties in the capillary endothelia [22]. Spatz suggested the capillary endothelium as an essential structure of the barrier in 1933 [23], and later, Crone also argued in favor of a peculiar tightness of the cerebral endothelium [24]. Nevertheless, only Reese and Karnovsky's 1967 experiment with horseradish peroxidase [25], followed by that of Brightman and Reese [26], provided a convincing proof for this idea.

20.2.2
Enzymatic and Transporter-Related Aspects

The BBB also has an additional, enzymatic aspect. Solutes crossing the cell membrane are exposed to degrading enzymes present in large numbers inside the endothelial cells that contain large densities of mitochondria, metabolically highly active organelles. Enzymes and receptors found in the BBB include, among others: adenylate cyclase, guanylate cyclase, Na$^+$–K$^+$ ATPase, alkaline phosphatase, catechol O-methyl transferase (COMT), monoamine oxidase (MAO), GABA transaminase, and DOPA decarboxylase [8, 27–30]. BBB enzymes also recognize and rapidly degrade most peptides, including naturally occurring neuropeptides [31, 32].

Active transporters also play an important BBB-role. While pointing to lipid solubility as an important factor determining access to the brain, in 1946 Krogh had already raised the possibility of some sort of active transport processes in the endothelium of the vessels in the CNS [33]. Today, it is well known that a few special solutes have access to specific, catalyzed transport, for example, the glucose transporter ensures that glucose, an essential nutrient, reaches the brain despite its low lipid solubility [34, 35]. Carriers are facilitative in nature and have been identified for hexoses, amino acids, peptides, monocarboxylic acids, organic cations, nucleosides and nucleoside bases, choline, and other biologically important molecules [36, 37]. These protein macromolecular systems are often characterized by saturability and molecular selectivity. Because this avenue provides for the movement of selected molecules, a number of attempts were made to exploit it for drug delivery purposes by coupling the therapeutic agent that has to be delivered to one of these molecules.

More recently, active efflux pumps have also been identified; and this recognition has considerably impacted CNS drug development [38]. The existence of probenecid-sensitive, active pumps for organic anions, such as β-lactam drugs or valproic acid, has been suspected for a while [39], and there is now considerable evidence for their existence [40, 41]. According to current knowledge [30, 38], three classes of transporters are implicated in the efflux of drugs from the brain: (1) monocarboxylic acid transporters (MCT), (2) organic ion transporters [organic anion transporters (OAT), organic cation transporters (OCT), organic anion transporter proteins (oatp), and organic cation/carnitine transporters (OCTN)], and (3) multidrug resistance (MDR) transporters [P-glycoprotein (P-gp), the prototypical MDR transporter, and MDR-associated protein (MRP)].

P-gp garnered considerable attention within the brain delivery field, as sufficient hydrophobicity is usually a target for CNS drugs, and classic P-gp substrates tend to be hydrophobic, amphipathic molecules (usually with a planar ring system) [42]. P-gp was first identified in 1976 as playing a role in cellular permeability (hence, P stands for permeability) [43], and it is a membrane protein belonging to the ATP-binding cassette (ABC) transport protein family, but contrary to most other members of this family, it appears to have wide substrate specificity. P-gp is a possible source of peculiarities in drug pharmacokinetics, including dose-dependent absorption, limited oral bioavailability, drug-drug interactions, intestinal secretion, and limited BBB permeability, and thus, it can considerably impact the therapeutic efficacy of its substrates [44]. The involvement of P-gp in active BBB transport has been confirmed by many studies [12, 45–51]. It seems that substrates, or at least some of them, are pumped out from the membrane itself [52].

20.3
Brain-Targeted Drug Delivery

Considering all these facts, it is not surprising that practically all drugs currently used for disorders that require CNS-delivery are lipid-soluble and can readily cross the BBB following oral administration. The delivery of other therapeutics to the brain requires some strategy to overcome the BBB, and therefore, general brain delivery approaches are of obvious therapeutic interest; and many related aspects have been reviewed [2, 17, 20, 32, 53–56]. Delivery of compounds such as neuropeptides or oligonucleotides is further complicated by their metabolic lability [32]. From a drug development perspective, one can aim either to modify existing drugs to increase their BBB penetration or to develop new chemical entities (NCEs) that already possess the desired permeability properties [57]. The spectrum of approaches intended to deliver potential therapeutic agents to the brain includes intracerebral delivery, intracerebroventricular delivery, intranasal delivery, BBB disruption, nanoparticles, receptor-mediated transport (vector-mediated transport or "chimeric" peptides), cell-penetrating peptides, prodrugs, and chemical delivery systems, which have been reviewed recently [56].

20.3.1
Lipophilicity and Its Role in CNS Entry

There has been a more or less continuous interest in defining lipophilicity/hydrophobicity and its role in biological activity in general and CNS activity in particular ever since Overton [58, 59] and Meyer [60, 61] discovered more than a century ago that narcotic potency in a set of congeners tends to increase as the oil/water partition coefficient P increases. Following its first use by Collander [62] and especially during the past 30 years, the log octanol/water partition coefficient, $\log P_{o/w}$, became a convenient standard measure for lipophilicity and one of the most informative physicochemical parameters used in the pharmaceutical and medicinal chemical fields [63]. The suitability of the octanol/water system to characterize CNS effects was underscored by a study of Franks and Lieb [64], who looked at anesthetic potency as a function of solvent/water partition coefficients and found that octanol gives an excellent correlation for all compounds and all animals over an extremely wide range.

$\log P_{o/w}$ alone seems to have a more limited performance in predicting brain/blood concentration ratios, but in combination with other parameters can still reasonably predict brain-blood partitioning [65–70]. In agreement with the structural aspects of the BBB described here, we also found a good correlation over a wide range of about eight log units between $\log P_{o/w}$ and in vivo log permeability data of rat brain capillaries ($\log \mathscr{P}_{BBB}$) for compounds that are not subject to active transport (Fig. 20.3) [63, 71]. However, increasing lipophilicity with the intent to improve membrane permeability might not only make chemical handling difficult, but might also increase the rate of oxidative metabolism by cytochromes P450 (CYPs) and other enzymes [72, 73]. In most cases, increasing lipophilicity increases potency and membrane permeability, but decreases dissolution and metabolic stability [74]. Hence, to improve bioavailability, the effects of lipophilicity on membrane permeability and first-pass metabolism have to be balanced [72, 75]. Furthermore, increasing lipophilicity tends to increase the volume of distribution [76] and tends to affect all other pharmacokinetic parame-

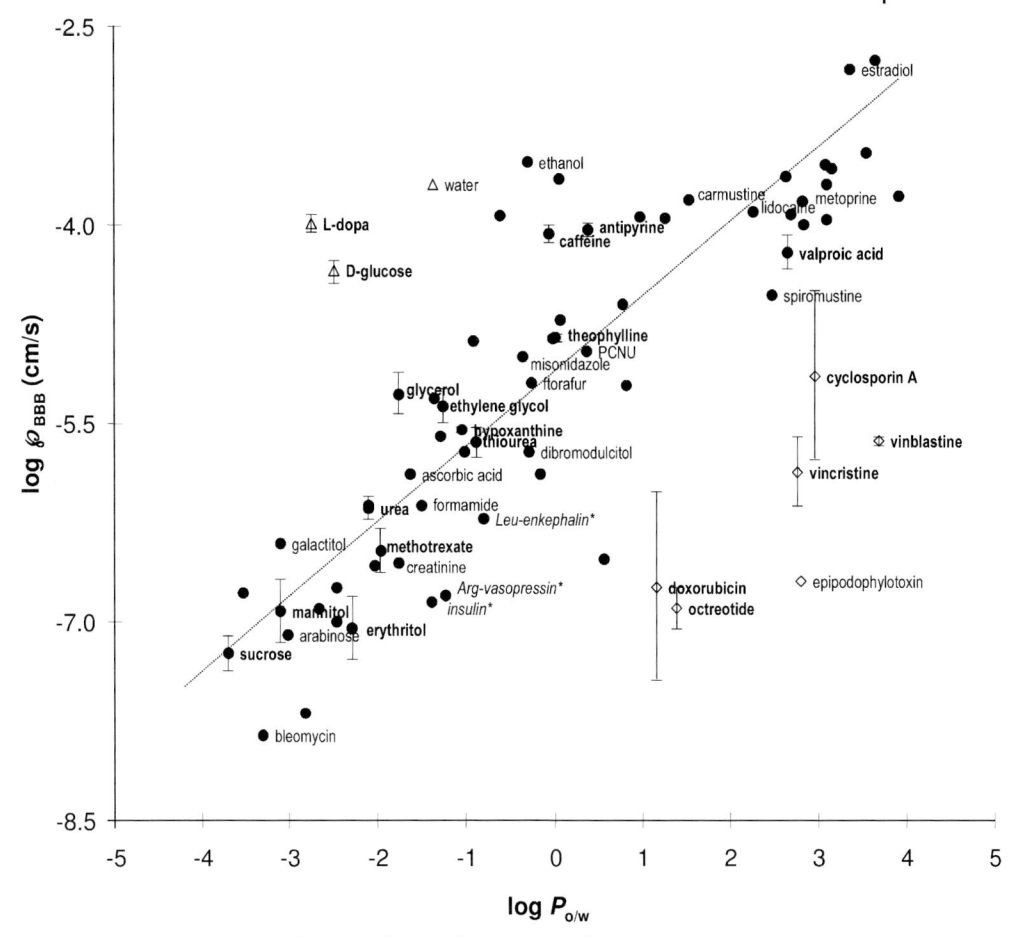

Fig. 20.3 In vivo log permeability coefficient of rat brain capillaries (log $\mathscr{P}_{BBB}$) as a function of log octanol/water partition coefficient (log$P_{o/w}$). Names are shown for a few well known compounds, values denoted with a star are for guinea pig, and values denoted in italics are for cases where the log distribution coefficient measured at physiological pH was used (log$D_{o/w}$). Strong deviants below the line, shown as diamonds, are known substrates for P-gp.

ters [72, 75]. Nonspecific toxicity is also well known to increase with log$P_{o/w}$ [77]; and King and Moffat have shown such a tendency even for approved hypnotics in humans [78]. It is difficult to formulate rules of general validity, but Hansch and co-workers [79] suggested that, when designing prospective new CNS drugs, the likelihood of success is better if one starts somewhere around an (apparent) calculated log$P_{o/w}=2$, and if CNS-related side-effects are to be avoided, it is probably better to avoid the $1<\log P_{o/w}<3$ region. This has been referred to as the "rule of two" [80, 81]. During the past two decades, a variety of in silico models [82] and in vitro permeability assays [83] have been developed in attempts to characterize and predict BBB permeability and integrate such prediction into the early phases of drug development, together with various other considerations [57, 72, 77, 84].

Obviously, brain exposure can be increased not only by enhancing influx, but also by restricting efflux through the BBB [85]. For example, coadministration of a P-gp inhibitor can increase not only oral absorption, but also BBB permeability for P-gp substrates [86, 87]. Coadministration of the P-gp blocker valspodar has been shown recently not only to increase the brain levels of paclitaxel, but also to considerably improve its therapeutic effect on tumor volume in mice [88]. At present, the investigation of such possibilities are in the early exploratory phase only, but they might become useful in the future, maybe in combination with other, influx-enhancing approaches if toxicity in nontarget tissues and other problems can be avoided.

20.3.2
Quantifying Brain-Targeting: Site-Targeting Index and Targeting Enhancement Factors

Being able to quantitatively assess the site-targeting effectiveness of drug delivery systems or drug targeting in general is very important, and is particularly so for CNS-targeting. Recently, we proposed the use of targeting enhancement factors in the pharmacokinetic assessment of drug delivery systems [56, 89]. A site-targeting index (STI) was defined as the ratio between the area under the concentration-time curve (AUC) for the drug itself at the targeted site and that at a systemic site, e.g., blood or plasma:

$$\text{STI} = \text{AUC}_{\text{target}}/\text{AUC}_{\text{blood}} \qquad (20.1)$$

This index gives an accurate measure of how effectively the active therapeutic agent is actually delivered to its intended site of action. For example, as data in Table 20.1 indicate, some drugs (e.g., AZT, benzylpenicillin) are very ineffective in penetrating the BBB and reaching brain tissues: the AUCs of their brain concentrations are less than 1% of the corresponding blood values.

For a drug delivery system, a site-exposure enhancement factor (SEF) was introduced to assess its effectiveness compared to the original drug itself. SEF measures the change in AUC of the active drug at the target site after administration of the delivery system, compared to administration of the drug alone:

$$\text{SEF} = \text{AUC}_{\text{target}}^{\text{Delivery System}}/\text{AUC}_{\text{target}}^{\text{Drug Alone}} \qquad (20.2)$$

Obviously, administration of equivalent (molar) doses is assumed for this definition. For slightly different doses, linearity between dose and AUC is a reasonable assumption; and the ratio of dose-normalized AUCs (AUC/dose) can be

Table 20.1 Site-targeting indices (STI), site-exposure factor (SEF) and targeting enhancement factor (TEF) of the brain-targeting CDSs with available in vivo data. AUC = area under the concentration-time curve; S = spacer; AZT = azidovudine (3′-azido-3′-deoxythymidine); AZdU = 3′-azido-2′,3′-dideoxyuridine; BP = benzylpenicillin (penicillin G).

Compound Species and reference	Dose (μmol kg^{-1})	Time (h)	D only AUC (μg h ml^{-1}) Brain	D only AUC (μg h ml^{-1}) Blood	STI (Eq. 20.1)	CDS AUC (μg h ml^{-1}) Brain	CDS AUC (μg h ml^{-1}) Blood	STI	SEF (Eq. 20.2)	TEF (Eq. 20.3)
AZdU Mouse[a] [167]	197	12	2.09	25.83	0.081	11.43	25.79	0.443	5.47	5.48
AZT Mouse[a] [167]	187	12	1.21	26.64	0.045	11.28	25.38	0.444	9.32	9.79
Rat [159]	130	6	0.23	26.33	0.009	7.09	9.67	0.733	31.51	85.84
Rabbit [156]	64	3	11.22	37.92	0.296	26.98	34.07	0.792	2.40	2.68
BP, S = Me Rat [141]	60	1	0.20	6.05	0.033	0.74	4.56	0.162	3.70	4.91
Dog[b] [142]	30	4	0.16	24.52	0.007	5.07	11.75	0.431	31.69	66.13
BP, S = Et Rat [141]	60	1	0.20	6.05	0.033	2.12	2.95	0.719	10.60	21.74
Dog[b] [142]	30	4	0.16	24.52	0.007	0.48	0.61	0.787	3.00	120.59
Dexamethasone Rat [115]	20	6	1.36	19.48	0.070	1.69	8.97	0.188	1.24	2.70
Ganciclovir Rat[c] [171]	80	6	0.66	10.46	0.063	3.61	1.42	2.542	5.44	40.05

a) Brain and serum concentrations monitored.
b) Cerebrospinal fluid and blood concentrations monitored.
c) Brain and plasma concentrations monitored.

used in the definition of Eq. (20.2). Because a good delivery system not only increases exposure to the active agent at the target site, but also decreases the corresponding systemic exposure, a targeting enhancement factor (TEF) was introduced to quantify the relative improvement in the STI produced by administration of the delivery system, compared to administration of the drug itself:

$$\text{TEF} = \text{STI}^{\text{Delivery System}} / \text{STI}^{\text{Drug Alone}} \qquad (20.3)$$

The TEF defined by Eq. (20.3) is the most rigorous measure that can be used to quantify the improvement in (pharmacokinetic) targeting produced by a delivery system. It compares not just concentrations, but concentrations along a time period, and it compares actual, active drug concentrations both at target and systemic sites. If: (1) the targeting or delivery ("carrier") moiety produces no toxicity of its own, (2) the therapeutic effect is linearly related to AUC_{target}, and (3) side-effects are linearly related to AUC_{blood} (assumptions that are to a good extent satisfied under most conditions), then TEF also represents the enhancement produced in the therapeutic index (TI), which is usually defined as the ratio between the median toxic dose (TD_{50}) and the median effective dose (ED_{50}), $TI = TD_{50}/ED_{50}$:

$$TEF = TI^{Delivery\ System}/TI^{Drug\ Alone} \qquad (20.4)$$

TEF, as defined here, represents the same ratio as a drug-targeting index used in a theoretical pharmacokinetic treatment [90]. A number of quantitative comparisons using these indices are presented in Table 20.1 for many of the CDSs discussed here.

20.4
Chemical Delivery Systems

CDS approaches provide novel, systematic methodologies for targeting active biological molecules to specific target sites or organs based on predictable enzymatic activation [71, 89, 91]. CDSs are inactive chemical derivatives of a drug obtained by one or more chemical modifications. They provide site-specific or site-enhanced delivery through sequential, multistep enzymatic and/or chemical transformations. The newly attached moieties are monomolecular units that are in general smaller or of similar size as the original molecule. Hence, the *chemical delivery system* term is used here in a stricter sense: they do not include systems in which the drug is covalently attached to large "carrier" molecules.

In a general formalism, the bioremovable moieties introduced can be classified into two categories. Targetor (T) moieties are responsible for targeting, site-specificity, and lock-in. Modifier functions ($F_1 \ldots F_n$) serve as lipophilizers, protect certain functions, or fine-tune the necessary molecular properties to prevent premature, unwanted metabolic conversions. CDSs are designed to undergo sequential metabolic conversions, first disengaging the modifier functions and then the targetor moiety, after it fulfills its site- or organ-targeting role. These transformations provide targeting (differential distribution) of the drug by either exploiting site-specific transport properties, such as those provided by the presence of the BBB, or by recognizing specific enzymes found primarily, exclusively, or at higher activity at the site of action. Three major CDS classes have been identified: (1) brain-targeting (enzymatic physical/chemical-based) CDSs, (2) site-specific enzyme-activated CDSs, and (3) receptor-based transient anchor-type CDSs. By relying on these general concepts, successful deliveries have been achieved to the brain, eye, and lung, as summarized recently [89]. However, brain-targeting CDSs represent the most developed class, and here, we will review only aspects related to their development.

20.5
Brain-Targeting CDSs

20.5.1
Design Principles

Brain-targeting chemical delivery systems represent the most developed CDS class. To obtain such a CDS, the drug is chemically modified to introduce the protective function(s) and the targetor (T) moiety. Upon administration, the resulting CDS is distributed throughout the body. Predictable enzymatic reactions convert the original CDS by removing some of the protective functions and modifying the T moiety, leading to a precursor form (T^+-D), which is still inactive, but has significantly different physicochemical properties (Fig. 20.4). Whereas these intermediates are continuously eliminated from the "rest of the body", at the targeted site, which has to be delimited by some specific membrane or other distribution barrier, efflux-influx processes are altered and provide a specific concentration. Consequently, release of the active drug essentially occurs only at the site of action.

In other words, such a system exploits the idea that if a lipophilic compound enters the brain and is converted there into a hydrophilic molecule, it will be-

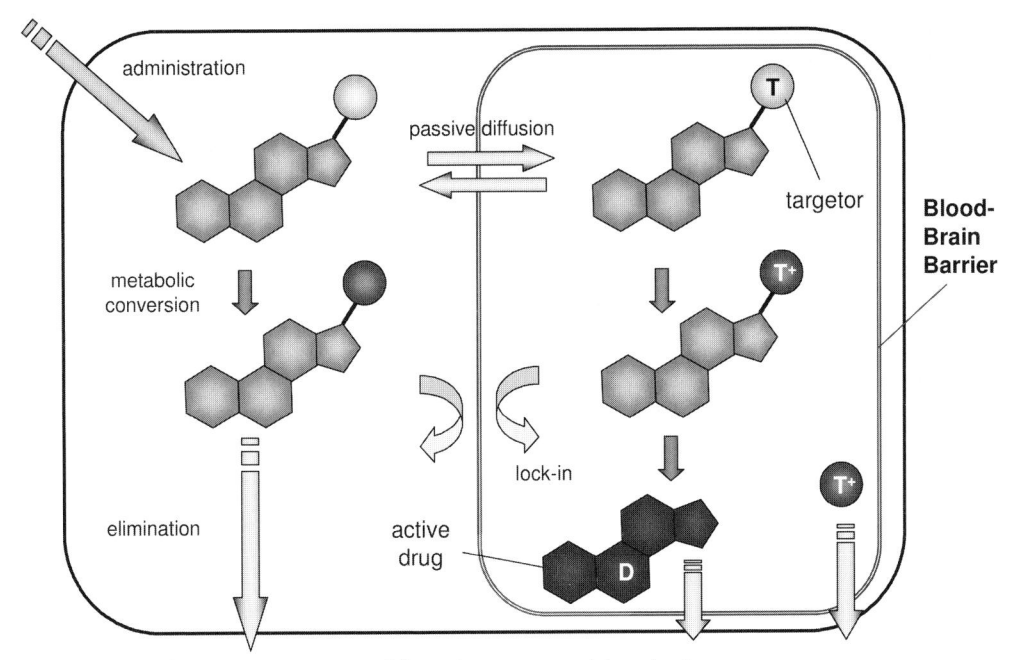

Fig. 20.4 Schematic representation of the multi-step sequential mechanism of brain-targeting CDSs designed to provide sustained and brain-specific activity.

come "locked-in": it will no longer be able to come out (Fig. 20.4). Targeting is assisted because the same conversion, as it takes place in the rest of the body, accelerates peripheral elimination, and further contributes to brain-targeting. In principle, many targetor moieties are possible for a general system of this kind [8, 92–96], but the one based on the 1,4-dihydrotrigonelline↔trigonelline (coffearine) system, in which the lipophilic 1,4-dihydro form (T) is converted in vivo to the hydrophilic quaternary form (T^+; Fig. 20.5), proved the most useful. This conversion takes place easily everywhere in the body, because it is closely related to the ubiquitous NADH↔NAD^+ coenzyme system associated with numerous oxidoreductases and cellular respiration [97, 98]. Because oxidation takes place with direct hydride transfer [99] and without generating highly active or reactive radical intermediates, it provides a nontoxic targetor system [100]. Furthermore, it was shown [101] that the trigonellinate ion formed after cleavage of the CDS undergoes rapid elimination from the brain, most likely by involvement of an active transport mechanism that eliminates small organic ions, and therefore, the T^+ moiety formed during the final release of the active drug D from the charged T^+-D form will not accumulate within the brain [101, 102].

Whereas the charged T^+-D form is locked behind the BBB into the brain, it is easily eliminated from the body as a result of the acquired positive charge, which enhances water solubility. After a relatively short time, the delivered drug D (as the inactive, locked-in T^+-D) is present essentially only in the brain, providing sustained and brain-specific release of the active drug. It has to be emphasized again that the system not only achieves delivery to the brain, but it

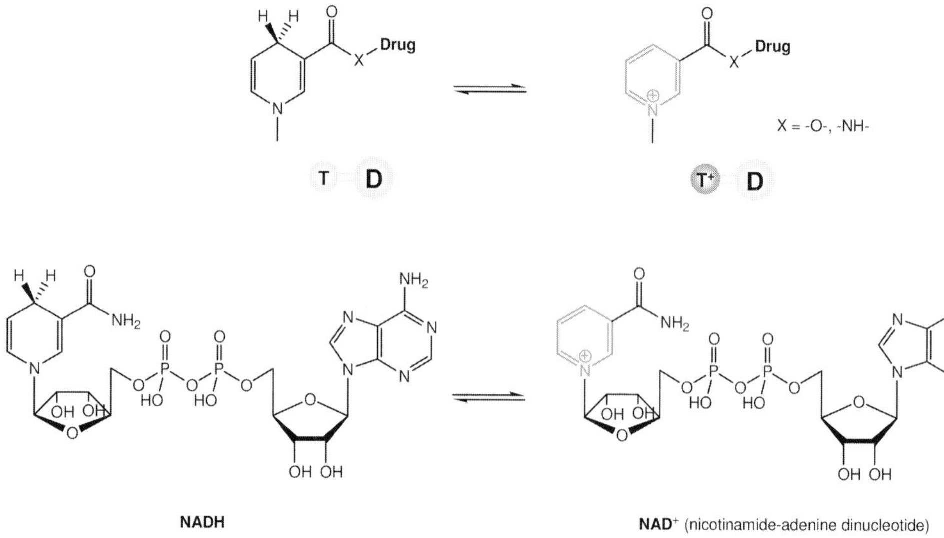

Fig. 20.5 The 1,4-dihydrotrigonellinate↔trigonellinate redox system used in brain-targeting CDSs, in analogy to the ubiquitous NADH↔NAD^+ coenzyme system.

achieves preferential delivery, which means brain-targeting. This should allow smaller doses and reduce peripheral side-effects. Furthermore, because the "lock-in" mechanism works against the concentration gradient, the system also provides more prolonged effects. Consequently, these CDSs can be used not only to deliver compounds that otherwise have no access to the brain, but also to retain lipophilic compounds within the brain, as has indeed been achieved, for example, with a variety of steroid hormones.

During the past two decades, the CDS approach has been explored with a wide variety of drug classes, e.g., biogenic amines: phenylethylamine [103–106], tryptamine [107, 108], steroids: testosterone [96, 109–111], progestins [112], progesterone [113], ethinyl estradiol [114], dexamethasone [115, 116], hydrocortisone [96], alfaxalone [117], estradiol [118–138], anti-infective agents: penicillins [139–142], sulfonamides [143], antivirals: acyclovir [91, 144], trifluorothymidine [145, 146], ribavirin [147–149], Ara-A [147], deoxyuridines [150–152], 2'-F-5-methylarabinosyluracil [153], antiretrovirals: zidovudine (AZT) [154–168], 2',3'-dideoxythymidine [102], 2',3'-dideoxycytidine [169, 170], ganciclovir [171], cytosinyl-oxathiolane [172], anticancer agents: lomustine (CCNU) [173, 174], HECNU [174], chlorambucil [175], neurotransmitters: dopamine [176–180], GABA [181, 182], nerve growth factor (NGF) inducers: catechol derivatives [183, 184], anticonvulsants: GABA [182], phenytoin [185], valproate [186], stiripentol [187], the calcium channel antagonist felodipine [188], MAO inhibitors: tranylcypromine [189, 190], anti-dementia drugs (cholinesterase inhibitors): tacrine [191, 192], antioxidants: LY231617 [193], chelating agents: ligands for technetium complexes [194], nonsteroidal anti-inflammatory drugs (NSAIDs): indomethacin [195], naproxen [195], anesthetics: propofol [196], nucleosides: adenosine [197], and peptides: tryptophan [198, 199], Leu-enkephalin analogs [200, 201], thyreotropin-releasing hormone (TRH) analogs [202–208], kyotorphin analogs [209, 210], and S-adenosyl-L-methionine (SAM) [211]. Selected examples will be briefly summarized together with representative physicochemical properties, metabolic pathways, and pharmacological data.

20.5.2
Zidovudine-CDS

Infection by human immunodeficiency virus type 1 (HIV-1), the causative agent of AIDS, has many devastating consequences, including encephalopathy with subsequent dementia. Many AIDS patients have demonstrable neurological damage related to this syndrome. Antiviral agents need to reach the CNS in therapeutically relevant concentrations to adequately treat AIDS dementia, but most potentially useful drugs cannot penetrate through the BBB. Zidovudine (3'-azido-3'-deoxythymidine, azidothymidine; AZT) was the first drug approved for the treatment of AIDS. This modified riboside has been shown to be useful in improving the neuropsychiatric course of AIDS encephalopathy in a few patients, but the doses required for this improvement usually precipitated severe anemia.

Following the increasing occurrence of AIDS, AZT-CDS was investigated in a number of laboratories to enhance the access of AZT to brain tissues [154–168]. AZT-CDS is relatively easily accessible synthetically; and it is a crystalline solid which is stable at room temperature for several months when protected from light and moisture. It is relatively stable in pH 7.4 phosphate buffer, but rapidly oxidizes in enzymatic media. In addition, T^+-AZT, the depot form of AZT, was shown to gradually release the parent compound.

The ability of AZT-CDS to provide increased brain AZT levels has been demonstrated in a number of different in vivo animal models. The corresponding AUC data for mice [167], rats [159], and rabbits [156] are presented in Table 20.1. In all species, AZT-CDS provided substantially increased brain-targeting of AZT. For example, the TEF was 86 in rats. It is interesting to note that the relatively large difference in the TEF values for different species is mainly due to the variability of the brain-STI of AZT itself, AZT-CDS gave very similar values in all three species.

In rabbits, the brain/blood ratio never exceeded 1 after AZT administration, indicating that the drug is always in higher concentration in the blood than in the brain, but it approached 3 at 1 h after AZT-CDS administration. In dogs, AZT levels were 46% lower in blood, but 175–330% higher in brain after AZT-CDS administration than after AZT administration [160]. Furthermore, in vitro experiments found AZT-CDS not only more effective in inhibiting HIV replication than AZT itself, presumably due to hydrolysis, but also less toxic to host lymphocytes [159, 162, 163]. In conclusion, a number of different studies found an improved delivery to the brain and a decreased potential for dose-limiting side-effects, suggesting that AZT-CDS may become useful in the treatment of AIDS encephalopathy and the related dementia.

20.5.3
Ganciclovir-CDS

A CDS approach also allowed enhanced brain-delivery for ganciclovir (Fig. 20.6), another antiviral agent that may be useful in treating human encephalitic cytomegalovirus (CMV) infection [171]. CMV infections are common and usually benign; however, human CMV occurs in 94% of all patients suffering with AIDS and it has been implicated as a deadly cofactor. Whereas positive results could be achieved in the treatment of the associated retinitis with ganciclovir, they were much harder to come by in the treatment of encephalitic cytomegalic disease. A brain-targeted CDS was therefore investigated; and improved organ selectivity was demonstrated. As shown in Fig. 20.6 and Table 20.1, a ca. 5-fold increase in the relative brain exposure to ganciclovir and a simultaneous 7-fold decrease in blood exposure was achieved with the CDS in rats.

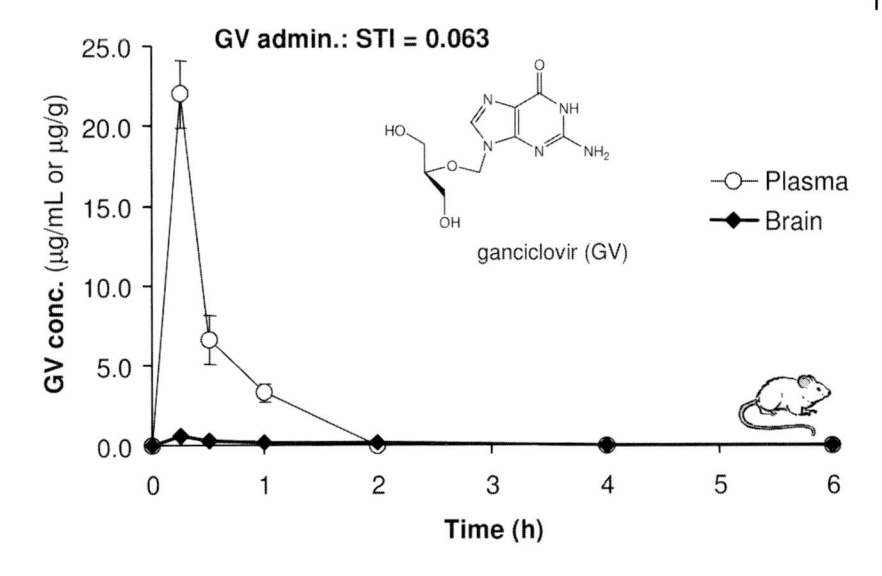

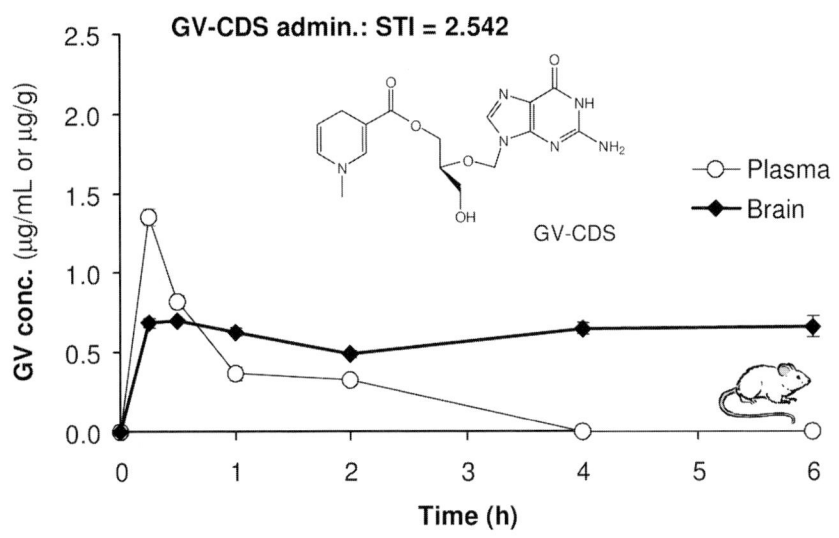

Fig. 20.6 Ganciclovir (GV) concentrations in brain and plasma as a function of time after an i.v. dose of 80 µmol kg^{-1} of GV or GV-CDS in rats. Data are means ± SEM.

20.5.4
Benzylpenicillin-CDS

The treatment of various bacterial infections localized in brain and cerebrospinal fluid is often hindered by the poor CNS penetration of β-lactam antibiotics, including penicillins and cephalosporins. In addition, because these antibiotics return back to the blood from the CNS (active transport may be involved), simple prodrug approaches could not provide real solutions. Consequently, a number of CDSs have been synthesized and investigated [139–142]. Because in these structures the targetor moiety had to be coupled to an acid moiety, a spacer function had to be inserted between the targetor and the drug. This renders additional flexibility to the approach, and several possibilities were investigated. Benzylpenicillin (BP)-CDS was one of the first cases where the influence of such functions on the stability and delivery properties of CDSs could be investigated. Systematic studies of the effects of various modifications have been performed since then [93, 201].

In rats, i.v. administration of BP-CDS with an ethylene 1,2-diol spacer (S = Et) gave an AUC-based brain-exposure enhancement factor for BP of around 11 (see Table 20.1) [141]. From the several coupling possibilities investigated, diesters of methylene diol and ethylene 1,2-diol proved worthy of further investigation, but ester-amide combinations did not prove successful. Consequently, the diesters have also been investigated in rabbits and dogs [142].

Both in rabbits and in dogs, i.v. administration of equimolar doses of CDSs provided BP levels in brain and cerebrospinal fluid (CSF) that were substantially higher and more prolonged than in the case of parent drug administration. In this study, the diester of methylene diol provided more significant increase in BP brain exposure (SEF = 32); and the diester of ethylene 1,2-diol gave better targeting, but released lower drug concentrations due to the formation of an inactive hydroxyethyl ester. Because dogs were not sacrificed, only blood and CSF data were collected (Table 20.1); and these indicated better CSF penetration in dogs than in rabbits.

Although parenteral injection of BPs may sometimes cause undesired side-effects, such as seizures, the high CNS levels obtained in dogs and rabbits after CDS administration were not accompanied by any toxic side-effects [142]. Similarly, although the delivered compound exhibited toxicity in the rotorod test at 0.5 h following a 300 mg kg^{-1} i.p. dose, felodipine-CDS displayed no toxicity at either 0.5 h or 4.0 h after a 100 mg kg^{-1} i.p. dose or 0.5 h for a 300 mg kg^{-1} i.p. dose [188]. Some toxicity, which is not surprising considering the toxicity of the delivered compound itself, was observed at 4.0 h for the 300 mg kg^{-1} CDS dose.

20.5.5
Estradiol-CDS

Estradiol-CDS (E_2-CDS; Estredox) is in the most advanced investigation stage of all CDSs (phase I/II clinical trials). Estrogens are endogenous hormones that produce numerous physiological actions and are among the most commonly prescribed drugs in the United States, mainly for hormone replacement therapy (HRT) in postmenopausal women and as a component of oral contraceptives [212]. They are lipophilic steroids that are not impeded in their entry to the CNS, and hence, they can readily penetrate the BBB and achieve high central levels after peripheral administration. However, estrogens are poorly retained within the brain. Therefore, frequent or sustained (e.g., transdermal) doses have to be administered to maintain therapeutically significant concentrations.

Menopause-related estrogen depletion is more than likely associated with a variety of symptoms that range the gamut from vasomotor complaints to cognitive deficits. Traditional HRT may alleviate many of these complications, but because of increased risks for cancer and other metabolic diseases, many women avoid treatment. Constant peripheral exposure to estrogens has been related, however, to a number of pathological conditions including cancer, hypertension, stroke, and altered metabolism [213–217]. The Heart and Estrogen/Progestin Replacement Study (HERS) has already suggested that, contrary to earlier expectations, HRT does not benefit the incidence of coronary heart disease (CHD) [218, 219]. As a very important development, two large-scale parallel randomized, double-blind, placebo-controlled clinical trials were undertaken within the Women's Health Initiative (WHI) to determine whether conjugated equine estrogen (CEE) alone (in women with prior hysterectomy) or in combination with progestin (medroxy-progesterone acetate) reduces cardiovascular events in postmenopausal women; and both were halted early because of unfavorable outcomes. The estrogen plus progestin component, which had a planned duration of 8.2 years and recruited 16 608 postmenopausal women, was stopped prematurely because it indicated an unacceptably increased risk for invasive breast cancer (26%) [220, 221]. It also indicated a 41% increase for stroke and a 29% increase for CHD. The estrogen alone component, which enrolled 10 739 postmenopausal women, was also ended early as the use of CEE was found to increase the risk of stroke and not affect the incidence of CHD (but decrease the risk of hip fracture) in postmenopausal women with prior hysterectomy [222]. Considering all this and that the approx. 30% increase in the risk of breast cancer caused by taking estrogen is suspected to rise to 50% if taken for more than 10 years [223], there is little justification now for systemic HRT. This is an important decision, since in industrialized nations, the average woman spends around a third of her life in the postmenopausal stage (menopause occurs at an average age of 51) [224]. As the CNS is the target site for many estrogenic actions [225–227], brain-targeted delivery may provide safer and more effective agents in many cases. With the recent unraveling of the many roles estrogen plays in males [228], there might be at least as many therapeutic possibilities in males as in females.

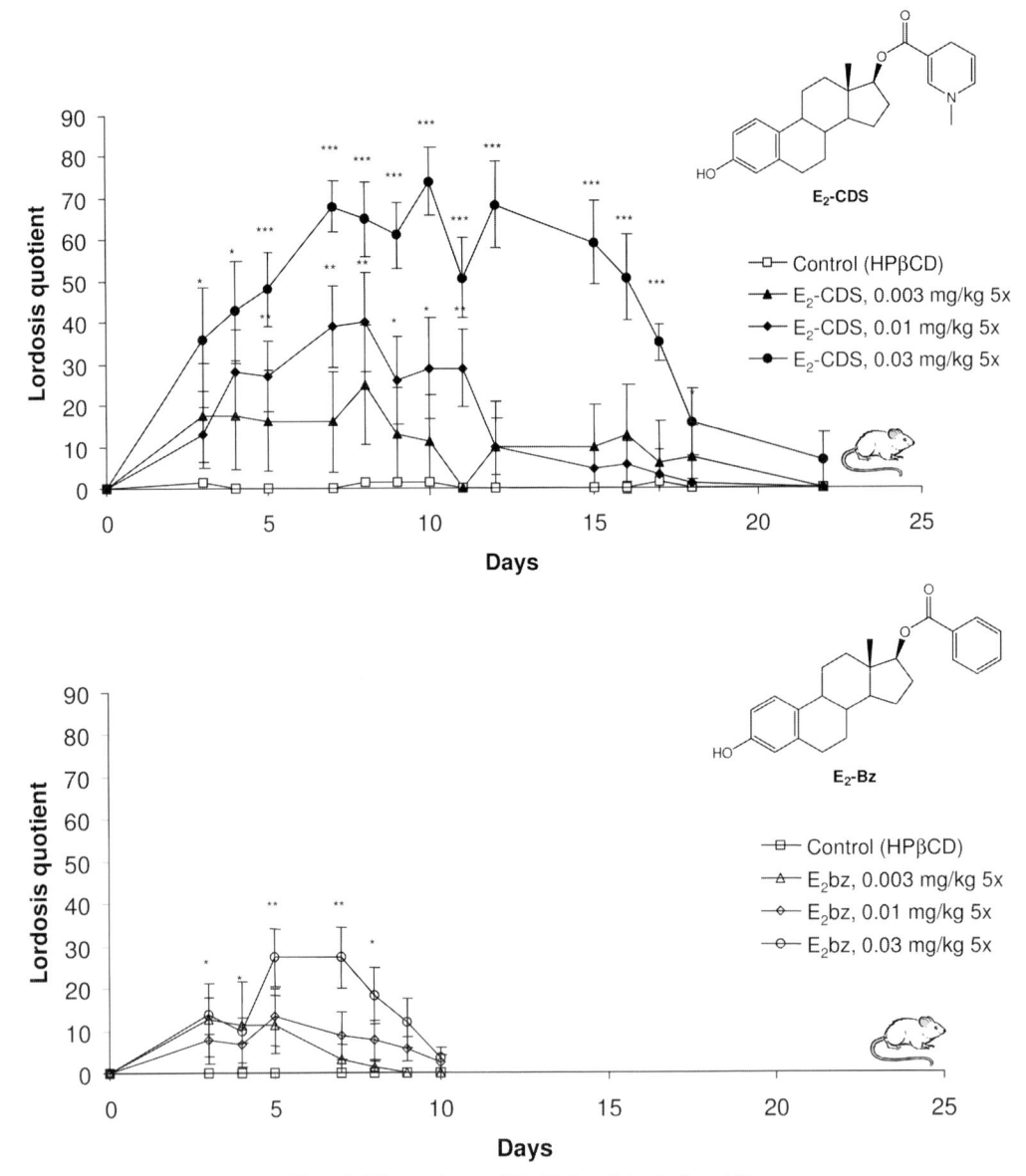

Fig. 20.7 Effect of different doses of E_2-CDS and the isolipophilic E_2-benzoate on the sexual behavior of ovariectomized female rats, as measured by the lordosis quotient. Lordosis is the vertebral dorsiflexion performed by female quadrupeds in response to adequate stimuli from a reproductively competent male; and the lordosis quotient is calculated as the number of lordosis per ten mounts (%). E_2-CDS was administered for 5 days i.v. dissolved in 27% HPβCD solution. Data are means ± SE for 8–12 animals per group (* $P<0.05$, ** $P<0.01$, *** $P<0.001$; using Mann-Whitney U test).

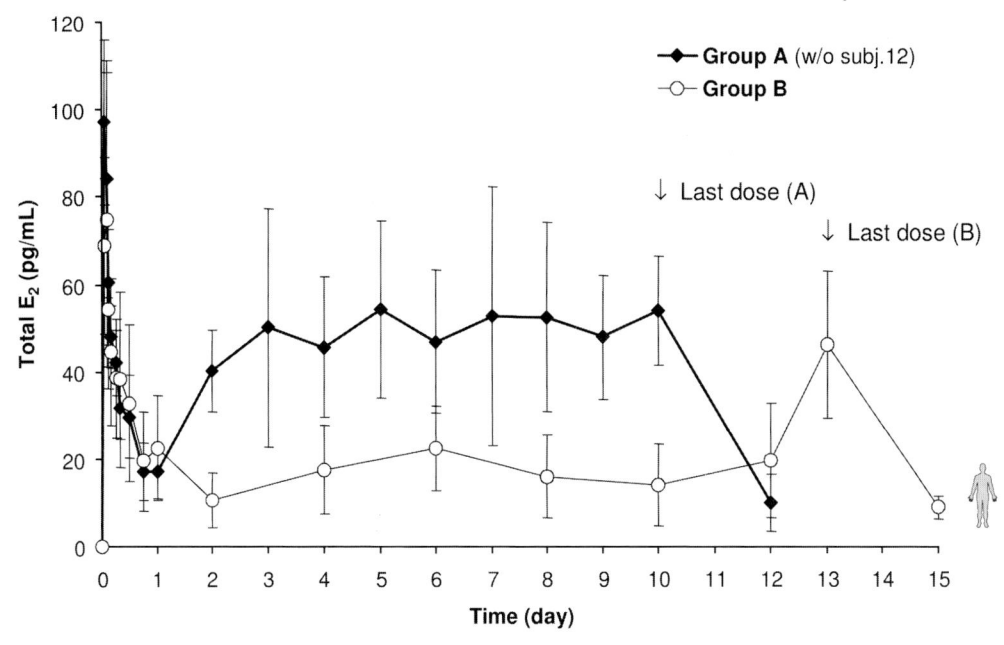

Fig. 20.8 Baseline-adjusted total E_2 concentrations for the whole study period following administration of 2.86 mg E_2-CDS as a buccal tablet once daily (group A) or once every other day (group B). Data are means±SD for six subjects in group B and five subjects in group A, where the data for one subject, which had about 10-fold higher E_2 levels than the other subjects, are not included in this graphic.

for 10 days (group A) or once every second day for 13 days (group B). Buccal E_2-CDS administered daily or every other day was well-tolerated. Serum E_2 levels were higher for the daily regimen (group A), and differences were significant (Fig. 20.8) [243]. Both dosing regimens caused significant LH- and FSH-suppression, providing evidence of CNS penetration. In conclusion, clinical evaluations suggested a potent central effect coupled with only marginal tendencies to elevate systemic estrogen levels. Furthermore, peripheral levels seem to be susceptible to easy manipulation through dosing amount and frequency, and therefore, this approach could represent the ideal estrogen replacement therapy. The data suggest that E_2-CDS may be a useful and safe therapy for menopausal symptoms, estrogen-dependent cognitive deficits, including Alzheimer's disease, male and female sexual dysfunction, and possibly even neuroprotection.

20.5.6
Cyclodextrin Complexes

Unfortunately, the same physicochemical characteristics that allow successful chemical delivery also complicate the development of acceptable pharmaceutical formulations. Increased lipophilicity allows partition into deep brain compartments, but also confers poor aqueous solubility. The oxidative lability, which is needed for the lock-in mechanism, and the hydrolytic instability, which releases the modifier functions or the active drug, combine to limit the shelf-life of the CDS. Cyclodextrins may provide a possible solution. They are torus-shaped oligosaccharides that contain various numbers of a-1,4-linked glucose units (6, 7, 8 for a-, β-, γ-cyclodextrin, respectively). The number of units determines the size of a cone-like cavity into which various compounds can enter and form stable complexes [244–247]. Formation of the host-guest inclusion complex generally involves only the simple spatial entrapment of the guest molecule without the formation of covalent bonds [248].

Complexation with 2-hydroxypropyl-β-cyclodextrin (HPβCD) provided important advances in the formulation of E_2-CDS [249]. HPβCD was selected based on its low toxicity observed using various administration routes and based on the fact that alkylation or hydroxyalkylation of the glucose oligomer can disrupt hydrogen bonding and provide increased water solubility for the compound and its inclusion complexes [250–253]. Indeed, the aqueous solubility of E_2-CDS was enhanced about 250 000-fold in a 40% (w/v) HPβCD solution (from 65.8 ng ml^{-1} to 16.36 mg ml^{-1}). The phase solubility diagram indicated that a 1:1 complex forms at low HPβCD concentration, but a 1:2 complex occurs at higher HPβCD concentration. The stability of E_2-CDS was also significantly increased, allowing formulation in acceptable form. The rate of ferricyanide-mediated oxidation, a good indicator of oxidative stability, was decreased about 10-fold and shelf life was increased about 4-fold, as indicated by t_{90} and t_{50} values in a temperature range of 23–80 °C [249]. The cyclodextrin complex even provided better distribution by preventing retention of the solid material precipitated in the lung. Promising results were also obtained for testosterone-CDS [111], lomustine-CDS [174], and BP-CDS [254]. For the latter, aqueous solubility was enhanced about 70 000-fold in a 20% (w/v) HPβCD solution (from 50–70 ng ml^{-1} to 4.2 mg ml^{-1}), and stability was also increased.

20.6
Molecular Packaging

The CDS approach also has been extended to achieve successful brain deliveries of a Leu-enkephalin analogs [200, 201], thyreotropin-releasing hormone (TRH) analogs [202–208], and kyotorphin analogs [209, 210]. Neuropeptides, peptides that act on the brain or spinal cord, represent the largest class of transmitter substance and have considerable therapeutic potential [209, 255, 256]. The number of identified endogenous peptides is growing exponentially, and recently, almost 600 sequences for neuropeptides and related peptides have been listed [257]. However, the delivery of peptides through the BBB is an even more complex problem than the delivery of other drugs, because they can be rapidly inactivated by ubiquitous peptidases [31, 32, 258–260]. Therefore, for a successful delivery, three issues have to be solved simultaneously: (1) enhance passive transport by increasing the lipophilicity, (2) ensure enzymatic stability to prevent premature degradation, and (3) exploit the lock-in mechanism to provide targeting.

The solution obtained as a generalization of the CDS approach is a complex molecular packaging strategy, in which the peptide unit is part of a bulky molecule dominated by lipophilic-modifying groups that direct BBB penetration and prevent recognition by peptidases [261]. Such a brain-targeted packaged peptide delivery system contains the following major components: the redox targetor (T), a spacer function (S), consisting of strategically used amino acids to ensure timely removal of the charged targetor from the peptide, the peptide itself (P), and a bulky lipophilic moiety (L) attached through an ester bond or sometimes through a C-terminal adjuster (A) at the carboxyl terminal to enhance lipid solubility and to disguise the peptide nature of the molecule.

Obviously, to achieve delivery and sustained activity with such complex systems, it is very important that the designated enzymatic reactions take place in a specific sequence. On delivery, the first step must be the conversion of the targetor to allow for lock-in. This must be followed by removal of the L function to form a direct precursor of the peptide that is still attached to the charged targetor. Subsequent cleavage of the targetor-spacer moiety finally releases the active peptide.

20.6.1
Leu-Enkephalin Analogs

Since the discovery of endogenous peptides with opiate activity [262], many potential roles have been identified for these substances [263, 264]. Possible therapeutic applications could extend from the treatment of opiate dependence to numerous other CNS-mediated dysfunctions. Opioid peptides are implicated in regulating neuroendocrine activity [265–267], motor behavior [268], seizure threshold [269, 270], feeding and drinking [271], mental disorders [263, 272], cognitive functions [263, 273], cardiovascular functions [274, 275], gastrointestinal functions [263, 276], and sexual behavior and development [264]. However, analgesia remains their best known role; and it is commonly used to evaluate endogenous opioid peptide activity. Whereas native opioid peptides could alter pain threshold after intracerebroventricular (i.c.v.) dosing, they were ineffective after systemic injection. It was reasonable, therefore, to attempt a brain-targeted CDS approach.

The first successful delivery and sustained release was achieved for Tyr-D-Ala-Gly-Phe-D-Leu (DADLE), an analog of Leu-enkephalin, a naturally occurring linear pentapeptide (Tyr-Gly-Gly-Phe-Leu) that binds to opioid receptors [200]. In rat brain tissue samples collected 15 min after i.v. CDS administration, electrospray ionization mass spectrometry clearly showed the presence of the locked-in T^+-D form at an estimated concentration of 600 pmol g^{-1} of tissue. The same ion was absent from the sample collected from the control animals treated with the vehicle solution only. To optimize this delivery strategy, an effective synthetic route was established for peptide CDSs and the role of the spacer and the lipophilic functions was investigated [201]. Four different CDSs were synthesized by a segment-coupling method. Their i.v. injection produced a significant and long-lasting (>5 h) response in rats monitored by classic [277] tail-flick latency test. Compared with the delivered peptide itself, the packaged peptide and the intermediates formed during the sequential metabolism had weak affinity to opioid receptors. The antinociceptive effect was naloxone-reversible and methylnaloxonium-irreversible. Because quaternary derivatives such as methylnaloxonium are unable to cross the BBB, these techniques [201, 278] are used to prove that central opiate receptors are responsible for mediating the induced analgesia. It could be concluded, therefore, that the peptide CDS successfully delivered, retained, and released the peptide in the brain.

The efficacy of the CDS package was strongly influenced by modifications of the spacer (S) and lipophilic (L) components, proving that they have important roles in determining molecular properties and the timing of the metabolic sequence. The bulkier cholesteryl group used as L showed a better efficacy than the smaller 1-adamantaneethyl, but the most important factor for manipulating the rate of peptide release and the pharmacological activity turned out to be the spacer (S) function: proline as spacer produced more potent analgesia than alanine.

20.6.2
TRH Analogs

A similar strategy was used [202–206] for the CNS delivery of a thyreotropin-releasing hormone (TRH; Glp-His-Pro) analog (Glp-Leu-Pro) in which histidine (His^2) was replaced with leucine (Leu^2) to dissociate CNS effects from thyreotropin-releasing activity [279]. Such compounds can increase extracellular acetylcholine (ACh) levels, accelerate ACh turnover, improve memory and learning, and reverse the reduction in high affinity choline uptake induced by lesions of the medial septal cholineric neurons [280, 281]. Therefore, as reviewed in the literature [282, 283], these peptides are potential agents for treating motor neuron diseases [284], spinal cord trauma [285], or neurodegenerative disorders such as Alzheimer's disease [286], and may also have a potential cytoprotective role [287].

Because the peptide that has to be delivered has no free -NH_2 and -COOH termini, a precursor sequence (Gln-Leu-Pro-Gly) that will ultimately produce the final TRH analog was packaged for delivery. Therefore, two additional steps had to be included in the metabolic sequence: one where the C-terminal adjuster glycine is cleaved by peptidyl glycine α-amidating monooxygenase (PAM) to form the ending prolinamide, and one where the N-terminal pyroglutamyl is formed from glutamine by glutaminyl cyclase [202]. In summary, the following sequential biotransformation has to take place after delivery to the brain: first, lock-in by oxidation of the dihydrotrigonellyl (T) to the corresponding pyridinium salt, then removal of cholesterol, oxidation of glycine by peptidyl glycine α-amidating monoxygenase to prolineamide, cleavage of the targetor-spacer combination, and finally, cyclization of glutamine by glutaminyl cyclase to pyroglutamate.

Selection of a suitable spacer proved also important for the efficacy of TRH-CDSs as measured by the decrease in the barbiturate-induced sleeping time in mice, an interesting and well documented effect of this neuropeptide [288, 289]. After i.v. administration of 15 µmol kg^{-1}, the Leu^2-TRH analog itself produced only a slight decrease of 17±7%, compared to the vehicle control. Equimolar doses of the CDS compounds with Pro, Pro-Pro, and Pro-Ala spacers produced statistically significant ($P<0.05$) decreases of 47±6%, 51±7%, and 56±4%, respectively [204]. A Nva^2-TRH analog, in which norvaline replaced leucine was also used in such an approach with a Pro-Pro spacer, and very similar results were obtained with a dose of 20.7 µmol kg^{-1} in Swiss-Webster mice [206]. Treatment with a Leu^2-TRH-CDS also significantly improved memory-related behavior in a passive avoidance paradigm in rats bearing bilateral fimbrial lesions without altering thyroid function [203].

Because the amide precursor was found to be susceptible to deamination by TRH-deaminase, a side-reaction process competitive with the designed-in cleavage of the spacer-Gln peptide bond, another analog of TRH was also examined [205]. The Pro-Gly C terminal was replaced by pipecolic acid (Pip) to obtain a TRH analog that is active in the carboxylate form and, hence, to eliminate the need for formation of the amide. The CDS was prepared by a 5+1 segment-coupling approach and the cholesteryl ester of pipecolic acid was prepared using either Fmoc (fluorenylmethyloxycarbonyl) or Boc (*tert*-butyloxycarbonyl) as protection. Pharmacological activity was assessed by antagonism of barbiturate-induced sleeping time in mice; and the pipecolic acid analog was found to be even somewhat more effective than the previous package.

20.6.3
Kyotorphin Analogs

Finally, a kyotorphin analog (Tyr-Lys; YK) that has activity similar to kyotorphin itself [290] was also successfully delivered [209, 210]. Kyotorphin (Tyr-Arg) is an endogenous neuropeptide that exhibits analgesic action through the release of endogenous enkephalin; and its analgesic activity is about four times larger than that of Met-enkephalin [291]. It also has other effects that may prove useful, such as seizure protective effects [292] or effects on nitric oxide synthase (NOS) [293–295]. Because this peptide contains a free amine residue, and because preliminary studies indicated that such ionizable functions might prevent

References

1 Regier, D. A., Boyd, J. H., Burke, J. D. Jr., Rae, D. S., Myers, J. K., Kramer, M., Robins, L. N., George, L. K., Karno, M., Locke, B. Z. **1988**, *Arch. Gen. Psychiat.* 45, 977–986.
2 Miller, G. **2002**, *Science* 297, 1116–1118.
3 Pardridge, W. M. **2002**, *Drug Discov. Today* 7, 5–7.
4 Crone, C., Thompson, A. M. **1970**, in *Capillary Permeability. The Transfer of Molecules and Ions between Capillary Blood and Tissue*, ed. C. Crone, N. A. Lassen, Munksgaard, Copenhagen, pp. 447–453.
5 Oldendorf, W. H. **1974**, *Proc. Soc. Exp. Biol. Med.* 147, 813–816.
6 Rapoport, S. I. **1976**, *Blood-Brain Barrier in Physiology and Medicine*, Raven Press, New York.
7 Bradbury, M. **1979**, *The Concept of a Blood-Brain Barrier*, Wiley, New York.
8 Bodor, N., Brewster, M. E. **1983**, *Pharmacol. Ther.* 19, 337–386.
9 Fenstermacher, J. D., Rapoport, S. I. **1984**, in *Microcirculation*, Part 2, Vol. 4, ed. E. M. Renkin, C. C. Michel, American Physiology Society, Bethesda, pp. 969–1000.
10 Goldstein, G. W., Betz, A. L. **1986**, *Sci. Am.* 255, 74–83.
11 Bradbury, M. W. B. (ed.) **1992**, *Physiology and Pharmacology of the Blood-Brain Barrier*, Springer, Berlin.
12 Begley, D. J. **1996**, *J. Pharm. Pharmacol.* 48, 136–146.
13 Schlossauer, B., Steuer, H. **2002**, *Curr. Med. Chem. CNS Agents* 2, 175–186.
14 Betz, A. L., Goldstein, G. W. **1984**, in *Structural Elements of the Nervous System*, vol. 7, ed. A. Lajtha, Plenum, New York, pp. 465–484.
15 Pardridge, W. M. **1991**, *Peptide Drug Delivery to the Brain*, Raven Press, New York.
16 Abbott, N. J., Bundgaard, M., Cserr, H. F. **1986**, In *The Blood-Brain Barrier in Health and Disease*, ed. A. J. Suckling, M. G. Rumsby, M. W. B. Bradbury, Ellis Horwood, Chichester, pp. 52–72.
17 Lo, E. H., Singhal, A. B., Torchilin, V. P., Abbott, N. J. **2001**, *Brain Res. Rev.* 38, 140–148.
18 Smith, Q. R. **1989**, In *Implications of the Blood-Brain Barrier and its Manipulation*, ed. E. A. Neuwelt, Plenum, New York, pp. 85–113.
19 Stewart, P. A., Tuor, U. I. **1994**, *J. Comp. Neurol.* 340, 566–576.
20 Siegal, T., Zylber-Katz, E. **2002**, *Clin. Pharmacokinet.* 41, 171–186.
21 Ehrlich, P. **1885**, *Das Sauerstoff Bedürfnis des Organismus. Eine farbenanalytische Studie*, Hirschwald, Berlin.
22 Janzer, R. C., Raff, M. C. **1987**, *Nature* 325, 253–257.
23 Spatz, H. **1933**, *Arch. Psychiat. Nervenkrank.* 101, 267–358.
24 Crone, C. **1961**, *The Diffusion of Some Organic Nonelectrolytes from Blood to Brain Tissue* (in Danish, English summary), Munksgaard, Copenhagen.
25 Reese, T. S., Karnovsky, M. J. **1967**, *J. Cell Biol.* 34, 207–217.
26 Brightman, M. W., Reese, T. S. **1969**, *J. Cell Biol.* 40, 648–677.

27 Crone, C. **1986**, in *The Blood-Brain Barrier in Health and Disease,* ed. A. J. Suckling, M. G. Rumsby, M. W. B. Bradbury, Ellis Horwood, Chichester, pp. 17–40.
28 Minn, A., Ghersi-Egea, J. F., Perrin, R., Leininger, B., Siest, G. **1991**, *Brain Res. Brain Res. Rev.* 16, 65–82.
29 Ghersi-Egea, J. F., Leninger-Muller, B., Suleman, G., Siest, G., Minn, A. **1994**, *J. Neurochem.* 62, 1089–1096.
30 Graff, C. L., Pollack, G. M. **2004**, *Curr. Drug Metab.* 5, 95–108.
31 Brownlees, J., Williams, C. H. **1993**, *J. Neurochem.* 60, 793–803.
32 Witt, K. A., Gillespie, T. J., Huber, J. D., Egleton, R. D., Davis, T. P. **2001**, *Peptides* 22, 2329–2343.
33 Krogh, A. **1946**, *Proc. R. Soc. Lond. B* 133, 140–200.
34 Crone, C. **1965**, *J. Physiol.* 181, 103–113.
35 Lund-Andersen, H. **1979**, *Physiol. Rev.* 59, 305–352.
36 Audus, K. L., Chikhale, P. J., Miller, D. W., Thompson, S. E., Borchardt, R. T. **1992**, *Adv. Drug Res.* 23, 1–64.
37 Tamai, I., Tsuji, A. **2000**, *J. Pharm. Sci.* 89, 1371–1388.
38 Taylor, E. M. **2002**, *Clin. Pharmacokinet.* 41, 81–92.
39 Barza, M. **1993**, *Eur. J. Clin. Microbiol. Infect. Dis.* 12, S31–S35.
40 Deguchi, Y., Nozawa, K., Yamada, S., Yokoyama, Y., Kimura, R. **1997**, *J. Pharmacol. Exp. Ther.* 280, 551–560.
41 Suzuki, H., Terasaki, T., Sugiyama, Y. **1997**, *Adv. Drug Deliv. Rev.* 25, 257–285.
42 Penzotti, J. E., Lamb, M. L., Evensen, E., Grootenhuis, P. D. **2002**, *J. Med. Chem.* 45, 1737–1740.
43 Juliano, R. L., Ling, V. **1976**, *Biochim. Biophys. Acta* 455, 152–162.
44 Lin, J. H., Yamazaki, M. **2003**, *Clin. Pharmacokinet.* 42, 59–98.
45 Cordon-Cardo, C., O'Brien, J. P., Casals, D., Rittman-Grauer, L., Biedler, J. L., Melamed, M. R., Bertino, J. R. **1989**, *Proc. Natl Acad. Sci. USA* 86, 695–698.
46 Tsuji, A., Tamai, I., Sakata, A., Tenda, Y., Terasaki, T. **1993**, *Biochem. Pharmacol.* 46, 1096–1099.
47 Schinkel, A. H., Smit, J. J. M., van Tellingen, O., Beijnen, J. H., Wagenaar, E., van Deemter, L., Mol, C. A. A. M., van der Walk, M. A., Robanus-Maadag, E. C., te Riele, H. P. J., et al. **1994**, *Cell* 77, 491–502.
48 Schinkel, A. H., Wagenaar, E., Mol, C. A. A. M., van Deemter, L. **1996**, *J. Clin. Invest.* 97, 2517–2524.
49 Tsuji, A., Tamai, I. **1997**, *Adv. Drug Deliv. Rev.* 25, 287–298.
50 van Asperen, J., Mayer, U., van Tellingen, O., Beijnen, J. H. **1997**, *J. Pharm. Sci.* 86, 881–884.
51 Schinkel, A. H. **1999**, *Adv. Drug Deliv. Rev.* 36, 179–194.
52 Stein, W. D. **1997**, *Physiol. Rev.* 77, 545–590.
53 Habgood, M. D., Begley, D. J., Abbott, N. J. **2000**, *Cell. Mol. Neurobiol.* 20, 231–253.
54 Thorne, R. G., Frey, W. H. 2nd **2001**, *Clin. Pharmacokinet.* 40, 907–946.
55 Filmore, D. **2002**, *Mod. Drug Discov.* 5, 22–27.
56 Bodor, N., Buchwald, P. **2003**, *Am. J. Drug Deliv.* 1, 13–26.

57 Mertsch, K., Maas, J. **2002**, *Curr. Med. Chem. CNS Agents* 2, 187–201.
58 Overton, E. **1897**, *Z. Phys. Chem.* 22, 189–209.
59 Overton, E. **1901**, *Studien über die Narkose, zugleich ein Beitrag zur allgemeiner Pharmakologie*, Gustav Fischer, Jena.
60 Meyer, H. **1899**, *Arch. Exp. Pathol. Pharmakol.* 42, 109–118.
61 Meyer, H. **1901**, *Arch. Exp. Pathol. Pharmakol.* 46, 338–346.
62 Collander, R. **1951**, *Acta Chem. Scand.* 5, 774–780.
63 Buchwald, P., Bodor, N. **1998**, *Curr. Med. Chem.* 5, 353–380.
64 Franks, N.P., Lieb, W.R. **1978**, *Nature* 274, 339–342.
65 Young, R.C., Mitchell, R.C., Brown, T.H., Ganellin, C.R., Griffiths, R., Jones, M., Rana, K.K., Saunders, D., Smith, I.R., Sore, N.E., et al. **1988**, *J. Med. Chem.* 31, 656–671.
66 Abraham, M.H., Chadha, H.S., Mitchell, R.C. **1994**, *J. Pharm. Sci.* 83, 1257–1268.
67 Lombardo, F., Blake, J.F., Curatolo, W.J. **1996**, *J. Med. Chem.* 39, 4750–4755.
68 Crivori, P., Cruciani, G., Carrupt, P.-A., Testa, B. **2000**, *J. Med. Chem.* 43, 2204–2216.
69 Iyer, M., Mishra, R., Han, Y., Hopfinger, A.J. **2002**, *Pharm. Res.* 19, 1611–1621.
70 Lobell, M., Molnár, L., Keserü, G.M. **2003**, *J. Pharm. Sci.* 92, 360–370.
71 Bodor, N., Buchwald, P. **1999**, *Adv. Drug Deliv. Rev.* 36, 229–254.
72 van de Waterbeemd, H., Smith, D.A., Beaumont, K., Walker, D.K. **2001**, *J. Med. Chem.* 44, 1313–1333.
73 Lewis, D.F.V., Dickins, M. **2002**, *Drug Discov. Today* 7, 918–925.
74 Smith, D., Schmid, E., Jones, B. **2002**, *Clin. Pharmacokinet.* 41, 1005–1019.
75 Lin, J.H., Lu, A.Y. **1997**, *Pharmacol. Rev.* 49, 403–449.
76 Lombardo, F., Obach, R.S., Shalaeva, M.Y., Gao, F. **2004**, *J. Med. Chem.* 47, 1242–1250.
77 Buchwald, P., Bodor, N., *Drugs Future* 27, 577–588.
78 King, L.A., Moffat, A.C. **1981**, *Lancet* 1, 387–388.
79 Hansch, C., Björkroth, J.P., Leo, A. **1987**, *J. Pharm. Sci.* 76, 663–687.
80 Gupta, S.P. **1989**, *Chem. Rev.* 89, 1765–1800.
81 Abraham, M.H. **2004**, *Eur. J. Med. Chem.* 39, 235–240.
82 Sippl, W. **2002**, *Curr. Med. Chem. CNS Agents* 2, 212–227.
83 de Boer, A.G., Gaillard, P.J. **2002**, *Curr. Med. Chem. CNS Agents* 2, 203–209.
84 Kerns, E.H. **2001**, *J. Pharm. Sci.* 90, 1838–1858.
85 Golden, P.L., Pollack, G.M. **1998**, *Biopharm. Drug Dispos.* 19, 263–272.
86 Sadeque, A.J., Wandel, C., He, H., Shah, S., Wood, A.J. **2000**, *Clin. Pharmacol. Ther.* 68, 231–237.
87 Savolainen, J., Edwards, J.E., Morgan, M.E., McNamara, P.J., Anderson, B.D. **2002**, *Drug Metab. Dispos.* 30, 479–482.
88 Fellner, S., Bauer, B., Miller, D.S., Schaffrik, M., Fankhänel, M., Spruss, T., Bernhardt, G., Graeff, C., Färber, L., Gschaidmeier, H., et al. **2002**, *J. Clin. Invest.* 110, 1309–1318.

89 Bodor, N., Buchwald, P. **2003**, in *Drug Discovery and Drug Development, Burger's Medicinal Chemistry and Drug Discovery*, Vol. 2, 6th edn, ed. D. J. Abraham, Wiley, New York, pp. 533–608.
90 Hunt, C. A., MacGregor, R. D., Siegel, R. A. **1986**, *Pharm. Res.* 3, 333–344.
91 Bodor, N., Brewster, M. E. **1991**, in *Targeted Drug Delivery*, Vol. 100, ed. R. L. Juliano, Springer, Berlin, pp. 231–284.
92 Ishikura, T., Senou, T., Ishihara, H., Kato, T., Ito, T. **1995**, *Int. J. Pharm.* 116, 51–63.
93 Pop, E. **1997**, *Curr. Med. Chem.* 4, 279–294.
94 Somogyi, G., Nishitani, S., Nomi, D., Buchwald, P., Prokai, L., Bodor, N. **1998**, *Int. J. Pharm.* 166, 15–26.
95 Somogyi, G., Buchwald, P., Nomi, D., Prokai, L., Bodor, N. **1998**, *Int. J. Pharm.* 166, 27–35.
96 Bodor, N., Farag, H. H., Barros, M. D. C., Wu, W.-M., Buchwald, P. **2002**, *J. Drug Target.* 10, 63–71.
97 Rydström, J., Hoek, J. B., Ernster, L. **1976**, in *The Enzymes*, Vol. 13, ed. P. D. Boyer, Academic Press, New York.
98 Hoek, J. B., Rydström, J. **1988**, *Biochem. J.* 254, 1–10.
99 Bodor, N., Brewster, M. E., Kaminski, J. J. **1990**, *J. Mol. Struct.* 206, 315–334.
100 Brewster, M. E., Estes, K. S., Perchalski, R., Bodor, N. **1988**, *Neurosci. Lett.* 87, 277–282.
101 Bodor, N., Roller, R. G., Selk, S. J. **1978**, *J. Pharm. Sci.* 67, 685–687.
102 Palomino, E., Kessel, D., Horwitz, J. P. **1989**, *J. Med. Chem.* 32, 622–625.
103 Bodor, N., Farag, H. H., Brewster, M. E. **1981**, *Science* 214, 1370–1372.
104 Bodor, N., Farag, H. H. **1983**, *J. Med. Chem.* 26, 313–318.
105 Bodor, N., Abdelalim, A. M. **1985**, *J. Pharm. Sci.* 74, 241–245.
106 Tedjamulia, M. L., Srivastava, P. C., Knapp, F. F. Jr. **1985**, *J. Med. Chem.* 28, 1574–1580.
107 Bodor, N., Nakamura, T., Brewster, M. E. **1986**, *Drug Des. Deliv.* 1, 51–64.
108 Bodor, N., Farag, H. H., Polgar, P. **2001**, *J. Pharm. Pharmacol.* 53, 889–894.
109 Bodor, N., Farag, H. H. **1984**, *J. Pharm. Sci.* 73, 385–389.
110 Bodor, N., Abdelalim, A. M. **1986**, *J. Pharm. Sci.* 75, 29–35.
111 Anderson, W. R., Simpkins, J. W., Brewster, M. E., Bodor, N. **1988**, *Drug Des. Del.* 2, 287–298.
112 Brewster, M. E., Estes, K. S., Bodor, N. **1986**, *Pharm. Res.* 3, 278–285.
113 Brewster, M. E., Deyrup, M., Czako, K., Bodor, N. **1990**, *J. Med. Chem.* 33, 2063–2065.
114 Brewster, M. E., Estes, K. S., Bodor, N. **1988**, *Ann. NY Acad. Sci.* 529, 298–300.
115 Anderson, W. R., Simpkins, J. W., Brewster, M. E., Bodor, N. **1989**, *Neuroendocrinology* 50, 9–16.
116 Siegal, T., Soti, F., Biegon, A., Pop, E., Brewster, M. E. **1997**, *Pharm. Res.* 14, 672–675.
117 Pop, E., Brewster, M. E., Prókai-Tátrai, K., Bodor, N. **1994**, *Org. Prep. Proced. Int.* 26, 379–382.

118 Bodor, N., McCornack, J., Brewster, M. E. **1987**, *Int. J. Pharm.* 35, 47–59.
119 Simpkins, J. W., McCornack, J., Estes, K. S., Brewster, M. E., Shek, E., Bodor, N. **1986**, *J. Med. Chem.* 29, 1809–1812.
120 Anderson, W. R., Simpkins, J. W., Brewster, M. E., Bodor, N. **1987**, *Pharmacol. Biochem. Behav.* 27, 265–271.
121 Estes, K. S., Brewster, M. E., Simpkins, J. W., Bodor, N. **1987**, *Life Sci.* 40, 1327–1334.
122 Estes, K. S., Brewster, M. E., Bodor, N. S. **1988**, *Life Sci.* 42, 1077–1084.
123 Anderson, W. R., Simpkins, J. W., Brewster, M. E., Bodor, N. **1988**, *Endocr. Res.* 14, 131–148.
124 Brewster, M. E., Estes, K. S., Bodor, N. **1988**, *J. Med. Chem.* 31, 244–249.
125 Mullersman, G., Derendorf, H., Brewster, M. E., Estes, K. S., Bodor, N. **1988**, *Pharm. Res.* 5, 172–177.
126 Howes, J., Bodor, N., Brewster, M. E., Estes, K., Eve, M. **1988**, Abstract 181, *J. Clin. Pharmacol.* 28, 951.
127 Simpkins, J. W., Anderson, W. R., Dawson, R., Jr., Seth, E., Brewster, M., Estes, K. S., Bodor, N. **1988**, *Physiol. Behav.* 44, 573–580.
128 Simpkins, J. W., Anderson, W. R., Dawson, R. Jr., Bodor, N. **1989**, *Pharm. Res.* 6, 592–600.
129 Sarkar, D. K., Friedman, S. J., Yen, S. S. C., Frautschy, S. A. **1989**, *Neuroendocrinology* 50, 204–210.
130 Millard, W. J., Romano, T. M., Bodor, N., Simpkins, J. W. **1990**, *Pharm. Res.* 7, 1011–1018.
131 Brewster, M. E., Simpkins, J. W., Bodor, N. **1990**, *Rev. Neurosci.* 2, 241–285.
132 Estes, K. S., Dewland, P. M., Brewster, M. E., Derendorf, H., Bodor, N. **1991**, *Pharm. Z. Wiss.* 136, 153–158.
133 Brewster, M. E., Bartruff, M. S. M., Anderson, W. R., Druzgala, P. J., Bodor, N., Pop, E. **1994**, *J. Med. Chem.* 37, 4237–4244.
134 Estes, K. S., Brewster, M. E., Bodor, N. **1994**, *Adv. Drug Deliv. Rev.* 14, 167–175.
135 Simpkins, J. W., Rajakumar, G., Zhang, Y.-Q., Simpkins, C. E., Greenwald, D., Yu, C. J., Bodor, N., Day, A. L. **1997**, *J. Neurosurg.* 87, 724–730.
136 Rabbani, O., Panickar, K. S., Rajakumar, G., King, M. A., Bodor, N., Meyer, E. M., Simpkins, J. W. **1997**, *Exp. Neurol.* 146, 179–186.
137 Bodor, N., Buchwald, P. **2002**, *Drug Discov. Today* 7, 766–774.
138 Tapfer, M. K., Sebestyen, L., Kurucz, I., Horvath, K., Szelenyi, I., Bodor, N. **2004**, *Pharmacol. Biochem. Behav.* 77, 423–429.
139 Pop, E., Wu, W.-M., Shek, E., Bodor, N. **1989**, *J. Med. Chem.* 32, 1774–1781.
140 Pop, E., Wu, W.-M., Bodor, N. **1989**, *J. Med. Chem.* 32, 1789–1795.
141 Wu, W.-M., Pop, E., Shek, E., Bodor, N. **1989**, *J. Med. Chem.* 32, 1782–1788.
142 Wu, W.-M., Pop, E., Shek, E., Clemmons, R., Bodor, N. **1990**, *Drug Des. Delivery* 7, 33–43.
143 Brewster, M. E., Deyrup, M., Seyda, K., Bodor, N. **1991**, *Int. J. Pharm.* 68, 215–229.

144 Venkatraghavan, V., Shek, E., Perchalski, R., Bodor, N. **1986**, *Pharmacologist* 28, 145.
145 Rand, K., Bodor, N., El-Koussi, A., Raad, I., Miyake, A., Houck, H., Gildersleeve, N. **1986**, *J. Med. Virol.* 20, 1–8.
146 El-Koussi, A., Bodor, N. **1987**, *Drug Des. Deliv.* 1, 275–283.
147 Canonico, P. G., Kende, M., Gabrielsen, B. **1988**, *Adv. Virus Res.* 35, 271–312.
148 Bhagrath, M., Sidwell, R., Czako, K., Seyda, K., Anderson, W., Bodor, N., Brewster, M. E. **1991**, *Antiviral Chem. Chemother.* 2, 265–286.
149 Deyrup, M., Sidwell, R., Little, R., Druzgala, P., Bodor, N., Brewster, M. E. **1991**, *Antiviral Chem. Chemother.* 2, 337–355.
150 Morin, K. W., Wiebe, L. I., Knaus, E. E. **1993**, *Carbohydr. Res.* 249, 109–116.
151 Morin, K. W., Knaus, E. E., Wiebe, L. I. **1994**, *J. Labelled Comp. Radiopharm.* 35, 205–207.
152 Balzarini, J., Morin, K. W., Knaus, E. E., Wiebe, L. I., De Clercq, E. **1995**, *Gene Therapy* 2, 317–322.
153 Pop, E., Anderson, W., Vlasak, J., Brewster, M. E., Bodor, N. **1992**, *Int. J. Pharm.* 84, 39–48.
154 Brewster, M. E., Little, R., Venkatraghavan, V., Bodor, N. **1988**, *Antiviral Res.* 9, 127.
155 Little, R., Bailey, D., Brewster, M. E., Estes, K. S., Clemmons, R. M., Saab, A., Bodor, N. **1990**, *J. Biopharm. Sci.* 1, 1–18.
156 Brewster, M. E., Anderson, W., Bodor, N. **1991**, *J. Pharm. Sci.* 80, 843–845.
157 Pop, E., Brewster, M. E., Anderson, W. R., Bodor, N. **1992**, *Med. Chem. Res.* 2, 457–466.
158 Pop, E., Liu, Z. Z., Vlasak, J., Anderson, W., Brewster, M. E., Bodor, N. **1993**, *Drug Deliv.* 1, 143–149.
159 Mizrachi, Y., Rubinstein, A., Harish, Z., Biegon, A., Anderson, W. R., Brewster, M. E. **1995**, *AIDS* 9, 153–158.
160 Brewster, M. E., Anderson, W. R., Webb, A. I., Pablo, L. M., Meinsma, D., Moreno, D., Derendorf, H., Bodor, N., Pop, E. **1997**, *Antimicrob. Agents Chemother.* 41, 122–128.
161 Torrence, P. T., Kinjo, J., Lesiak, K., Balzarini, J., DeClerq, E. **1988**, *FEBS Lett.* 234, 135–140.
162 Gogu, S. R., Aggarwal, S. K., Rangan, S. R. S., Agrawal, K. C. **1989**, *Biochem. Biophys. Res. Commun.* 160, 656–661.
163 Aggarwal, S. K., Gogu, S. R., Rangan, S. R. S., Agrawal, K. C. **1990**, *J. Med. Chem.* 33, 1505–1510.
164 Lupia, R. H., Ferencz, N., Aggarwal, S. K., Agrawal, K. C., Lertora, J. J. L. **1990**, *Clin. Res.* 38, 15A.
165 Lupia, R. H., Ferencz, N., Lertora, J. J. L., Aggarwal, S. K., George, W. J., Agrawal, K. C. **1993**, *Antimicrob. Agents Chemother.* 37, 818–824.
166 Gallo, J., Boubinot, F., Doshi, D., Etse, J., Bhandti, V., Schinazi, R., Chu, C. K. **1989**, *Pharm. Res.* 6, S161.

167 Chu, C. K., Bhadti, V. S., Doshi, K. J., Etse, J. T., Gallo, J. M., Boudinot, F. D., Schinazi, R. F. **1990**, *J. Med. Chem.* 33, 2188–2192.

168 Gallo, J. M., Etse, J. T., Doshi, K. J., Boudinot, F. D., Chu, C. K. **1991**, *Pharm. Res.* 8, 247–253.

169 Palomino, E., Kessel, D., Horwitz, J. P. **1992**, *Nucleosides Nucleotides* 11, 1639–1649.

170 Torrence, P. F., Kinjo, J., Khamnei, S., Greig, N. H. **1993**, *J. Med. Chem.* 36, 529–537.

171 Brewster, M. E., Raghavan, K., Pop, E., Bodor, N. **1994**, *Antimicrob. Agents Chemother.* 38, 817–823.

172 Camplo, M., Charvet Faury, A. S., Borel, C., Turin, F., Hantz, O., Trabaud, C., Niddam, V., Mourier, N., Graciet, J. C., Chermann, J. C., et al. **1996**, *Eur. J. Med. Chem.* 31, 539–546.

173 Raghavan, K., Shek, E., Bodor, N. **1987**, *Anticancer Drug Des.* 2, 25–36.

174 Raghavan, K., Loftsson, T., Brewster, M. E., Bodor, N. **1992**, *Pharm. Res.* 9, 743–749.

175 Bodor, N., Venkatraghavan, V., Winwood, D., Estes, K., Brewster, M. E. **1989**, *Int. J. Pharm.* 53, 195–208.

176 Bodor, N., Simpkins, J. W. **1983**, *Science* 221, 65–67.

177 Bodor, N., Farag, H. H. **1983**, *J. Med. Chem.* 26, 528–534.

178 Simpkins, J. W., Bodor, N., Enz, A. **1985**, *J. Pharm. Sci.* 74, 1033–1036.

179 Omar, F. A., Farag, H. H., Bodor, N. **1994**, *J. Drug Target.* 2, 309–316.

180 Carelli, V., Liberatore, F., Scipione, L., Impicciatore, M., Barocelli, E., Cardellini, M., Giorgioni, G. **1996**, *J. Controlled Release* 42, 209–216.

181 Anderson, W. R., Simpkins, J. W., Woodard, P. A., Winwood, D., Stern, W. C., Bodor, N. **1987**, *Psychopharmacology* 92, 157–163.

182 Woodard, P. A., Winwood, D., Brewster, M. E., Estes, K. S., Bodor, N. **1990**, *Drug Des. Deliv.* 6, 15–28.

183 Kourounakis, A., Bodor, N., Simpkins, J. **1996**, *Int. J. Pharm.* 141, 239–250.

184 Kourounakis, A., Bodor, N., Simpkins, J. **1997**, *J. Pharm. Pharmacol.* 49, 1–9.

185 Shek, E., Murakami, T., Nath, C., Pop, E., Bodor, N. S. **1989**, *J. Pharm. Sci.* 78, 837–843.

186 Pop, E., Bodor, N. **1992**, *Epilepsy Res.* 13, 1–16.

187 Boddy, A. V., Zhang, K., Lepage, F., Tombret, F., Slatter, J. G., Baillie, T. A., Levy, R. H. **1991**, *Pharm. Res.* 8, 690–697.

188 Yiu, S. H., Knaus, E. E. **1996**, *J. Med. Chem.* 39, 4576–4582.

189 Pop, E., Prókai-Tátrai, K., Anderson, W., Lin, J.-L., Brewster, M. E., Bodor, N. **1990**, *Eur. J. Pharmacol.* 183, 1909–1919.

190 Prókai-Tátrai, K., Pop, E., Anderson, W., Lin, J.-L., Brewster, M. E., Bodor, N. **1991**, *J. Pharm. Sci.* 80, 255–261.

191 Brewster, M. E., Robledo-Luiggi, C., Miyakeb, A., Pop, E., Bodor, N. **1989**, in *Novel Approaches to the Treatment of Alzheimer's Disease*, ed. E. M. Meyer, J. W. Simpkins, J. Yamamoto, Plenum Press, New York, pp. 173–183.

192 Pop, E., Prókai-Tátrai, K., Scott, J.D., Brewster, M.E., Bodor, N. **1990**, *Pharm. Res.* 7, 658–664.
193 Pop, E., Soti, F., Anderson, W.R., Panetta, J.A., Estes, K.S., Bodor, N.S., Brewster, M.E. **1996**, *Int. J. Pharm.* 140, 33–44.
194 Bailey, D., Perchalski, R., Bhagrath, M., Shek, E., Winwood, D., Bodor, N. **1991**, *J. Biopharm. Sci.* 2, 205–218.
195 Phelan, M.J., Bodor, N. **1989**, *Pharm. Res.* 6, 667–676.
196 Pop, E., Anderson, W., Prókai-Tátrai, K., Vlasak, J., Brewster, M.E., Bodor, N. **1992**, *Med. Chem. Res.* 2, 16–21.
197 Anderson, W., Pop, E., Lee, S.-K., Bodor, N., Brewster, M. **1991**, *Med. Chem. Res.* 1, 74–79.
198 Pop, E., Anderson, W., Prókai-Tátrai, K., Brewster, M.E., Fregly, M., Bodor, N. **1990**, *J. Med. Chem.* 33, 2216–2221.
199 Pop, E., Prókai-Tátrai, K., Brewster, M.E., Bodor, N. **1994**, *Org. Prep. Proced. Int.* 26, 687–690.
200 Bodor, N., Prokai, L., Wu, W.-M., Farag, H.H., Jonnalagadda, S., Kawamura, M., Simpkins, J. **1992**, *Science* 257, 1698–1700.
201 Prokai-Tatrai, K., Prokai, L., Bodor, N. **1996**, *J. Med. Chem.* 39, 4775–4782.
202 Prokai, L., Ouyang, X.-D., Wu, W.-M., Bodor, N. **1994**, *J. Am. Chem. Soc.* 116, 2643–2644.
203 Prokai, L., Ouyang, X., Prokai-Tatrai, K., Simpkins, J.W., Bodor, N. **1998**, *Eur. J. Med. Chem.* 33, 879–886.
204 Prokai, L., Prokai-Tatrai, K., Ouyang, X., Kim, H.-S., Wu, W.-M., Zharikova, A., Bodor, N. **1999**, *J. Med. Chem.* 42, 4563–4571.
205 Yoon, S.-H., Wu, J., Wu, W.-M., Prokai, L., Bodor, N. **2000**, *Bioorg. Med. Chem.* 9, 1059–1063.
206 Wu, J., Yoon, S.-H., Wu, W.-M., Bodor, N. **2002**, *J. Pharm. Pharmacol.* 54, 945–950.
207 Prokai-Tatrai, K., Perjési, P., Zharikova, A.D., Li, X., Prokai, L. **2002**, *Bioorg. Med. Chem. Lett.* 12, 2171–2174.
208 Prokai, L., Prokai-Tatrai, K., Zharikova, A.D., Nguyen, V., Perjesi, P., Stevens, S.M. Jr. **2004**, *J. Med. Chem.* 47, 6025–6033.
209 Bodor, N., Buchwald, P. **1998**, *Chem. Ber.* 34, 36–40.
210 Chen, P., Bodor, N., Wu, W.-M., Prokai, L. **1998**, *J. Med. Chem.* 41, 3773–3781.
211 Prokai, L., Prokai-Tatrai, K., Bodor, N. **2000**, *Med. Res. Rev.* 20, 367–416.
212 Williams, C.L., Stancel, G.M. **1996**, in *Goodman & Gilman's Pharmacological Basis of Therapeutics*, ed. J.G. Hardman, L.E. Limbird, McGraw-Hill, New York, pp. 1411–1440.
213 Kaplan, N.M. **1978**, *Annu. Rev. Med.* 29, 31–40.
214 Fotherby, K. **1985**, *Contraception* 31, 367–394.
215 Lobo, R.A. **1995**, *Am. J. Obstet. Gynecol.* 173, 982–989.
216 Yager, J.D., Liehr, J.G. **1996**, *Annu. Rev. Pharmacol. Toxicol.* 36, 203–232.
217 Beral, V., Banks, E., Reeves, G., Appleby, P. **1999**, *J. Epidemiol. Biostat.* 4, 191–210.

218 Wells, G., Herrington, D. M. **1999**, *Drugs Aging* 15, 419–422.
219 Meade, T. W., Vickers, M. R. **1999**, *J. Epidemiol. Biostat.* 4, 165–190.
220 Rossouw, J. E., Anderson, G. L., Prentice, R. L., LaCroix, A. Z., Kooperberg, C., Stefanick, M. L., Jackson, R. D., Beresford, S. A. A., Howard, B. V., Johnson, K. C., et al. **2002**, *JAMA* 288, 321–333.
221 Hays, J., Ockene, J. K., Brunner, R. L., Kotchen, J. M., Manson, J. E., Patterson, R. E., Aragaki, A. K., Shumaker, S. A., Brzyski, R. G., LaCroix, A. Z., et al. **2003**, *N. Engl. J. Med.* 348, 1839–1854.
222 Anderson, G. L., Limacher, M., Assaf, A. R., Bassford, T., Beresford, S. A., Black, H., Bonds, D., Brunner, R., Brzyski, R., Caan, B., et al. **2004**, *JAMA* 291, 1701–1712.
223 Aldridge, S. **1998**, *Magic Molecules: How Drugs Work*, Cambridge University Press, Cambridge.
224 Greendale, G. A., Lee, N. P., Arriola, E. R. **1999**, *Lancet* 353, 571–580.
225 McEwen, B. S., Alves, S. E. **1999**, *Endocr. Rev.* 20, 279–307.
226 Cyr, M., Calon, F., Morissette, M., Grandbois, M., Callier, S., Di Paolo, T. **2000**, *Curr. Pharm. Des.* 6, 1287–1312.
227 Shepherd, J. E. **2001**, *J. Am. Pharm. Assoc.* 41, 221–228.
228 Sharpe, R. M. **1998**, *Trends Endocrin. Metabol.* 9, 371–377.
229 Upton, G. V. **1984**, *J. Reprod. Med.* 29, 71–79.
230 Maggi, A., Perez, J. **1985**, *Life Sci.* 37, 893–906.
231 García-Segura, L. M., Chowen, J. A., Párducz, A., Naftolin, F. **1994**, *Progr. Neurobiol.* 44, 279–307.
232 Ohkura, T., Isse, K., Akazawa, K., Hamamoto, M., Yaoi, Y., Hagino, N. **1994**, *Endocr. J.* 41, 361–371.
233 Tang, M.-X., Jacobs, D., Stern, Y., Marder, K., Schofield, P., Gurland, B., Andrews, H., Mayeux, R. **1996**, *Lancet* 346, 429–432.
234 Jacobs, D. M., Tang, M.-X., Stern, Y., Sano, M., Marder, K., Bell, K. L., Schofield, P., Dooneief, G., Gurland, B., Mayeux, R. **1998**, *Neurology* 50, 368–373.
235 Yaffe, K., Sawaya, G., Lieberburg, I., Grady, D. **1998**, *JAMA* 279, 688–695.
236 Asthana, S., Craft, S., Baker, L. D., Raskind, M. A., Birnbaum, R. S., Lofgreen, C. P., Veith, R. C., Plymate, S. R. **1999**, *Psychoneuroendocrinology* 24, 657–677.
237 Costa, M. M., Reus, V. I., Wolkowitz, O. M., Manfredi, F., Lieberman, M. **1999**, *Biol. Psychiatry* 46, 182–188.
238 Janocko, L., Lamer, J. M., Hochberg, R. B. **1984**, *Endocrinology* 114, 1180–1186.
239 Shepherd, J. E. **2002**, *J. Am. Pharm. Assoc.* 42, 479–488.
240 Laumann, E. O., Paik, A., Rosen, R. C. **1999**, *JAMA* 281, 537–544.
241 Dennerstein, L., Randolph, J., Taffe, J., Dudley, E., Burger, H. **2002**, *Fertil. Steril.* 77[Suppl. 4], S42–S48.
242 Juhász, A., Howes, J., Mantelle, L., Halabi, A., Bodor, N. **2003**, *Proc. N. Am. Menopause Soc.* 14.
243 Juhász, A., Howes, J., Halabi, A., Bodor, N. **2004**, *Proc. N. Am. Menopause Soc.* 15.

244 Saenger, W. **1980**, *Angew. Chem. Int. Ed. Engl.* 19, 344–362.
245 Szejtli, J. **1982**, *Cyclodextrins and Their Inclusion Complexes*, Akadémiai Kiadó, Budapest.
246 Pagington, J. S. **1987**, *Chem. Ber.* 23, 455–458.
247 Szejtli, J. **1998**, *Chem. Rev.* 98, 1743–1753.
248 Buchwald, P. **2002**, *J. Phys. Chem. B* 106, 6864–6870.
249 Brewster, M. E., Estes, K. E., Loftsson, T., Perchalski, R., Derendorf, H., Mullersman, G., Bodor, N. **1988**, *J. Pharm. Sci.* 77, 981–985.
250 Pitha, J., Pitha, J. **1985**, *J. Pharm. Sci.* 74, 987–990.
251 Pitha, J., Milecki, J., Fales, H., Pannell, L., Uekama, K. **1986**, *Int. J. Pharm.* 29, 73–82.
252 Yoshida, A., Arima, H., Uekama, K., Pitha, J. **1988**, *Int. J. Pharm.* 46, 217–222.
253 Brewster, M. E., Estes, K. S., Bodor, N. **1990**, *Int. J. Pharm.* 59, 231–243.
254 Pop, E., Loftsson, T., Bodor, N. **1991**, *Pharm. Res.* 8, 1044–1049.
255 Kastin, A. J., Ehrensing, R. H., Banks, W. A., Zadina, J. E. **1987**, in *Neuropeptides and Brain Function*, vol. 72, ed. E. R. de Kloet, V. M. Wiegant, D. de Wied, Elsevier, Amsterdam, pp. 223–234.
256 Nemeroff, C. B. E. **1988**, *Neuropeptides in Psychiatric and Neurological Disorders*, John Hopkins University Press, Baltimore.
257 Hoyle, C. H. V. **1996**, *Neuropeptides. Essential Data*, Wiley, Chichester.
258 Banks, W. A., Kastin, A. J. **1990**, *Adv. Exp. Med. Biol.* 274, 59–69.
259 Ermisch, A., Brust, P., Kretzschmar, R., Rühle, H.-J. **1993**, *Physiol. Rev.* 73, 489–527.
260 Oliyai, R., Stella, V. J. **1993**, *Annu. Rev. Pharmacol. Toxicol.* 33, 521–544.
261 Bodor, N., Prokai, L. **1995**, in *Peptide-Based Drug Design: Controlling Transport and Metabolism*, ed. M. Taylor, G. Amidon, American Chemical Society, Washington, D.C., pp. 317–337.
262 Hughes, J., Smith, T. V., Kosterlitz, H. W., Fothergill, L., Morgan, B. A., Morris, H. R. **1975**, *Nature* 258, 577–579.
263 Olson, G. A., Olson, R. D., Kastin, A. J. **1985**, *Peptides* 6, 769–791.
264 Olson, G. A., Olson, R. D., Kastin, A. J. **1992**, *Peptides* 13, 1247–1287.
265 Bicknell, R. J. **1985**, *J. Endocrinol.* 107, 437–446.
266 Millan, M. J., Herz, A. **1985**, *Int. Rev. Neurobiol.* 26, 1–83.
267 Yen, S. S. C., Quigley, M. E., Reid, R. L., Ropert, J. F., Cetel, N. S. **1985**, *Am. J. Obstet. Gynecol.* 152, 485–493.
268 Sandyk, R. **1985**, *Life Sci.* 37, 1655–1663.
269 Frenk, H. **1983**, *Brain Res. Rev.* 6, 197–210.
270 Tortella, F. C., Long, J. B., Holaday, J. W. **1985**, *Brain Res.* 332, 174–178.
271 Baile, C. A., McLaughlin, C. L., Della-Fera, M. A. **1986**, *Physiol. Rev.* 66, 172–234.
272 Schmauss, C., Emrich, H. M. **1985**, *Biol. Psychiatry* 20, 1211–1231.
273 Izquierdo, I., Netto, C. A. **1985**, *Ann. NY Acad. Sci.* 444, 162–177.
274 Holaday, J. W. **1983**, *Annu. Rev. Pharmacol. Toxicol.* 23, 541–594.
275 Johnson, M. W., Mitch, W. E., Wilcox, C. S. **1985**, *Prog. Cardiovasc. Dis.* 27, 435–450.

276 Porreca, F., Burks, T. F. **1983**, *J. Pharmacol. Exp. Ther.* 227, 22–27.
277 D'Amour, F. E., Smith, D. L. **1941**, *J. Pharmacol. Exp. Ther.* 72, 74–79.
278 Kastin, A. J., Pearson, M. A., Banks, W. A. **1991**, *Pharmacol. Biochem. Behav.* 40, 771–774.
279 Szirtes, T., Kisfaludy, L., Pálosi, É., Szporny, L. **1984**, *J. Med. Chem.* 27, 741–745.
280 Santori, E. M., Schmidt, D. E. **1980**, *Regul. Pept.* 1, 69–74.
281 Itoh, Y., Ogasawara, T., Mushiroi, T., Yamazaki, A., Ukai, Y., Kimura, K. **1994**, *J. Pharmacol. Exp. Therap.* 271, 884–890.
282 Kelly, J. A. **1995**, *Essays Biochem.* 30, 133–149.
283 Bennett, G. W., Ballard, T. M., Watson, C. D., Fone, K. C. **1997**, *Exp. Gerontol.* 32, 451–469.
284 Yarbrough, G. G. **1983**, *Life Sci.* 33, 111–118.
285 Faden, A. I., Vink, R., McIntosh, T. K. **1989**, *Ann. NY Acad. Sci.* 553, 380–384.
286 Mellow, A. H., Sunderland, T., Cohen, R. M., Lawlor, B. A., Hill, J. L., Newhouse, P. A., Cohen, M. R., Murphy, D. L. **1989**, *Psychopharmacology* 98, 403–407.
287 Horita, A., Carino, M. A., Zabawska, J., Lai, H. **1989**, *Peptides* 10, 121–124.
288 Horita, A., Carino, M. A., Smith, J. R. **1976**, *Pharmacol. Biochem. Behav.* 5[Suppl. 1], 111–116.
289 Horita, A., Carino, M. A., Lai, H. **1986**, *Fed. Proc. Fed. Am. Soc. Exp. Biol.* 45, 795.
290 Rolka, K., Oslisok, E., Krupa, J., Kruszinsky, M., Baran, L., Przegalinska, E., Kupryszewski, G. **1983**, *Pol. J. Pharmacol. Pharm.* 35, 473–480.
291 Takagi, H., Shiomi, H., Ueda, H., Amano, H. **1979**, *Nature* 282, 410–412.
292 Godlevsky, L. S., Shandra, A. A., Mikhaleva, I. I., Vastyanov, R. S., Mazarati, A. M. **1995**, *Brain Res. Bull.* 37, 223–226.
293 Arima, T., Kitamura, Y., Nishiya, T., Takagi, H., Nomura, Y. **1996**, *Neurosci. Lett.* 212, 1–4.
294 Arima, T., Kitamura, Y., Nishiya, T., Taniguchi, T., Takagi, H., Nomura, Y. **1997**, *Neurochem. Int.* 30, 605–611.
295 Summy-Long, J. Y., Bui, V., Gestl, S., Koehler-Stec, E., Liu, H., Terrell, M. L., Kadekaro, M. **1998**, *Brain Res. Bull.* 45, 395–403.
296 Zincke, T. **1903**, *Ann. Chem.* 330, 361–374.
297 Lettré, H., Haede, W., Ruhbaum, E. **1953**, *Ann. Chem.* 579, 123–132.
298 Génisson, Y., Marazano, C., Mehmandoust, M., Gnecco, D., Das, B. C. **1992**, *Synlett* 1992, 431–434.
299 Rivier, J., Vale, W., Monahan, M., Ling, N., Burgus, R. **1972**, *J. Med. Chem.* 15, 479–482.
300 Wallace, B. M., Lasker, J. S. **1993**, *Science* 260, 912–913.

21
Drug Delivery to the Brain by Internalizing Receptors at the Blood-Brain Barrier

Pieter J. Gaillard, Corine C. Visser, and Albertus (Bert) G. de Boer

21.1
Introduction

Drug delivery to the brain is still a major obstacle for the effective treatment of brain diseases. However, effective and safe ways of targeting drugs to the entire brain have been developed by applying delivery systems directed to endogenous receptor-mediated uptake mechanisms present at the cerebral capillaries. Such systems have been shown to be effective up to primates, but not yet in humans. In this chapter we focus on the well characterized transferrin system and the more recently described systems that use the low-density lipoprotein receptor-related protein 1 (LRP1) receptor, the LRP2 receptor (also known as megalin and glycoprotein 330) or the diphtheria toxin receptor (DTR, which is the membrane-bound precursor of heparin-binding epidermal growth factor-like growth factor, HB-EGF). The possibilities and limitations of these systems together with their future human application are discussed.

For many diseases of the brain, such as Alzheimer's disease, Parkinson's disease, stroke, depression, schizophrenia, epilepsy and migraine headache, the drugs on the market are far from being ideal, let alone curative, drugs. A significant part of the problem is the poor blood-brain barrier penetration of most of the drugs in development against neuronal targets for treatment of these disorders. This includes approximately 98% of the small molecules and nearly 100% of large molecules, such as recombinant proteins or gene-based medicines [1]. Therefore, much effort is put towards delivery and targeting of drugs to the brain. As a result, various drug delivery and targeting strategies are currently being developed. We discuss in this chapter the possibilities to deliver drugs to the brain via the vascular route by receptor-mediated uptake/transcytosis into the (human) brain.

21.2
Blood-Brain Barrier Transport Opportunities

The development and the anatomy of the blood-brain barrier have been treated elsewhere in this book (see Chapters 1 to 3). Here we want to state that the blood-brain barrier, located in the brain capillaries, can be considered as an interface between blood and brain regulating the homeostasis of the brain and limiting the transport of many compounds into the brain (Fig. 21.1). Next to the blood-brain barrier, there is the blood-cerebrospinal fluid barrier (B-CSFB) that is located in the choroid plexus. Although choroid plexus epithelial cells have microvilli, it can be stated that, with respect to intravascular administration of drugs, the surface area of the blood-brain barrier is several decades larger than the B-CSFB. For an extensive review of the concept and the properties of the blood-brain barrier, we refer to preceeding chapters in this book (Chapters 4 to 9) and the literature [2–9]. In addition, important contributions have been made to the topic of drug delivery to the brain (see Chapters 18 to 20); [10–12]. Here we want to focus on receptor-mediated drug delivery to the brain.

One of the most important properties of the brain capillary endothelial cells (BCEC) of the blood-brain barrier with respect to drug transport to the brain is that they have specific characteristics, such as tight junctions, which prevent paracellular transport of small and large (water-soluble) compounds from blood to the brain [9, 13, 14]. Furthermore, transcellular transport from blood to brain is limited as a result of low vesicular transport, high metabolic activity and a

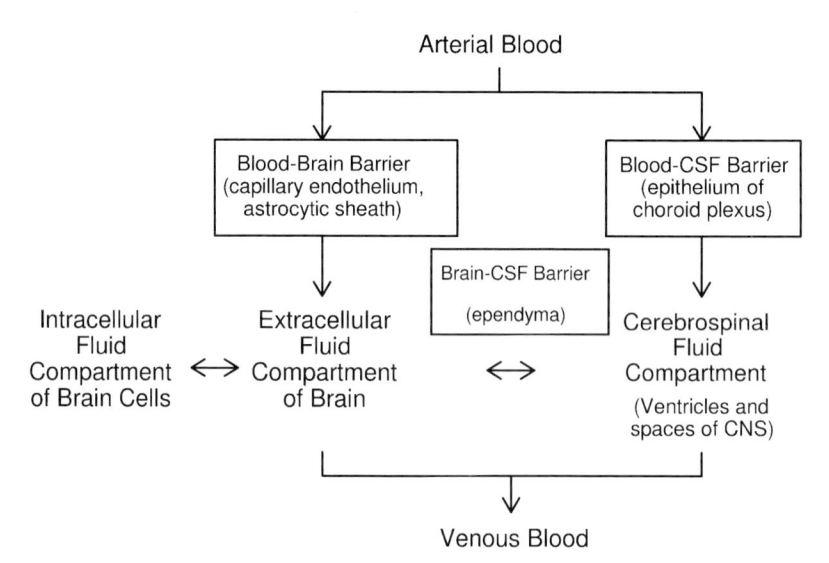

Fig. 21.1 Schematic representation of the blood-brain barrier, the blood-cerebrospinal fluid (CSF) barrier and the brain-CSF barrier. The blood-brain barrier has the largest surface area and is, therefore, considered to be the most important influx barrier for solutes to reach the brain.

lack of fenestrae [5]. These specific characteristics of the blood-brain barrier are induced and maintained by the (endfeet of) astrocytes (see Chapters 8 and 9), surrounding the BCEC [9, 15], as well as by neuronal endings, which can directly innervate the BCEC [5, 16]. Pericytes also play a role at the blood-brain barrier, as they share the continuous capillary basement membrane with the BCEC. Pericytes cover about 20–30% of the cerebral capillary surface, so they do not really constitute a physical barrier for the movement of solutes across the BCEC. However, their phagocytotic activity does form an additional blood-brain barrier property [8, 17]. Because of these complex interactions between cell types, as well as the dynamic regulation of the blood-brain barrier properties (e.g. receptor expression, formation of tight junctions) this barrier is considered to be an organ protecting the brain [18].

Potential neurotoxic endogenous and exogenous compounds are excluded from the brain by the blood-brain barrier. In addition, the blood-brain barrier maintains ion homeostasis and transports essential nutrients, such as glucose, amino acids, purines, nucleosides, peptides and proteins [4, 6]. There are several influx mechanisms which can be divided into active or passive blood-brain barrier transport mechanisms. Passive diffusion depends on lipophilicity and molecular weight [7]. Furthermore, the ability of a compound to form hydrogen bonds limits its diffusion through the blood-brain barrier [19]. In general, Lipinski's "rule of five" and the Abraham's equation can be used to predict the passive transport of a drug molecule across the blood-brain barrier (see Chapters 18 and 20) [20, 21]. Transport of hydrophilic compounds via the paracellular route is limited, while lipophilic drugs smaller than 400–600 Da may freely enter the brain via the transcellular route. Other blood-brain barrier transport systems can be divided into adsorptive-, carrier- or receptor-mediated uptake/transcytosis.

Adsorptive-mediated transcytosis is initiated by the binding of polycationic substances (such as most cell-penetrating peptides) to negative charges on the plasma membrane [22, 23]. This process does not involve specific plasma membrane receptors. Upon binding of the cationic compound to the plasma membrane, endocytosis occurs, followed by the formation of endosomes. Indeed, several drugs have been described to enter the brain via this mechanism [24, 25]. However, vesicular transport is actively down-regulated in the blood-brain barrier to protect the brain from nonspecific exposure to polycationic compounds. Therefore, forcing drugs to enter the brain by adsorptive-mediated transcytosis may go against the neuroprotective barrier function, as was recently illustrated for anionic and cationic nanoparticles, which were shown to disrupt the blood-brain barrier [26].

Carrier-mediated transcytosis is used for the transcytosis of nutrients, such as glucose, amino acids and purine bases [27–29]. At least eight different nutrient transport systems have been identified, which each transport a group of nutrients of the same structure. Carrier-mediated transcytosis is substrate-selective and the transport rate is dependent on the degree of occupation of the carrier [29]. In addition, carriers stay in the membrane of the cell, which means that the size of the drugs that can be transported must closely mimic the endogenous ligand if they will be taken up and transported into the brain.

The focus of this chapter is on receptor-mediated transcytosis by the transferrin receptor [31, 32], the insulin receptor [30] and the transporters for low-density lipoprotein [33], leptin [34] and insulin-like growth factors [35]. These systems enable larger molecules, such as peptides, proteins and genes to specifically enter the brain. Besides many influx mechanisms, several efflux mechanisms exist at the blood-brain barrier as well. The best known is P-glycoprotein (Pgp). Pgp is a transmembrane protein commonly regarded to be located at the apical membrane of the BCEC, although some recent publications describe (conditional) Pgp expression on astrocytes and astrocyte endfeet as well [36]. It has a high affinity for a wide range of compounds, including cytotoxic anticancer drugs, antibiotics, hormones and HIV protease inhibitors [37]. Other multidrug resistance (MDR) efflux mechanisms at the blood-brain barrier include the MDR-related protein (MRP), such as MRP1, -2, -5 and -6 [38]. In addition, many other transporters are present at the blood-brain barrier, such as the organic anion transporter (influx, efflux), the organic cation transport system (influx) and the nucleoside transporter system (influx) [18, 39].

In conclusion, research over the years has shown that the blood-brain barrier is a dynamic organ, which combines restricted diffusion to the brain for endogenous and exogenous compounds with specialized transport mechanisms for essential nutrients.

21.3
Drug Delivery and Targeting Strategies to the Brain

Drug delivery to the brain can be achieved via several methods, including invasive, pharmaco-chemical or physiological strategies.

Invasive brain drug delivery strategies, such as direct intracerebral injections of, e.g., slow-release products allow only for local delivery (Fig. 21.2 B). This may be effective for drug delivery to localized brain tumors, but not for the administration of drugs against more widespread diseases. Another invasive method is intracerebroventricular (ICV) or intrathecal drug infusion, in which a drug is directly injected into the CSF. However, the CSF is completely drained into the venous circulation and drugs still have to cross the ependymal brain-CSF barrier. As a result, the infused drug has minimal access to the parenchyma by diffusion (Fig. 21.2 C) [40]. In general, invasive strategies are not effective for drug delivery to the whole brain, but only to a localized part of the brain.

The advantage of the vascular route is the widespread diffusion of the infused drug across the whole brain [11]. This can be explained by the large surface area of the human blood-brain barrier (approx. 20 m^2). In addition, approximately each neuron has its own brain capillary for oxygen supply as well as the supply of other nutrients (see also Fig. 21.2 A). In fact, every cubic centimeter of cortex comprises the amazing sum of 1 km of blood vessel. This means that the vascular route is a very promising one for drug targeting and delivery to the brain. However, drug delivery through blood-brain barrier disruption by osmotic im-

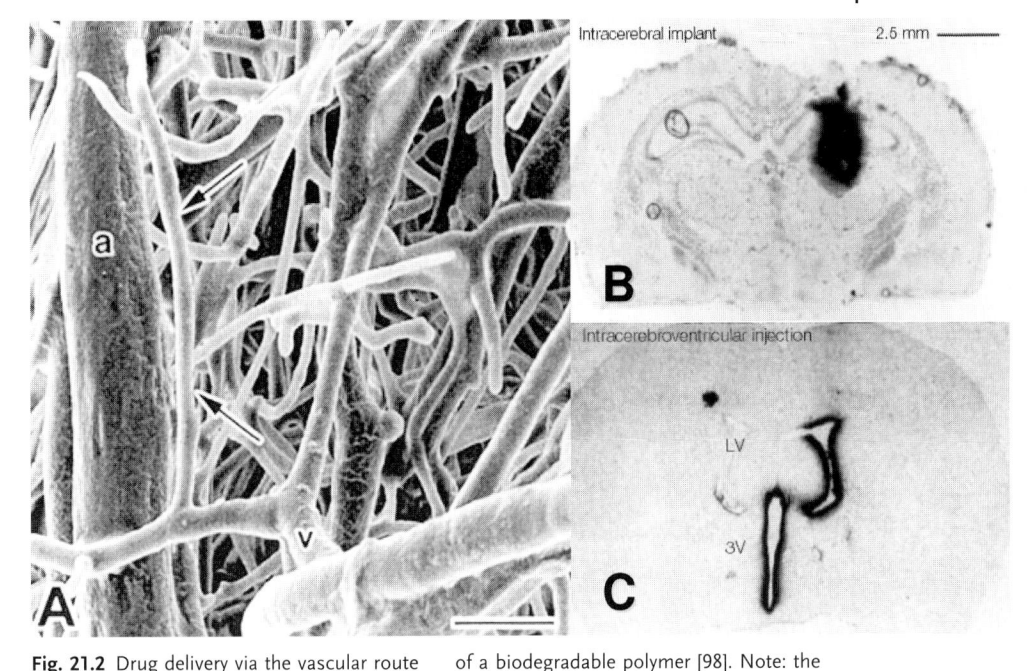

Fig. 21.2 Drug delivery via the vascular route will enable widespread distribution of the drug to each single neuron within the brain. (A) Scanning electron micrograph of a vascular cast of a mouse brain [97]. a = artery; v = vein. Note: the bar indicates 25 μm, which is about the size of a single neuron. (Reprinted with permission from Lippincott Williams & Wilkins). (B) Minimal diffusion of [^{125}I]-nerve growth factor (NGF) after intracerebral implantation of a biodegradable polymer [98]. Note: the bar indicates 2.5 mm, which is the size of the implant. (Reprinted with permission from Elsevier). (C) ICV injection of [^{125}I]-brain-derived neurotrophic factor (BDNF) [99]. Note that the neurotrophin does not distribute into the brain beyond the ipsilateral ependymal surface. (Reprinted with permission from Elsevier).

balance or vaso-active compounds, though effective in reaching the entire brain, has the disadvantage that neurons may be damaged permanently due to unwanted blood components entering the brain [41]. In contrast, physiological drug delivery strategies rather aim to use endogenous transport mechanisms at the blood-brain barrier, such as carrier- or receptor-mediated transcytosis. The focus of this chapter is on receptor-mediated transcytosis technologies that have successfully been employed to target drugs to the brain.

21.4
Receptor-Mediated Drug Delivery to the Brain

In general receptor-mediated transcytosis occurs in three steps: receptor-mediated endocytosis of the compound at the luminal (blood) side, movement through the endothelial cytoplasm and exocytosis of the drug at the abluminal (brain) side of the brain capillary endothelium [8]. Upon receptor-ligand internalization, clathrin-coated vesicles are formed, which are approximately 120 nm in diameter [42, 43]. Therefore, receptor-mediated transcytosis allows the specific delivery of larger drug molecules or drug carrying particles (like liposomes or nanoparticles) to the brain.

21.5
Transferrin Receptor

The most widely characterized receptor-mediated transcytosis system for the targeting of drugs to the brain is the transferrin receptor (TfR). The TfR is a transmembrane glycoprotein consisting of two 90-kDa subunits. A disulfide bridge links these subunits and each subunit can bind one transferrin molecule [31]. The TfR is expressed mainly on hepatocytes, erythrocytes, intestinal cells, monocytes and endothelial cells of the blood-brain barrier [44, 45]. Furthermore, in the brain the TfR is expressed on choroid plexus epithelial cells and neurons [31]. The TfR mediates cellular uptake of iron bound to transferrin.

Drug targeting to the TfR can be achieved either by using the endogenous ligand, transferrin or by using an antibody directed against the TfR (OX-26 anti-rat TfR). Each of these targeting vectors has its advantages and disadvantages. It was shown that transferrin conjugated to horseradish peroxidase (HRP) was very efficiently taken up by blood-brain barrier endothelial cells in vitro (Fig. 21.3) [46]. However, for the TfR, in vivo application is limited due to high endogenous concentrations of transferrin in plasma and the likely overdosing with iron when one tries to displace the endogenous transferrin with exogenously applied transferrin-containing systems. Nevertheless, recent studies in our group have shown that liposomes tagged with transferrin are suitable for drug delivery to blood-brain barrier endothelial cells in vitro, even in the presence of serum (Visser et al., submitted). OX-26 does not bind to the transferrin-binding site and is therefore not displaced by endogenous transferrin.

The TfR is responsible for iron transport to the brain. So far, the intracellular trafficking of transferrin and OX-26 upon internalization via the TfR has not yet been elucidated. Some literature reports suggest transcytosis of transferrin across the BCEC, while others claim endocytosis of transferrin, followed by an intracellular release of iron and a subsequent return of apo-transferrin to the apical side of the BCEC [32, 47, 48]. Moos and Morgan [31] have shown that the transcytosis of iron exceeds the transcytosis of transferrin across the blood-brain barrier, supporting the second theory. Furthermore, these authors have proposed

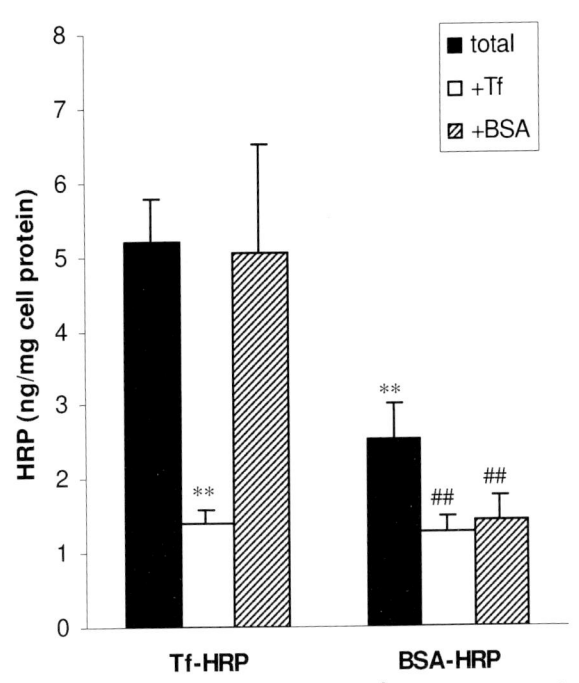

Fig. 21.3 Association of Tf-HRP (3 µg ml^{-1}, 2 h, 37 °C) was reduced by a 500-fold excess of nonconjugated Tf, but not by an equal concentration of BSA. Total association of BSA-HRP (3 µg ml^{-1}, 2 h, 37 °C) was approximately 50% of the association of Tf-HRP and was inhibited by both Tf and BSA. ** $P<0.01$, Tf-HRP + Tf, or total BSA-HRP vs total Tf-HRP. ## $P<0.01$, BSA-HRP + Tf or BSA-HRP + BSA vs total BSA-HRP, one-way ANOVA. (From [46], with permission from Elsevier).

a new theory in which the TfR-transferrin complex is transcytosed to the basolateral side of the BCEC, where transferrin remains bound to the TfR, but iron is released into the brain extracellular fluid [49]. Subsequently, apo-transferrin bound to the TfR will recycle back to the apical side of the blood-brain barrier. This theory is supported by data from Zhang and Pardridge [47], who found a 3.5-fold faster efflux from brain to blood of apo-transferrin than holo-transferrin. In addition, Deane et al. [50] recently illustrated that free iron is also rapidly taken up by brain capillaries and subsequently released into the brain extracellular fluid and CSF at controlled moderate to slow rates.

The mechanism of transcytosis of OX-26 is not yet fully elucidated either. Pardridge and colleagues have shown efficient drug targeting and delivery to the brain in vivo by applying OX-26, e.g. [32, 51–53]. In contrast, Broadwell et al. [54] have shown that both transferrin and OX-26 are able to cross the blood-brain barrier, but that the transcytosis of transferrin is more efficient. Furthermore, Moos and Morgan [55] have shown that OX-26 mainly accumulates in the BCEC and not in the postcapillary compartment. In addition, iron deficiency did not increase OX-26 uptake in rats. Both our data and literature reports show that iron deficiency causes an increase in TfR expression [44, 56, 57]. Therefore, it is expected that the uptake of OX-26 would also increase. The data by Moos and Morgan [55] suggest that OX-26 transcytosis might result from a high-affinity accumulation by the BCEC, followed by a nonspecific exocytosis at the basolateral side of the BCEC. In addition, these authors found a periventricular localisation of OX-26, which suggests that OX-26 probably is also transported across the blood-CSF barrier.

Although the mechanism of transcytosis of transferrin and OX-26 may not yet be fully elucidated, it is important to realize that drug delivery to the brain via the TfR is efficient. By these means, vasoactive intestinal peptide (VIP), brain derived neurotrophic factor (BDNF), basic fibroblast growth factor (bFGF), epidermal growth factor (EGF), peptide nucleic acids and pegylated immunoliposomes containing plasmid DNA encoding beta-galactosidase, tyrosine hydroxylase and short hairpin RNAs have all been made available to the brain [58]. However, OX-26 is an antibody against the rat TfR and does not bind to the human TfR, making it impossible to transfer this technology to the clinic. Moreover, rat antibodies cause immunogenic reactions in humans, unless they are humanized. The preparation of humanized or chimeric antibodies is difficult and in some cases may lead to a loss of affinity for the target receptor [59]. In addition, one can argue that the administration of antibodies directed against such an important uptake mechanism involved in iron homeostasis poses a risk for human application.

Still, preferably, a targeting vector directed to the TfR would be small, nonimmunogenic and should initialize internalization of the TfR upon binding. Xu et al. [60] used a single-chain antibody Fv fragment against the human TfR, which was tagged with a lipid anchor for insertion into a liposomal bilayer. The molecular weight of this antibody fragment, including the lipid anchor was approximately 30 kDa. In addition, Lee et al. [61] used a phage display technique to find small peptide ligands for the human TfR. They obtained a 7- and a 12-mer peptide that bind to a different binding site than transferrin and are intenalized by the TfR. Although these small peptides can also exert immunogenic reactions in humans, they are promising ligands for drug targeting to the human TfR on the blood-brain barrier.

21.6
Insulin Receptor

Another widely characterized, classic, receptor-mediated transcytosis system for the targeting of drugs to the brain is the insulin receptor. Again, just as for the TfR system, Pardridge and colleagues have predominantly documented the use of the insulin receptor for the targeted delivery of drugs to the brain.

The insulin receptor is a large 300-kDa protein and is a heterotetramer of two extracellular alpha and two transmembrane beta subunits. Each beta chain contains a tyrosine kinase activity in its cytosolic extension. The alpha and beta subunits are coded by a single gene and are joined by disulfide bonds to form a cylindrical structure. Primarily, insulin binds and changes the shape of the receptor to form a tunnel, to allow entry of molecules such as glucose into the cells. The insulin receptor is a tyrosine kinase receptor and induces a complex cellular response by phosphorylating proteins on their tyrosine residues. The binding of a single insulin molecule into a pocket created by the two alpha chains effects a conformational change in the insulin receptor so that the beta chains approximate one another and carry out transphosphorylation on tyrosine residues. This autophosphorylation is necessary for the receptor to internalize into endosomes. The endosomal system has been shown to be a site where insulin signaling is regulated, but where the degradation of endosomal insulin also occurs. Most of the insulin is degraded, but less so in endothelial cells [62], whereas the receptors are largely recycled to the cell surface. Endocytosis is not necessary for insulin action, but probably is important for removing insulin from the cell surface so that the target cell for insulin responds to the hormone in a time-limited fashion. This endocytotic mechanism of the insulin receptor has been exploited for the targeting of drugs to the brain.

As for transferrin, the in vivo application of insulin as the carrier protein is limited, mainly due to the high insulin concentrations needed and the resulting lethal overdosing with insulin. Therefore, drug or gene delivery to, for instance, rhesus monkeys is performed with the mouse 83–14 mAb that binds to the exofacial epitope on the alpha subunit of the human insulin receptor. The mAb has a blood-brain barrier permeability surface area (PS) product in the primate that is 9-fold greater than mouse mAbs to the human TfR [63]. By using this mAb, Pardridge and coworkers have successfully made radiolabeled amyloid-β peptide$_{1-40}$ (Aβ $_{1-40}$), serving as a diagnostic probe for Alzheimer's disease, and pegylated immunoliposomes containing plasmid DNA encoding β-galactosidase available to the brain of primates [58].

Unfortunately, the 83–14 mAb cannot be used in humans owing to immunogenic reactions to this mouse protein. However, genetically engineered, effective, forms of the mAb have now been produced, which may allow for drug and gene delivery to the human brain [64]. Still, one can argue that the administration of antibodies directed against such an important mechanism involved in glucose homeostasis poses a risk for human application.

21.7
LRP1 and LRP2 Receptors

During the past few years, the low-density lipoprotein receptor related protein-1 (LRP1) and LRP2 (also known as megalin or glycoprotein 330) receptors have been exploited to target drugs to the brain in a fashion similar to that used for the transferrin and insulin receptors. Both LRP1 and LRP2 receptors belong to the structurally closely related cell surface LDL receptor gene family. Both receptors are multifunctional, multiligand scavenger and signaling receptors. A large number of substrates are shared between the two receptors, like lipoprotein lipase (LPL), a_2-macroglobulin (a_2M), receptor-associated protein (RAP), lactoferrin, tissue- and urokinase-type plasminogen activator (tPA/uPA), plasminogen activator inhibitor (PAI-1) and tPA/uPA:PAI-1 complexes. More specific ligands for the LRP1 receptor are described to be, e.g., melanotransferrin (or P97), thrombospondin 1 and 2, hepatic lipase, factor VIIa/tissue factor pathway inhibitor (TFPI), factor VIIIa, factor IXa, $A\beta_{1-40}$, amyloid-β precursor protein (APP), C1 inhibitor, complement C3, apolipoproteinE (apoE), *Pseudomonas* exotoxin A, HIV-1 Tat protein, rhinovirus, matrix metalloproteinase 9 (MMP-9), MMP-13 (collagenase-3), sphingolipid activator protein (SAP), pregnancy zone protein, antithrombin III, heparin cofactor II, a_1-antitrypsin, heat-shock protein 96 (HSP-96) and platelet-derived growth factor (PDGF, mainly involved in signaling) [65–69], where apolipoproteinJ (apoJ, or clusterin), $A\beta$ bound to apoJ and apoE, aprotinin and very low-density lipoprotein (VLDL) are more specific for the LRP2 receptor [70, 71].

The group of Richard Béliveau first reported that melanotransferrin/P97 was actively transcytosed across the blood-brain barrier and suggested that this was mediated by the LRP1 receptor [69]. Melanotransferrin is a membrane-bound transferrin homolog that can also exist in a soluble form and is highly expressed on melanoma cells compared to normal melanocytes. Intravenously applied melanotransferrin delivers the majority of its bound iron to the liver and kidney, where only a small part is taken up by the brain [72]. After conjugation to melanotransferrin, the group of Béliveau was able to successfully deliver doxorubicin to brain tumors in animal studies [73, 74]. This melanotransferrin-mediated drug targeting technology (now designated NeuroTrans) is currently being further developed by BioMarin Pharmaceuticals Inc. (Novato, Calif., USA) for the delivery of enzyme replacement therapies to the brain. Interestingly, together with researchers from BioMarin, Pan et al. [75] recently reported on the efficient transfer of RAP across the blood-brain barrier by means of the LRP1/LRP2 receptors, suggesting the discovery of a novel means of protein-based drug delivery to the brain. RAP is a 39-kDa protein that functions as a specialized endoplasmic reticulum chaperone assisting in the folding and trafficking of members of the LDL receptor family. In yet unpublished results, the group of Béliveau have now filed a patent application on the use of the LRP2-specific ligand aprotinin, and more specifically on functional derivatives thereof (e.g. angio-pep1), thereby providing a noninvasive and flexible method and a carrier for transporting a compound or drug across the blood-brain barrier [76]. Aprotinin (Trasylol) is known as a potent inhibitor of serine proteases such as trypsin, plasmin, tissue and plasma kallikrein and is the only pharmacologic treatment approved by the United States Food and Drug Administration to reduce blood transfusion in coronary artery bypass grafting [77].

In addition to being a tumor marker protein, melanotransferrin is also associated with brain lesions in Alzheimer's disease and is a potential marker of the disorder [78]. In addition, the proposed receptor for melanotransferrin, LRP1, has genetically been linked to Alzheimer's disease and may influence APP processing and metabolism and Aβ uptake by neurons through α_2M (α_2M is one of the Aβ carrier proteins next to, e.g., apoE, apoJ, transthyretin, albumin) [70, 79]. Furthermore, a close relationship with RAGE (receptor for advanced glycation end products), in shuttling Aβ across the blood-brain barrier has been described [70, 79]. In addition, the LRP2 receptor has also been described to mediate the uptake of Aβ complexed to apoJ and apoE across the blood-brain barrier [71, 80, 81]. This complex interaction with Alzheimer's disease makes it difficult to predict the safety of the use of the LRP1/LRP2 receptors for the targeting of drugs to the brain in human application. This is especially so when the complex signaling function of these receptors is included in the assessment (including e.g. control of permeability of the blood-brain barrier, vascular tone, expression of MMPs) [66], as well as the fact that both the receptors are critically involved in the coagulation-fibrinolysis system. On top of which, melanotransferrin is also reported to be directly involved in the activation of plasminogen [82]; and high plasma concentrations of melanotransferrin are needed to deliver drugs to the brain, resulting perhaps in dose limitations because of the high iron load in the body.

The same line of reasoning for the interactions at the level of the uptake receptors may apply to the use of RAP and aprotinin (derivatives). However, the latter has already been successfully applied to humans usually without severe side-effects, indeed making the peptide derivatives potentially safe drug carriers. As for RAP, though, no results on the efficacy or capacity of the aprotinin peptides as carrier for drugs have been made available yet.

21.8
Diphtheria Toxin Receptor

Recently, our group identified a novel human-applicable carrier protein (known as CRM197) for the targeted delivery of conjugated proteins across the blood-brain barrier [83, 84]. Uniquely, CRM197 has already been used as a safe and effective carrier protein in human vaccines for a long time [85] and recently also as a systemically active therapeutic protein in anti-cancer trials [86]. This has resulted in a large body of prior knowledge on the carrier protein, including its transport receptor and mechanism of action, receptor binding domain, conjugation and manufacturing process and its kinetic and safety profiles, both in animals and humans. CRM197 delivers drugs across the blood-brain barrier by the well characterized, safe and effective mechanism called receptor-mediated transcytosis. From the literature, it was already known that CRM197 uses the membrane-bound precursor of heparin-binding epidermal growth factor-like growth factor (HB-EGF) as its transport receptor [87]. This precursor is also known as the diphtheria toxin receptor (DTR). In fact, CRM197 is a nontoxic mutant of diphtheria toxin. Membrane-bound HB-EGF is constitutively expressed on the blood-brain barrier, neurons and glial cells [88]. Moreover, HB-EGF expression

60 Xu L, Tang WH, Huang CC, et al. **2001**, Systemic p53 gene therapy of cancer with immunolipoplexes targeted by anti-transferrin receptor scFv, *Mol Med* 7, 723–734.

61 Lee JH, Engler JA, Collawn JF, Moore BA **2001**, Receptor mediated uptake of peptides that bind the human transferrin receptor, *Eur J Biochem* 268, 2004–2012.

62 Bottaro DP, Bonner-Weir S, King GL **1989**, Insulin receptor recycling in vascular endothelial cells, regulation by insulin and phorbol ester, *J Biol Chem* 264, 5916–5923.

63 Wu D, Yang J, Pardridge WM **1997**, Drug targeting of a peptide radiopharmaceutical through the primate blood-brain barrier in vivo with a monoclonal antibody to the human insulin receptor, *J Clin Invest* 100, 1804–1812.

64 Coloma MJ, Lee HJ, Kurihara A, et al. **2000**, Transport across the primate blood-brain barrier of a genetically engineered chimeric monoclonal antibody to the human insulin receptor, *Pharm Res* 17, 266–274.

65 Herz J, Strickland DK **2001**, LRP: a multifunctional scavenger and signaling receptor, *J Clin Invest* 108, 779–784.

66 Herz J **2003**, LRP: a bright beacon at the blood-brain barrier, *J Clin Invest* 112, 1483–1485.

67 Boucher P, Gotthardt M, Li WP, Anderson RG, Herz J **2003**, LRP: role in vascular wall integrity and protection from atherosclerosis, *Science* 300, 329–332.

68 Yepes M, Sandkvist M, Moore EG, et al. **2003**, Tissue-type plasminogen activator induces opening of the blood-brain barrier via the LDL receptor-related protein, *J Clin Invest* 112, 1533–1540.

69 Demeule M, Poirier J, Jodoin J, et al. **2002**, High transcytosis of melanotransferrin (P97) across the blood-brain barrier, *J Neurochem* 83, 924–933.

70 Deane R, Wu Z, Zlokovic BV **2004**, RAGE (yin) versus LRP (yang) balance regulates Alzheimer amyloid beta-peptide clearance through transport across the blood-brain barrier, *Stroke* 35[Suppl 1], 2628–2631.

71 Chun JT, Wang L, Pasinetti GM, Finch CE, Zlokovic BV **1999**, Glycoprotein 330/megalin (LRP-2) has low prevalence as mRNA and protein in brain microvessels and choroid plexus, *Exp Neurol* 157, 194–201.

72 Gabathuler R, Kolaitis G, Brooks RC, et al. **2002**, Patent WO0213843.

73 Richardson DR, Morgan EH **2004**, The transferrin homologue, melanotransferrin (p97), is rapidly catabolized by the liver of the rat and does not effectively donate iron to the brain, *Biochim Biophys Acta* 1690, 124–133.

74 Gabathuler R, Arthur G, Kennard M, et al. **2005**, Development of a potential vector (NeuroTrans) to deliver drugs across the blood-brain barrier, *Int Congr Ser* 1277 (in press).

75 Pan W, Kastin AJ, Zankel TC, et al. **2004**, Efficient transfer of receptor-associated protein (RAP) across the blood-brain barrier, *J Cell Sci* 117, 5071–5078.

76 Béliveau R, Demeule M **2004**, Patent WO2004060403.

77 Sedrakyan A, Treasure T, Elefteriades JA **2004**, Effect of aprotinin on clinical outcomes in coronary artery bypass graft surgery: a systematic review and meta-analysis of randomized clinical trials, *J Thorac Cardiovasc Surg* 128, 442–448.

78 Jefferies WA, Food MR, Gabathuler R, et al. **1996**, Reactive microglia specifically associated with amyloid plaques in Alzheimer's disease brain tissue express melanotransferrin, *Brain Res* 712, 122–126.

79 Deane R, Du Yan S, Submamaryan RK, et al. **2003**, RAGE mediates amyloid-beta peptide transport across the blood-brain barrier and accumulation in brain, *Nat Med* 9, 907–913.

80 Zlokovic BV **1996**, Cerebrovascular transport of Alzheimer's amyloid beta and apolipoproteins J and E: possible anti-amyloidogenic role of the blood-brain barrier, *Life Sci* 59, 1483–1497.

81 Zlokovic BV, Martel CL, Matsubara E, et al. **1996**, Glycoprotein 330/megalin: probable role in receptor-mediated transport of apolipoprotein J alone and in a complex with Alzheimer disease amyloid beta at the blood-brain and blood-cerebrospinal fluid barriers, *Proc Natl Acad Sci USA* 93, 4229–4234.

82 Demeule M, Bertrand Y, Michaud-Levesque J, et al. **2003**, Regulation of plasminogen activation: a role for melanotransferrin (p97) in cell migration, *Blood* 102, 1723–1731.

83 Gaillard PJ, Brink A, De Boer AG **2005**, Diphtheria toxin receptor-targeted brain drug delivery, *Int Congr Ser* 1277 (in press).

84 Gaillard PJ, Brink A, De Boer AG **2004**, Patent WO2004069870.

85 Anderson P **1983**, Antibody responses to *Haemophilus influenzae* type b and diphtheria toxin induced by conjugates of oligosaccharides of the type b capsule with the nontoxic protein CRM197, *Infect Immun* 39, 233–238.

86 Buzzi S, Rubboli D, Buzzi G, et al. **2004**, CRM197 (nontoxic diphtheria toxin): effects on advanced cancer patients, *Cancer Immunol Immunother* 53, 1041–1048.

87 Raab G, Klagsbrun M **1997**, Heparin-binding EGF-like growth factor, *Biochim Biophys Acta* 1333, F179–F199.

88 Mishima K, Higashiyama S, Nagashima Y, et al. **1996**, Regional distribution of heparin-binding epidermal growth factor-like growth factor mRNA and protein in adult rat forebrain, *Neurosci Lett* 213, 153–156.

89 Mishima K, Higashiyama S, Asai A, et al. **1998**, Heparin-binding epidermal growth factor-like growth factor stimulates mitogenic signaling and is highly expressed in human malignant gliomas, *Acta Neuropathol (Berl)* 96, 322–328.

90 Tanaka N, Sasahara M, Ohno M, et al. **1999**, Heparin-binding epidermal growth factor-like growth factor mRNA expression in neonatal rat brain with hypoxic/ischemic injury, *Brain Res* 827, 130–138.

91 Gaillard PJ, Voorwinden LH, Nielsen JL, et al. **2001**, Establishment and functional characterization of an in vitro model of the blood-brain barrier, comprising a co-culture of brain capillary endothelial cells and astrocytes, *Eur J Pharm Sci* 12, 215–222.

92 Mitamura T, Higashiyama S, Taniguchi N, Klagsbrun M, Mekada E **1995**, Diphtheria toxin binds to the epidermal growth factor (EGF)-like domain of human heparin-binding EGF-like growth factor/diphtheria toxin receptor and inhibits specifically its mitogenic activity, *J Biol Chem* 270, 1015–1019.

93 Cha JH, Chang MY, Richardson JA, Eidels L **2003**, Transgenic mice expressing the diphtheria toxin receptor are sensitive to the toxin, *Mol. Microbiol.* 49, 235–240.

94 Bendell JC, Domchek SM, Burstein HJ, et al. **2003**, Central nervous system metastases in women who receive trastuzumab-based therapy for metastatic breast carcinoma, *Cancer* 97, 2972–2977.

95 Broadwell RD, Baker-Cairns BJ, Friden PM, Oliver C, Villegas JC **1996**, Transcytosis of protein through the mammalian cerebral epithelium and endothelium III, Receptor-mediated transcytosis through the blood-brain barrier of blood-borne transferrin and antibody against the transferrin receptor, *Exp. Neurol.* 142, 47–65.

96 Gosk S, Vermehren C, Storm G, Moos T **2004**, Targeting anti-transferrin receptor antibody (OX26) and OX26-conjugated liposomes to brain capillary endothelial cells using in situ perfusion, *J Cereb Blood Flow Metab* 24, 1193–1204.

97 Satomi et al. **2003**, Cerebral vascular abnormalities in a mouse model of hereditary hemorrhagic telangiectasia, *Stroke* 34, 783–789.

98 Krewson et al. **1995**, Distribution of nerve growth factor following direct delivery to brain interstitium, *Brain Res.* 680, 196–206.

99 Yan et al. **1994**, Distribution of intracerebral ventricularly administered neurotrophins in rat brain and its correlation with trk receptor expression, *Exp Neurol* 127, 23–36.

**Part VI
Vascular Perfusion**

22
Blood-Brain Transfer and Metabolism of Oxygen

Albert Gjedde

22.1
Introduction

Oxygen is essential to brain function. Disruption of oxygen delivery to brain results in complete cessation of functional activity in the human brain in 6 s (Hansen 1984, 1985). The reason for this dramatic effect is not known with certainty. Increased potassium conductance in neuronal membranes plays a role, but the effect also suggests that oxygen reserves in brain tissue are very low in relation to the steady-state turnover, and it implies that oxygen turnover is intimately related to the mechanisms that subserve consciousness, because the effect cannot be explained by depletion of intermediary metabolites.

The delivery of oxygen to brain tissue differs in major respects from the delivery of oxygen to other tissues. In the classic description of Krogh (1919), the standard capillary bed is a system of parallel tubes serving uniform cylinders of tissue (known as Krogh cylinders). This simple arrangement yields a quantitative expression of oxygen delivery to the tissue, suggestive of specific mechanisms of regulation. However, in brain tissue, the vascular bed differs from the vascular beds of other tissues in two important ways. Both the absence of the recruitment of capillaries to neurons in a state of activation and the general principle of the topographic arrangement of the vessels are features unlike those of other vascular beds, but the importance of these differences has eluded students of the vascular physiology of the brain.

The absence of recruitment suggests that the work of the brain, measured as energy turnover, varies little over time; and the random arrangement of capillaries suggests that the geometry of the vascular network is fixed competitively by the work of the nerve cells, rather than by the need to make frequent, orderly and substantive changes to the diffusion capacity, as in skeletal and cardiac muscles. If the geometry is always the same, there is no need for any particular order.

Thus, in brain, the vascular bed is so poorly organized that no single geometric description has general validity and hence no single prediction of oxygen distribution in the tissue is possible (Wang and Bassingthwaighte 2001). To predict the general properties of oxygen delivery and consumption, the remaining valid approach is the averaging of properties across large volumes of tissue, on the strength of observations that have wide applicability, always with the provision that local and regional properties may differ significantly from the global average.

Only two claims of the capillary bed in brain appear to be indisputable, i.e., that the capillaries have multiple arterial sources but a common venous terminus and that the density of capillaries is proportional to the average regional rate of metabolism at the steady-state (also known as the "default" state). From these claims arises a revised distribution of oxygen in brain tissue that is different from the classic distribution in Krogh cylinders. The revision is made possible by the simple realization, first introduced by Erwin R. Weibel (1984) in "*The Pathway for Oxygen*", that every segment of the capillary bed exists to supply an equivalent amount of brain tissue with oxygen, i.e., that every fraction of the tissue is served by commensurate fractions of a capillary bed, having the appropriate oxygen diffusibility and accounting for equivalent fractions of the total oxygen consumption.

This approach considers the random delivery of oxygen molecules from the source in arbitrarily distributed capillaries to the sink in equally arbitrarily distributed mitochondria. The only discernable regulation is the need to make enough oxygen molecules available to the cells served by the capillaries. The approach assumes that suitable mechanisms originally served to place the capillary bed and the cells in relation to each other in such a way that this requirement would be fulfilled. It is an intriguing possibility that the astrocytic projections, in addition to other services, could provide a structural restraint on the anatomic relationship of capillaries and neurons, established to achieve the optimal match of randomly placed capillaries and cells. The current approach is not limited to capillaries per se. It is suspected that oxygen exchange with the tissue may begin in arteriolar microvessels (Duling and Berne 1970). For the purposes of this analysis, arteriolar microvessels are therefore part of the effective capillary bed.

The variable organization can be reduced to compartments of predictable behavior by making the simple assumption that the structure of the vascular bed evolved to satisfy the existing needs for oxygen, in contrast to the more traditional view that the needs for oxygen can be deduced from a specific structure of the vascular bed. It turns out that the resulting temporal and anatomical averages of the random processes of oxygen delivery achieve an appearance of orderliness that is illusory.

22.2
Blood-Brain Transfer of Oxygen

22.2.1
Capillary Model of Oxygen Transfer

Capillary models of blood-brain transfer focus on the distribution of molecules inside microvessels. In the case of molecular oxygen, the extraction from the capillary to the tissue establishes an oxygen profile in the capillary bed that ranges from the tension associated with the more or less fully saturated hemoglobin at the arterial inlet to the tension of the more or less fully desaturated hemoglobin at the venous outlet. Between these extremes, the tension depends on the extraction of oxygen from the network of capillaries serving the tissue. The discussion above leads to the conclusion that it is impossible to make a specific prediction of the profile in individual capillaries but it is possible to look at the ensemble of capillaries as a unit that provides the tissue with the oxygen it needs. This simple assumption leads to a solution that is close, but not identical, to the general claim that the oxygen tension declines monoexponentially from the arterial inlet to the venous outlet.

The solution is obtained by letting the oxygen extraction from the blood flowing in the general direction from the arterial inlet to the venous outlet be proportional to the fraction of the capillary bed cumulatively served by the blood flow. The fractional segment of the capillary bed is denoted by a variable z that ranges from 0 to 1. In every segment, an equivalent amount of oxygen is extracted by the cells served by their respective segments of the capillary bed:

$$E_{O_2}(z) = \frac{J_{O_2}(z)}{FC_{O_2}^{art}} = \frac{z\bar{J}_{O_2}}{FC_{O_2}^{art}} \tag{22.1}$$

where the oxygen extraction fraction is $E_{O_2}(z)$, $J_{O_2}(z)$ is the segmental rate of cerebral metabolism of oxygen (CMR_{O_2}) at the segment z of the capillary bed in question, $\bar{J}_{O_2}$ is the measured CMR_{O_2} value per unit mass or volume of the brain region served by this particular capillary bed, F is the measured blood

Table 22.1 Average properties of oxygen compartments in human brain. (Modified from Gjedde et al. 2005).

Variable	Unit	Arterial	Capillary	Venous	Tissue	Mitochondrial
C_{O_2}	mM	8.5	7.0	5.3		
P_{O_2}	mmHg	114.0	53.0	40.0	20–30	8.5
P_{50}	mmHg	26.0	35.0	35.0		0.5
h		2.84	3.5	3.5		
S_{O_2}	Ratio	0.985	0.81	0.61		0.94

Capillary Model

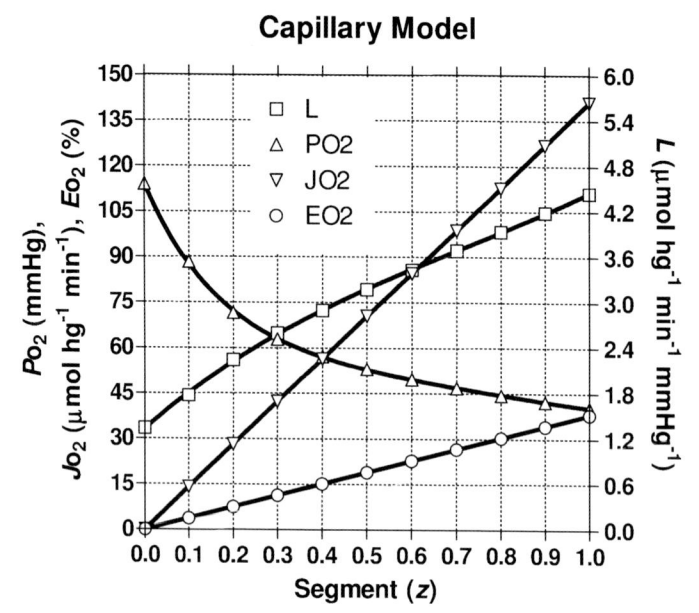

Fig. 22.1 Oxygen tension, extraction fraction, and oxygen diffusibility as function of axial segmentation of cortical capillary bed (z direction), calculated for baseline condition of human brain. Abscissa: axial segments of capillary bed (z). Ordinate: capillary oxygen extraction fraction $[E_{O_2}(z)]$, tension $[P_{O_2}^{cap}(z)]$, and diffusibility $[L(z)]$ as functions of axial segment. The variables were calculated with Eqs. (22.7) to (22.9) under normal conditions of whole human brain, as listed in Tables 22.1 and 22.2: normal CMR_{O_2} (141 µmol hg^{-1} min^{-1} at the baseline), normal blood flow (F, 43 ml hg^{-1} min^{-1}), normal arterial oxygen concentration (8.7 mM), normal P_{50}^{cap}, the half-saturation tension of oxygen (35 mmHg when corrected for the Bohr shift), and normal $\bar{h}$, the Hill coefficient (3.5 when corrected for Bohr shift). $P_{O_2}^{cap}(0)$ was assumed to equal the arterial oxygen tension $P_{O_2}^{art}$ (114 mmHg).

flow to the region, and $C_{O_2}^{art}$ is the measured arterial oxygen concentration at the arterial inlet. Standard properties of arterial blood are given in Table 22.1.

The extraction of oxygen from brain capillaries causes the capillary oxygen tension to decline from the arterial oxygen tension at the arterial end of the capillary bed to the venous oxygen tension at the venous end. The decline of the oxygen tension as a function of z is determined by the oxygen remaining after the extraction. The loss of oxygen establishes a declining oxygen saturation of hemoglobin according to the formula:

$$S_{O_2}^{cap}(z) = S_{O_2}^{art}[1 - E_{O_2}(z)] \tag{22.2}$$

where $S_{O_2}^{art}$ is the saturation of hemoglobin in arterial blood and $S_{O_2}^{cap}$ is the saturation in capillary blood at the segmental position z. The amount of oxygen remaining in the capillary bed also satisfies the Hill equation of the relationship between oxygen tension and hemoglobin saturation:

$$S_{O_2}^{cap}(z) = \frac{1}{1 + \left(\frac{P_{50}^{cap}(z)}{P_{O_2}^{cap}(z)}\right)^{h(z)}} \qquad (22.3)$$

where $P_{O_2}^{cap}(z)$ is the oxygen tension as a function of position in the capillary, $P_{50}^{cap}(z)$ is the half-saturation tension of oxygen at the position z, and $h(z)$ is the value of the Hill coefficient at that position. The Hill equation then is solved for the capillary oxygen tension associated with the particular extraction fraction:

$$P_{O_2}^{cap}(z) = P_{50}^{cap}(z)^{h(z)}\sqrt{\frac{S_{O_2}^{art}(1 - E_{O_2}(z))}{1 - S_{O_2}^{art}(1 - E_{O_2}(z))}} \qquad (22.4)$$

which, on the assumption that the saturation of hemoglobin in arterial blood is negligibly different from unity, reduces to the expression:

$$P_{O_2}^{cap}(z) = P_{50}^{cap}(z)^{h(z)}\sqrt{\frac{1}{E_{O_2}(z)} - 1} \qquad (22.5)$$

where $P_{O_2}^{cap}(z)$ is the pressure head of oxygen diffusion when arterial oxygen saturation is 100%.[1] The third determinant of oxygen delivery is the oxygen diffusibility, which depends on the diffusion coefficient, the solubility, and the position of the oxygen sink or sinks. The oxygen diffusibility is given by the ratio between the oxygen flux and the difference between the pressures of oxygen in the z segment of the capillary bed and in the mitochondria of the tissue segment served by the segment of capillary bed:

$$L(z) = \frac{\bar{J}_{O_2}}{P_{O_2}^{cap}(z) - P_{O_2}^{mit}(z)} \qquad (22.6)$$

where the oxygen diffusibility $L(z)$ is a function of the value of the segmental position z and $P_{O_2}^{mit}(z)$ is the operative mitochondrial oxygen tension at the site, which a priori is not known with certainty in brain tissue, although attempts have been made to calculate the value (Gjedde et al. 2005).

---

1) As an alternative to the assumption of unity for $S_{O_2}^{art}$, it is possible to lump the effect of arterial desaturation into an apparent change of the half-saturation constant of hemoglobin, $P_{50}^{app}(z)$. The apparent half-saturation constant is then given by:

$$P_{50}^{app}(z) = \frac{P_{50}^{cap}(z)}{\sqrt[h(z)]{\frac{1 - S_{O_2}^{ven}}{S_{O_2}^{art} - S_{O_2}^{ven}}}} \text{ in}$$

which P_{50}^{app} is the half-saturation tension corrected for arterial desaturation of hemoglobin and $S_{O_2}^{ven}$ is the venous oxygen saturation. The denominator assumes a value within a fewpercentage points of unity for ordinary extraction fractions at arterial oxygen tensions above 100 mmHg. At lower oxygen tensions, the value of the denominator depends on the position in the capillary bed. Since the physically dissolved oxygen represents no more than 1–2% in normoxia (although less in hypoxia), the physically dissolved oxygen "spares" some of the bound oxygen during extraction. There is indication that the half-saturation tension of arterial blood is adjusted in hypoxia to counter the effects of the low arterial hemoglobin saturation (Gjedde 2002).

The diffusibility depends on the surface area of the capillary segment. The diffusion gradient also varies with the distribution of mitochondria in the tissue, which is not known with certainty for any specific location and hence must be inferred from consideration of averages across larger regions.

If the sinks are placed at the end of the diffusion path, $L(z)$ is an estimate of the diffusion coefficient, corrected for solubility, that reflects the average distance from the z segment to the mitochondria served by the segment. The distinction between even and uneven distribution of metabolic sinks for oxygen will be discussed in greater detail in the subsequent section on tissue oxygen tensions.

Equations (22.1), (22.5), and (22.6) can be solved in sequence, beginning with the extraction fraction, as functions of the capillary segment z. The results are shown in Fig. 22.1. The extraction rises linearly as shown in Eq. (22.1); and the capillary oxygen tension declines accordingly as shown in Eq. (22.5). The decline is approximately monoexponential as predicted by conventional capillary models (Crone 1963), but the reason is not per se the simple pressure gradient between capillary and tissue, but is the more complex satisfaction of a constant oxygen delivery, organized as a combination of the linearly increasing extraction fraction and the sigmoidal relationship between oxygen tension and hemoglobin saturation.

The diffusibility also rises in response to the needs of the tissue segment served, to maintain the constant flux of oxygen; and the increase is not linear. A nonlinearly rising diffusibility can be achieved in several ways, including a declining distance between adjacent microvessels as well as by branching of vessels.

22.2.2
Compartment Model of Oxygen Transfer

A compartment is an idealization of kinetic relationships in which all concentration gradients are placed at the interfaces among the compartments. Inside the compartments, in contravention of the true nature of gradients, concentrations are considered to be uniform. The linearity of the extraction fraction means that the capillary bed can be reduced to a single microvascular compartment with a uniform oxygen tension, $\bar{P}_{O_2}^{cap}$, derived from the linearly averaged extraction fraction. The linearity of the extraction fraction as function of the capillary segmentation further implies that the entire axial dimension can be projected onto a single transaxial plane at the position $z = 0.5$, where the extraction fraction is half of the totally extracted fraction $[E_{O_2}(0.5) = \bar{E}_{O_2}/2]$. Provided the arterial saturation of hemoglobin is close to unity, the weighted average capillary oxygen tension associated with the oxygen present in all segments of the capillary bed (Vafaee and Gjedde 2000) is then:

$$\bar{P}_{O_2}^{cap} = P_{50}^{cap} \sqrt[h]{\frac{2}{\bar{E}_{O_2}} - 1} \qquad (22.7)$$

where $\bar{P}_{50}^{cap}$ is the average capillary half-saturation tension of oxygen binding to hemoglobin[2]) and $\bar{h}$ is the average value of the Hill coefficient in the capillary. When $E_{O_2}(z)$ rises linearly, the corresponding weighted average oxygen diffusibility is:

$$\bar{L} = \frac{\bar{J}_{O_2}}{\bar{P}_{O_2}^{cap} - \bar{P}_{O_2}^{mit}} \tag{22.8}$$

where $\bar{P}_{O_2}^{mit}$ is the tissue's minimum oxygen tension, and $\bar{L}$ is the mean oxygen diffusibility, equal to $L(z)$ at $z=0.5$. The magnitude of $\bar{L}$ depends on the unknown value of $\bar{P}_{O_2}^{mit}$, the oxygen tension at cytochrome c oxidase. Equations (22.7) and (22.8) can be used to estimate $\bar{L}$ by identifying an extraction fraction at which the oxygen reserve of the mitochondria has been exhausted and calculating the corresponding value of $\bar{L}$ by means of Eq. (22.8). It follows from Eq. (22.8) that the oxygen consumption is:

$$\bar{J}_{O_2} = \bar{L}\left[P_{50}^{cap}\,\bar{h}\sqrt{\frac{2}{\bar{E}_{O_2}} - 1} - \bar{P}_{O_2}^{mit}\right] \tag{22.9}$$

where the average value of h is $\bar{h}$. Solutions to these equations are shown in Figs. 22.1 and 22.2 for normal values of human brain.

22.3 Oxygen in Brain Tissue

22.3.1 Cytochrome Oxidation

The solutions to the equations above depend on the magnitude of the mitochondrial oxygen tension in Eqs. (22.8) and (22.9). This tension, together with the affinity of cytochrome c oxidase towards oxygen, the maximum reaction rate of each mitochondrion, and the number of mitochondria in a given segment of tissue, determines the oxygen consumption according to the simplest formulation of the Michaelis-Menten equation (Gnaiger et al. 1995, 1998; Guilivi 1998):

---

2) As an alternative to the assumption of arterial hemoglobin saturation of unity, it is again possible to define an apparent average half-saturation tension that corrects for the degree of desaturation of hemoglobin that exists prior to the entry into the capillary. In the compartmental model, in agreement with Eq. (4), $\bar{P}_{50}^{app}$ is given by the expression

$$\bar{P}_{50}^{app} = \frac{P_{50}^{cap}}{\sqrt[h]{\frac{2 - S_{O_2}^{art} - S_{O_2}^{ven}}{S_{O_2}^{art} - S_{O_2}^{ven}}}}$$

which has the value 1.02 for an extraction of 40% at an arterial oxygen tension of 114 mmHg (standard). At a tension of 52 mmHg, the denominator has the values 1.16 at an extraction of 40% and 1.12 at 60% extraction.

Capillary Oxygen Profiles

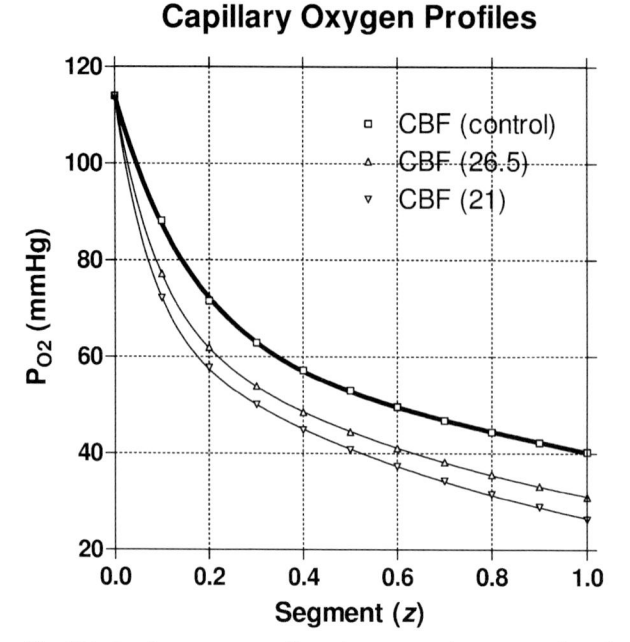

Fig. 22.2 Capillary oxygen profiles. Abscissa: axial segment of capillary bed (z). Ordinate: capillary oxygen tensions $[P^{cap}_{O_2}(z)]$ as a function of axial segment, calculated as explained in Fig. 22.1 for baseline (CBF 43 ml kg^{-1} min^{-1}), and two oligemia (low blood flow) states (CBF 26.5 ml hg^{-1} min^{-1}, 21 ml hg^{-1} min^{-1}), chosen to be just at and just below the limit of adequate oxygen delivery. Numerical solution of the three equations with the three unknowns $E_{O_2}(z)$, $P^{cap}_{O_2}(z)$, and $x_0(z)$ yielded the capillary oxygen tension profile for insufficient oxygen supply (see Section 22.5).

$$J_{O_2} = \frac{J_{max} P^{mit}_{O_2}}{P^{cyt}_{50} + P^{mit}_{O_2}} \tag{22.10}$$

where P^{cyt}_{50} is the apparent half-saturation tension of the oxygen reaction with cytochrome oxidase, $P^{mit}_{O_2}$ is the mitochondrial oxygen tension, and J_{max} is the maximum velocity of the cytochrome oxidase reaction. This equation is applicable to all mitochondria, but it is not known to what extent the magnitudes of each of the parameters of the equation, as well as the resulting oxygen consumption, apply generally to the tissue as a whole. Therefore it is necessary to operate with the concept of an average mitochondrion. In vivo, the affinity is considerably (an order of magnitude) below the inherent affinity observed under in vitro circumstances. The expected saturation of the enzyme therefore is not as pronounced in vivo as in vitro. The standard properties of this enzyme are given in Table 22.1.

The actual flux of oxygen in each tissue volume element depends on the regional oxygen consumption, which is often assumed to be the same per unit tissue mass or volume everywhere but may in fact vary considerably from site to site. The flux is the difference between the average rate of oxygen consumption of the tissue segment and the regionally (but not locally) invariant rate of oxygen delivery to each unit of tissue. The mitochondrial oxygen tension in turn links the oxygen consumption with the oxygen delivery. The mitochondrial oxygen tension represents the balance between the diffusion of oxygen to the site of metabolism and the reaction of oxygen with cytochrome c oxidase, according to the simple Michaelis-Menten expression. The local variability of mitochondrial oxygen tension in brain tissue is therefore the key to the understanding of oxygen homeostasis.

22.3.2
Mitochondrial Oxygen Tension

22.3.2.1 Distributed Model of Tissue and Mitochondrial Oxygen

In the classic model of nonuniformly distributed oxygen, the mitochondria are evenly scattered in the tissue but the oxygen tension varies spatially as oxygen is consumed. This situation is reminiscent of the conditions existing in the Krogh cylinder, in which the local mitochondrial oxygen tension varies with the distance of the mitochondria from the microvessels, while the oxygen consumption rate is assumed to be the same for each mitochondrion, because affinity or maximum reaction rate or both are adjusted to maintain the same oxygen consumption rate in every tissue element. This situation is consistent with a one-dimensional diffusion equation of the form:

$$j_{O_2}(z, x) = -KA(z)\frac{\partial p_{O_2}(z, x)}{\partial x} = (1 - x)J_{O_2}(z) \tag{22.11}$$

where $j_{O_2}(z, x)$ is the flux of oxygen from the capillary segment z to the tissue segment x, $J_{O_2}(z)$ is the regionally invariant oxygen consumption of each unit of tissue mass or volume, K is Krogh's diffusion coefficient for a unit surface area, and $A(z)$ is the magnitude of that area. The sites of oxygen consumption generate an extravascularly directed gradient of oxygen tension from the microvascular compartment to the tissue mass or volume fraction x served by the microvascular compartment:

$$\frac{dP_{O_2}(x)}{dx} = -\left(\frac{1-x}{K\bar{A}}\right)\bar{J}_{O_2} \tag{22.12}$$

where $\bar{A}$ is the diffusion area of a segmental surface. Integration with respect to the tissue mass or volume fraction x yields the pressure as a function of the tissue mass or volume fraction served by this segment of the capillary:

$$P_{O_2}(x) = P_{O_2}(1) + \left[\frac{\bar{J}_{O_2}}{2K\bar{A}}\right](1-x)^2 \qquad (22.13)$$

where $P_{O_2}(1)$ is the minimum oxygen tension in the tissue. For evenly distributed (i.e., equally abundant more and less "distant") sites of oxygen consumption in the one-dimensional tissue, obtained by projecting the higher dimensions onto the segment at $z=0.5$, the relationship between Krogh's coefficient K and the weighted average oxygen diffusibility is:

$$\bar{L} = 2K\bar{A} \qquad (22.14)$$

such that substitution of $\bar{L}/2$ for $K\bar{A}$ yields the weighted average oxygen tension as a function of the weighted average distance from the capillary:

$$P_{O_2}(x) = \bar{P}_{O_2}^{cap} - \frac{\bar{J}_{O_2}}{\bar{L}}[2x - x^2] \qquad (22.15)$$

where $\bar{L}$ is the weighted average oxygen diffusibility defined in Eq. (22.5), and $\bar{J}_{O_2}$ is the average rate of oxygen consumption of the entire tissue. This solution is shown in Fig. 22.3 for the cases of one normal and two low levels of blood flow to the human cerebral cortex.

For $x=1.0$, Eq. (22.15) yields the minimum mitochondrial oxygen tension reached at the end of the diffusion path:

$$\bar{P}_{O_2}^{mit} = \bar{P}_{O_2}^{cap} - \frac{\bar{J}_{O_2}}{\bar{L}} \qquad (22.16)$$

which is identical to the evenly distributed oxygen tension associated with the linear diffusion case above. Integration and normalization of Eq. (22.15) yields the weighted tissue average of the oxygen tension:

$$\bar{P}_{O_2}^{br} = \bar{P}_{O_2}^{cap} - \frac{2\bar{J}_{O_2}}{3\bar{L}} \qquad (22.17)$$

where $\bar{P}_{O_2}^{br}$ is the weighted average tissue tension.

Equation (22.15) is in fact a special solution to a more general formulation in which the local mitochondrial oxygen tension varies with the distance from the microvessels, depending on the local oxygen consumption rate, which in turn varies with the oxygen tension according to the Michaelis-Menten equation for cytochrome c oxidase, provided the enzyme has the same affinity and maximum reaction rate everywhere. In that case, the diffusion equation assumes the form:

$$K\bar{A}\frac{d^2 P_{O_2}(x)}{dx^2} = J_{max}\left(\frac{P_{O_2}(x)}{P_{50}^{cyt} + P_{O_2}(x)}\right) \qquad (22.18)$$

Tissue Oxygen Tension Profiles

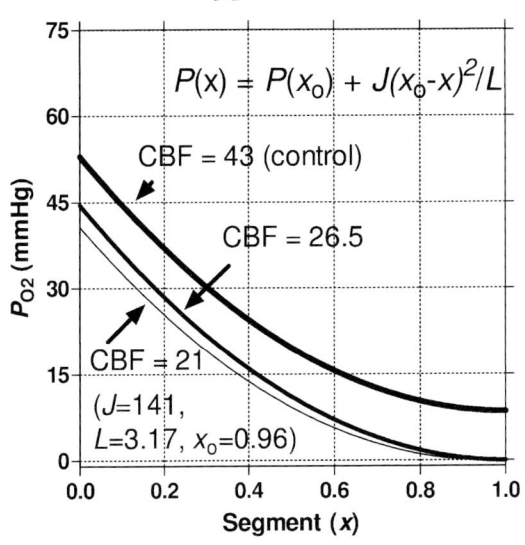

$$P(x) = P(x_0) + J(x_0'-x)^2/L$$

CBF = 43 (control)

CBF = 26.5

CBF = 21
(J=141,
L=3.17, x_o=0.96)

Fig. 22.3 Tissue oxygen tensions of distributed tissue model. Abscissa: tissue volume fraction (x). Ordinate: oxygen tension profiles [$P_{O_2}(x)$] as a function of tissue volume fraction for the three cases in Fig. 22.2, using the equations indicated in the figure, identical to Eqs. (22.13) and (22.24) in the text. The normal end-path oxygen tension was assumed to be 8.5 mmHg, as listed in Table 22.1.

which equation in principle is soluble only by numerical means. When $P_{O_2}(x) = P_{50}^{cyt}$, the equation reduces to the special case of spatially uniform oxygen consumption discussed above. Whether this happens depends on the tension as well as on the affinity.

22.3.2.2 Compartment Models of Tissue and Mitochondrial Oxygen

In the simplest case of a compartment model, the local mitochondrial oxygen tensions are uniformly negligible with respect to the capillary oxygen tension and the oxygen diffusion from the capillary compartment, although they remain significant with respect to the cytochrome c oxidase reaction. In this case, the local oxygen consumption still varies with the negligible tension but has little influence on the delivery of oxygen to the tissue, which is then entirely flow-limited. This is the basis for the original explanation of the apparent uncoupling of blood flow from oxygen consumption during functional activation (Buxton and Frank 1997; Gjedde 1997), which also cannot be sustained in light of the maintenance of normal oxygen consumption rates during substantial flow decreases, for example in hypocapnia (Gjedde et al. 2005). In this case, the oxygen diffusibility is simply the ratio between oxygen consumption and the capillary oxygen tension.

The negligible mitochondrial oxygen tension is in reality a special case of a more general situation in which the local tension is maintained at a constant magnitude with uniform distribution, implying that the majority of mitochondria reside at the same distance from the relevant microvessels and have about the same rate of oxygen consumption. In this case, the oxygen is delivered by the simplest diffusion relation established by the tension difference between the microvessels and the mitochondria. The fixed distance from the relevant microvessels and the mitochondria represents a diffusion barrier, for which the tension at the far end is given by

$$\bar{P}_{O_2}^{mit} = \bar{P}_{O_2}^{cap} - \frac{\bar{J}_{O_2}}{\bar{L}} \tag{22.19}$$

where $\bar{P}_{O_2}^{mit}$ is the tension of the uniformly distributed mitochondrial oxygen. The resulting compartmental model is shown in Fig. 22.4. The gradient associated with the interface between the capillary and mitochondrial compartments allows the approximation of an average tissue oxygen diffusibility according to Eq. (22.16).

It is of course possible that distributed and compartmental oxygen states coexist in the tissue. The local mitochondrial oxygen tension, affinity, maximum reaction rate, and distance from microvessels may vary unpredictably for all mitochondria, the density of which may also vary to maintain the appropriate oxygen consumption for each tissue element. However, since every arrangement discussed above yields the same or similar expressions of the average oxygen diffu-

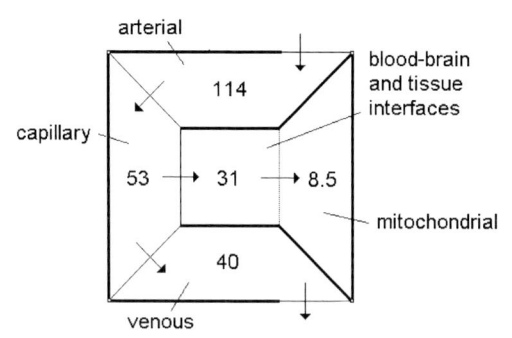

Fig. 22.4 Combined compartment model of oxygen tensions in brain. The compartments include arterial, capillary, venous, and mitochondrial spaces. The interface between the capillary and mitochondrial compartments is a diffusion barrier, the exact position of which is not known with certainty, although it may be dominated by the capillary endothelium. The numbers refer to normal oxygen tensions (mmHg), calculated from Eq. (22.7) for capillary oxygen tension, Eq. (22.5) for venous oxygen tension ($z=1$), and Eq. (22.19) for mitochondrial oxygen tension ($x=1$). The oxygen tension of the average tissue compartment or capillary-mitochondrial diffusion interface is a simple linear average. Note that the term capillary bed is used for the entire portion of the vascular bed that interacts with the tissue. This portion may include elements of arterial microvessels (Duling and Berne 1970).

sibility, it appears generally and for most practical purposes valid to estimate the lowest oxygen tension in the tissue (which may also be the tension of the uniformly distributed oxygen, of course) by the formula

$$\bar{P}_{O_2}^{\text{mit}} = \bar{P}_{O_2}^{\text{cap}} - \frac{\bar{J}_{O_2}}{\bar{L}} \tag{22.20}$$

in which $\bar{P}_{O_2}^{\text{mit}}$ is either the uniform tissue oxygen tension in the linear case, or the minimum tissue oxygen tension in the nonlinear case.

The average tissue oxygen tension, in contrast, depends on the actual oxygen distribution but generally must observe the following inequalities:

$$\bar{P}_{O_2}^{\text{cap}} - \frac{2\bar{J}_{O_2}}{3\bar{L}} < \bar{P}_{O_2}^{\text{br}} < \bar{P}_{O_2}^{\text{cap}} - \frac{\bar{J}_{O_2}}{2\bar{L}} \tag{22.21}$$

22.4
Flow-Metabolism Coupling of Oxygen

The expressions for the blood-brain transfer and metabolism of oxygen must be combined to obtain a general expression of the flow-metabolism coupling of oxygen. In this coupling, the mitochondrial oxygen tension represents the balance between the diffusion of oxygen to the site of metabolism and the reaction of oxygen with cytochrome c oxidase, known to follow the simple Michaelis-Menten expression given above. Solution of Eq. (22.9) for the cerebral blood flow yields:

$$F = \frac{\bar{J}_{O_2}}{2C_{O_2}^{\text{art}}} \left(1 + \left[\frac{\bar{J}_{O_2} + \bar{L}\bar{P}_{O_2}^{\text{mit}}}{\bar{L}\bar{P}_{50}^{\text{cap}}} \right]^{\bar{h}} \right) \tag{22.22}$$

where F is the blood flow to the tissue (per unit weight or volume), $\bar{J}_{O_2}$ is the average oxygen metabolism of the tissue, $C_{O_2}^{\text{art}}$ is the arterial oxygen concentration, $\bar{L}$ is the oxygen diffusibility, $\bar{P}_{50}^{\text{cap}}$ is the half-saturation oxygen tension of hemoglobin in capillaries, corrected for the Bohr shift and prior arterial desaturation, and $\bar{h}$ is the weighted average of Hill's coefficient for the capillary bed.[3] When the two equations involving the uniform or minimum oxygen tension are linked (Eqs. (22.10) and (22.22))9, $\bar{P}_{O_2}^{\text{mit}}$ can be eliminated from the rela-

---

[3] Adjustment of $\bar{L}$ may compensate for the degree of hypoxemic desaturation of arterial blood, provided the product $\bar{L}\bar{P}_{50}^{\text{app}}$ can be considered a constant. This is the case if the magnitude of L is adjusted to compensate for the decline of $\bar{P}_{50}^{\text{app}}$ by desaturation, by a process of recruitment. Equation (4) shows that the com-bined desaturation prior to capillary entry and a commensurate adjustment of the blood-brain oxygen diffusion capacity together could serve to render the baseline magnitude of the product valid for all degrees of desaturation.

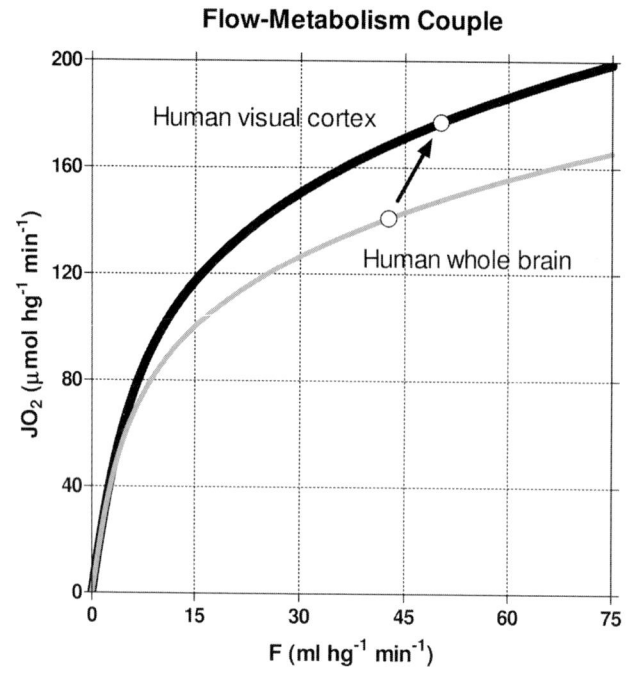

Fig. 22.5 Flow-metabolism couple predicted by Eqs. (22.22) and (22.23) for human whole brain and visual cortex. Parameters used to generate curves are given in Tables 22.1 and 22.2 for whole human brain and human cerebral cortex. Abscissa: blood flow (ml hg^{-1} min^{-1}). Ordinate: oxygen consumption (μmol hg^{-1} min^{-1}). Points indicate normal values listed in Table 22.2 for the two tissues. Change from whole-brain to cortex is associated with slight decline of the oxygen extraction fraction (E_{O_2} declines from 0.39 to 0.38).

tionship. The resulting formula prescribes the rate of blood flow that is associated with given properties of cytochrome c oxidase:

$$F = \frac{\bar{J}_{O_2}}{2 C_{O_2}^{art}} \left(1 + \left(\left[1 + \frac{\bar{L} P_{50}^{cyt}}{J_{max} - \bar{J}_{O_2}} \right] \left[\frac{\bar{J}_{O_2}}{\bar{L} P_{50}^{cap}} \right] \right)^{\bar{h}} \right) \tag{22.23}$$

where all the parameters in principle are variables under homeostatic control. The variables have all been defined above. The equation describes a mechanism of nonlinear flow-metabolism coupling, cast in terms of the nominally independent variables or constants F, $C_{O_2}^{art}$, P_{50}^{cyt}, $\bar{L}$, J_{max}, P_{50}^{cap}, and $\bar{h}$, and the nominally dependent variable $\bar{J}_{O_2}$. If independent variables other than flow remain constant, oxygen consumption must follow flow nonlinearly. As oxygen metabolism is known not to depend rigidly on blood flow, for example in hypocapnia, it is safe to conclude that several independent variables undergo simultaneous

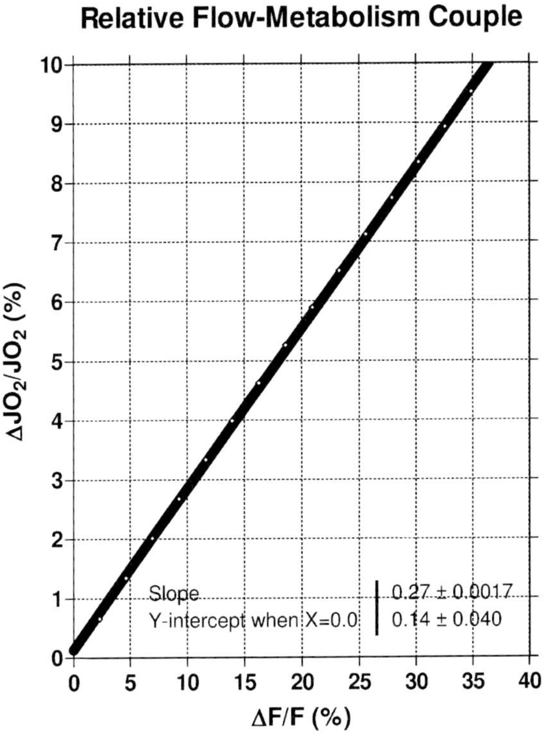

Relative Flow-Metabolism Couple

[Graph showing ΔJO₂/JO₂ (%) vs ΔF/F (%), linear relationship from origin to approximately (40, 10). Slope: 0.27 ± 0.0017. Y-intercept when X=0.0: 0.14 ± 0.040.]

Fig. 22.6 Changes of oxygen consumption and blood flow, relative to respective baselines for human whole-brain, calculated from human whole-brain curve shown in Fig. 22.5. Abscissa: change in blood flow, relative to baseline (%). Ordinate: change in oxygen consumption, relative to baseline (%). Note approximate linearity, the slope of which is close to the reciprocal of magnitude of the average capillary Hill coefficient $(1/\bar{h})$, as indicated by Eqs. (22.22) and (22.23).

change. One possible adjustment is the compensation for low oxygen saturation of arterial blood suggested by the substitution $\bar{P}_{50}^{app}$ for $\bar{P}_{50}^{cap}$ for as argued above.[4]

In principle, the relationship between oxygen consumption and blood flow described by the flow-metabolism couple is flow-limited. It is shown in Fig. 22.5 for normal values of the independent variables and constants for human brain. Assuming these parameters to be invariant, the relationship is highly nonlinear and fixed. The changes of oxygen consumption, relative to baseline, in contrast,

---

[4] The complete flow-metabolism equation shows that the effect of arterial hypoxemia can also be counteracted by maintaining a constant $P_{O_2}^{mit}/P_{50}^{cyt}$ ratio as $P_{O_2}^{mit}$ declines. It is of interest that experimental observations conform less well to the equation when it is correctly modified for the prior arterial desaturation of hemoglobin. This suggests either that reserves of oxygen are available in the tissue for use by the mitochondria, or that "recruitment" may reset $\bar{L}$ to compensate for the prior arterial desaturation of hemoglobin.

and the accompanying changes of blood flow, also relative to baseline, are linearly linked, as shown in Fig. 22.6, indicating that the assumption of invariant variables and constants other than blood flow and oxygen consumption is valid under these conditions.

It is increasingly clear, however, that the independent variables as well as the constants must undergo substantial changes in some circumstances. Independent variables or constants that may undergo simultaneous adjustment include the cytochrome oxidase maximum reaction rate or affinity, or both. There is evidence that they change in the same direction during the activation of cytochrome oxidase in vitro, while there is other evidence that the affinity of cytochrome oxidase is subject to inhibition by nitric oxide, which is also likely to play a major role in flow adjustment (Gjedde et al. 2005).

22.5
Limits to Oxygen Supply

Oxygen consumption is secured by an adequate blood flow and the possible adjustments of independent variables in Eq. (22.23). However, these adjustments eventually are exhausted if the flow declines to a certain threshold, below which the compulsatory flow-limitation of oxygen consumption intervenes. At that threshold, it is no longer possible for the circulation to maintain normal oxygen metabolism of all parts of the tissue.

The imbalance may be due to insufficient capillary driving pressure or increased maximum reaction rate, or both, and the effect of the imbalance is different for the different arrangements and regulations of mitochondria and cytochrome oxidase discussed above. However, in all cases, the oxygen tension will decline until no mismatch exists between the metabolism and the delivery, but the needs of tissue will not be satisfied in part or in whole.

22.5.1
Distributed Model of Insufficient Oxygen Delivery

The issue of when the needs of the tissue are not satisfied can be resolved by consideration of the tissue oxygen distribution. Equations (22.11) to (22.15) apply to the situation in which the capillary oxygen supply is not sufficient to supply the entire tissue at the necessary rate. When that happens, the segmental fraction x served with oxygen will not reach unity ($x<1$).

In the case of nonuniform oxygen distribution, the effect depends on the distribution of metabolism. It is possible that some parts of the tissue exhaust the oxygen supply, leaving other parts without oxygen. The exhaustion of the oxygen supply by a given fraction (x) of the tissue mass or volume can be predicted from Eq. (22.15) by introduction of a limit, x_0, at which the oxygen supply is exhausted ($x_0<1$):

$$P_{O_2}(x) = P_{O_2}^{cap} - \left(2\left[\frac{x}{x_0}\right] - \left[\frac{x}{x_0}\right]^2\right)\frac{\bar{J}_{O_2}}{L} \tag{22.24}$$

where $\bar{J}_{O_2}$ is the normal rate of oxygen consumption of the tissue, excluding the part that is not supplied with oxygen (equal to $1 - x_0$), and x_0 is the fraction consuming oxygen at the normal rate. The oxygen tensions accompanying the depletion of the oxygen supply are given by Eq. (22.24), which yields the value of $x(x_0)$ at which the tension reaches zero, when the supply does not match the demand. The actual oxygen consumption average of the tissue is then $J'_{O_2} = x_0 \bar{J}_{O_2}$.

In this case, Eq. (22.1) is no longer valid as written but must be modified to account for the nonlinear rise of the extraction fraction caused by the exhaustion of the oxygen supply. Equation (22.1) modifies to

$$E_{O_2}(z) = \frac{z\bar{J}_{O_2}}{FC_{O_2}^{art}} \int_0^z x_0(z)dz \tag{22.25}$$

where as above $\bar{J}_{O_2}$ is the normal oxygen consumption per unit mass or volume of tissue. To solve Eq. (22.25) for $x_0(z)$, it is necessary to establish the capillary oxygen tension as a function of the extraction fraction reached in segment (z) serving the volume fraction x according to Eq. (22.5):

$$P_{O_2}^{cap}(z) = P_{50}^{app}(z)^{h(z)}\sqrt{\frac{1}{E_{O_2}(z)} - 1} \tag{22.26}$$

as well as the fraction of the tissue supplied with oxygen at that tension according to Eq. (22.6) for $\bar{P}_{O_2}^{mit} = 0 \ (x_0 \leq 1)$:

$$x_0(z) = \sqrt{\frac{L(z)P_{O_2}^{cap}(z)}{\bar{J}_{O_2}}} \tag{22.27}$$

where $L(z)$ and $\bar{J}_{O_2}$ are assumed to be known from the normal cases of adequate oxygen supply. Thus, when $L(z)$ and $\bar{J}_{O_2}$ are known, numerical solution of the three equations (Eqs. 22.25 to 22.27) with the three unknowns $E_{O_2}(z)$, $P_{O_2}^{cap}(z)$ and $x_0(z)$ yields both the capillary oxygen tension profile for a case of insufficient oxygen supply and the resulting oxygen consumption $J'_{O_2}(z) = x_0(z)\bar{J}_{O_2}$.

The solution can be further illustrated by consideration of the flux capacity by differentiation of Eq. (22.24), which yields a description of the oxygen metabolism in each segment of the tissue (x) at each segment of the capillary bed (z):

$$j_{O_2}(z, x) = (x_0(z) - x)\bar{J}_{O_2} = \frac{L(z)P_{O_2}^{cap}(z)}{x_0(z)} - x\bar{J}_{O_2} \tag{22.28}$$

Tissue Oxygen Flux Reserve

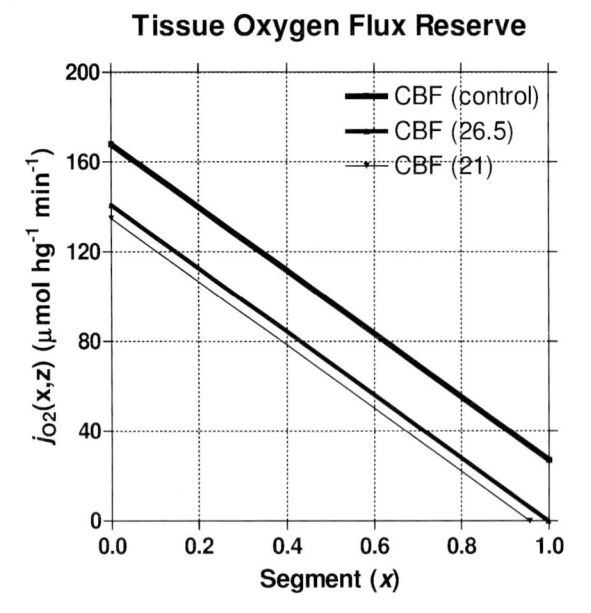

Fig. 22.7 Flux capacitor lines for distributed model of tissue oxygen, predicted from normal values shown in Tables 22.1 and 22.2, and in conditions of reduced blood flow as indicated in the graph. Dependent variable was calculated by means of Eqs. (22.11) and (22.28). Abscissa: tissue segment (x) served by average capillary compartment ($z=0.5$). Ordinate: oxygen flux as a function of x. Flux reserve at normal blood flow is consistent with minimum mitochondrial oxygen tension of 8.5 mmHg. At the lowest blood flow rate, only 96% of tissue ($x_0=0.96$) can consume oxygen at the normal rate.

which is a line of slope $-\bar{J}_{O_2}$ and ordinate intercept $J'_{O_2}(z)$. The ordinate intercept is the actual oxygen flux from that part of the capillary bed, equal to the product of the capillary oxygen pressure and oxygen diffusibility, relative to the tissue fraction served at that point $[x_0(z) \leq 1]$.

The linearly declining flux capacity of Eq. (22.28) is illustrated in Fig. 22.7 for normal human brain as well as for lowered blood flow rates, including a blood flow rate just below the ischemic threshold. For values of z at which $x_0 < 1$, $E_{O_2}(z)$ fails to rise linearly and the compartmental simplification is then possible only by numerical solution of Eqs. (22.25) to (22.27).

22.5.2
Compartment Model of Insufficient Oxygen Delivery

In the case of uniform oxygen distribution, the reaction rate declines in every mitochondrion until the consumption again matches the supply, as dictated by the flow. The extraction of oxygen at which this happens can be evaluated by means of Eq. (22.9) by setting $\bar{P}_{O_2}^{mit}$ equal to zero and solving for $\bar{E}_{O_2}$, assuming diffusibility and hemoglobin saturability to remain constant. The threshold and the declining oxygen consumption below the threshold can also be illustrated as a contraction of the tissue fraction served by the capillary compartment. This is shown in Fig. 22.8 as a diminution of the capillary segmental fraction z served by the capillary compartment, beginning at the blood flow rate at which x declines below unity before the end of the capillary bed has been reached.

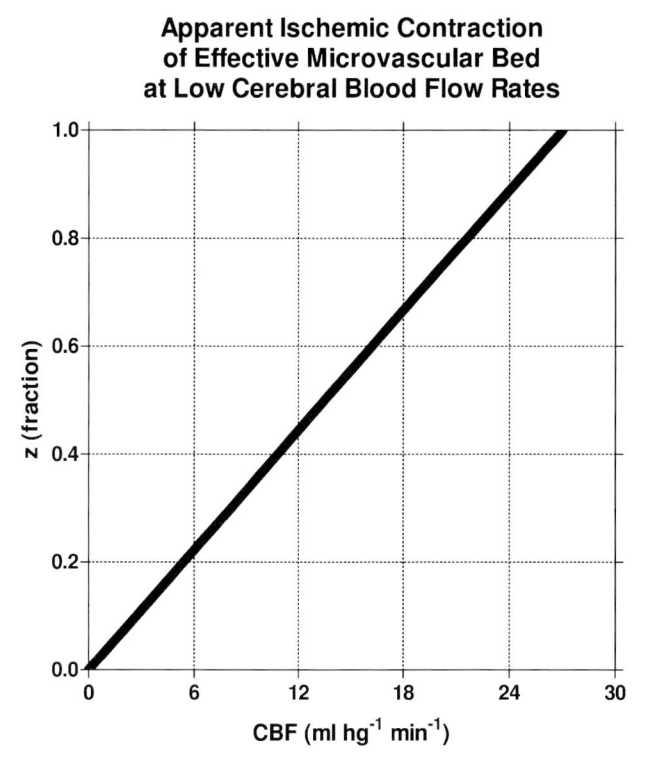

Fig. 22.8 Prediction from Eq. (22.1) of apparent ischemic contraction of magnitude of effective capillary bed serving brain tissue, using standard values listed in Tables 22.1 and 22.2. Abscissa: cerebral blood flow (ml hg^{-1} min^{-1}). Ordinate: capillary bed segments serving entire tissue segment ($x_0 = 1$). Effective capillary bed starts contracting at blood flow rates below 27 ml hg^{-1} min^{-1}.

The actual oxygen flux is less than the arterial supply of oxygen, because it depends on the average oxygen tension in the capillary, which in turn depends on the extraction of oxygen. The discrepancy between the arterial delivery of oxygen and the actually available oxygen supply is a result of the observed diffusion-limitation of oxygen delivery to brain tissue (Gjedde et al. 1991; Kassissia et al. 1995), as expressed in the magnitude of $\bar{L}$. Thus, while oxygen consumption is flow-limited, oxygen delivery is diffusion-limited.

22.6
Experimental Results

22.6.1
Brain Tissue and Mitochondrial Oxygen Tensions

Using values listed in Tables 22.1 and 22.2, the inequality given as Eq. (22.21) makes the prediction that average oxygen tensions in human brain tissue range over 20–30 mmHg, depending on the distribution and properties of oxygen consumption sites. Measurements of oxygen tensions date back many decades for animal brain, but are not particularly well described in human brain. As expected from the treatment leading to inequality (Eq. 22.21), most authors found that values in human brain tissue in vivo range from 100 mmHg close to arterioles to less than 30 mmHg further from the microvessels. However, the majority of values lie in the range of 20–40 mmHg (Zauner et al. 2002), which is completely consistent with the tissue compartmental average of 30 mmHg summarized in Fig. 22.4.

Mitochondrial oxygen tension in human brain tissue in vivo can only be inferred indirectly from the oxygen-binding properties of hemoglobin and the oxygen-metabolizing properties of cytochrome c oxidase. These are summarized in Table 22.1. Using these values and the well known ischemic threshold of oxygen extraction of 60% (see below), Gjedde et al. (2005) employed Eq. (22.19) to estimate an average mitochondrial oxygen tension of 8.5 mmHg in human brain.

Table 22.2 Average properties of human whole brain and cerebral cortex. (Modified from Kuwabara et al. 1992), Vafaee et al. 1999), and Gjedde et al. 2005).

Variable (unit)	Whole brain	Cerebral cortex
CMR_{glc} ($\mu mol\ hg^{-1}\ min^{-1}$)	0.25	0.30
CMR_{O_2} ($\mu mol\ hg^{-1}\ min^{-1}$)	1.40	1.60
CBF ($ml\ hg^{-1}\ min^{-1}$)	0.43	0.50
Oxygen extraction fraction (E_{O_2}; ratio)	0.39	0.38
Oxygen diffusibility (L; $\mu mol\ hg^{-1}\ min^{-1}\ mmHg^{-1}$)	0.032	0.041
Cytochrome oxidase activity (J_{max}; $\mu mol\ hg^{-1}\ min^{-1}$)	1.59	1.70

22.6.2
Flow-Metabolism Coupling

Equation (22.23) is the quantitative expression of the flow-metabolism couple in compartmental kinetics. The equation considers only the steady-state, although it could possibly be used to predict some transient events, provided the rate constants of the transients were of the appropriate magnitude. The foundations of the equation is shown in Figs. 22.1 to 22.3. Figure 22.1 shows that the oxygen extraction as a function of capillary segment rises linearly.

Fundamentally similar models of non-linear flow-metabolism coupling have been formulated by several authors (Weibel 1984; Buxton and Frank 1997; Gjedde 1997; Vafaee and Gjedde 2000; Aubert and Costalat 2002). The inverse relation between the oxygen extraction fraction and the oxygen tension of the capillary bed is central to these revisions of the conventional coupling of flow to

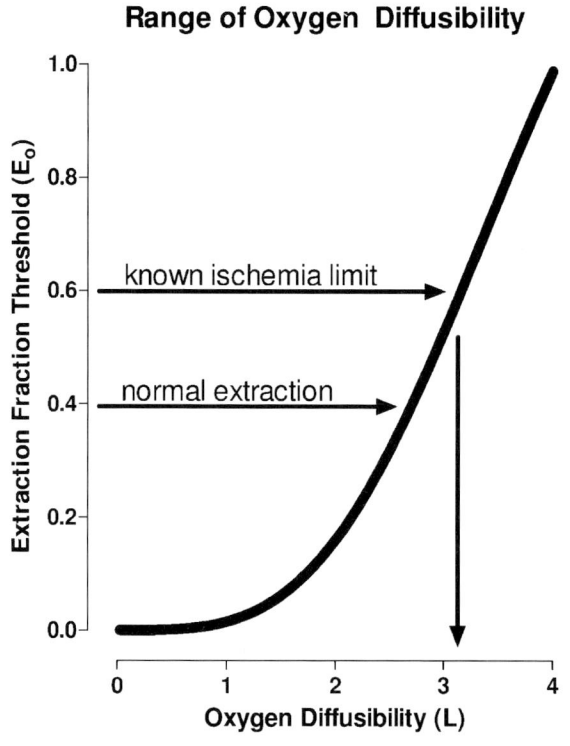

Fig. 22.9 Estimation of whole-brain oxygen diffusibility ($\bar{L}$) from Eqs. (22.7) and (22.8), using standard values listed in Tables 22.1 and 22.2. Abscissa: possible magnitudes of oxygen diffusibility (μmol hg^{-1} min^{-1} mmHg^{-1}). Ordinate: predicted oxygen extraction fraction threshold as a function of possible oxygen diffusibility. The known ischemic oxygen extraction limit (E_0) of 60% is consistent with a magnitude of $\bar{L}$ of 3.17 μmol hg^{-1} min^{-1} mmHg^{-1}.

Visual Stimulation

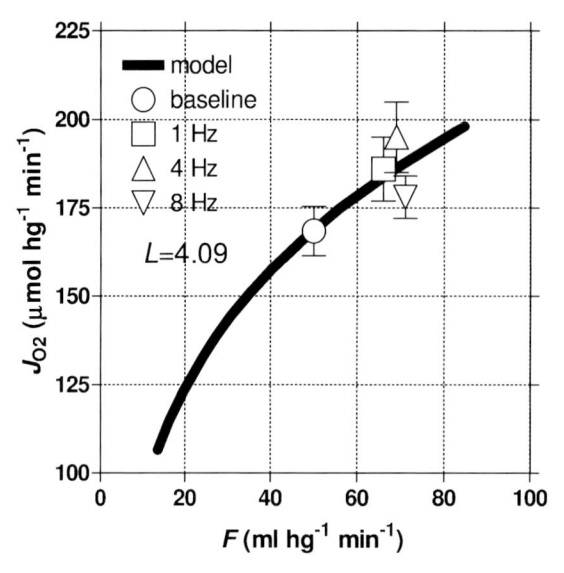

Fig. 22.10 Flow-metabolism coupling during stimulation of visual cortex of human brain, measured with positron emission tomography of labeled oxygen uptake. Curve shows segment of cortical flow-metabolism couple illustrated in Fig. 22.5 for cerebral cortex. Abscissa: cortical blood flow (ml hg^{-1} min^{-1}). Ordinate: oxygen consumption in human visual cortex (μmol hg^{-1} min^{-1}). (Modified from Vafaee et al. 1999).

metabolism, which is assumed to maintain a normal oxygen extraction fraction. The revisions arise from the insight that an inverse relation is necessary because every increase in the extraction fraction represents a reduction in the partial pressure of oxygen in the capillary compartment.

The formulation of the steady-state flow-metabolism couple is consistent with evidence obtained in visual cortex activation (Vafaee et al. 1999; Gjedde and Marrett 2001), motor cortex activation (Kastrup et al. 2002; Vafaee and Gjedde 2004), and hypothermia (Sakoh and Gjedde 2003), as shown in Figs. 22.9 to 22.13. The evidence shown in Fig. 22.10 is consistent with human cortical gray matter having an oxygen diffusibility of 4 μmol kg^{-1} min^{-1} mmHg^{-1}, while the evidence shown in Fig. 22.11 is based on the slightly higher whole-brain oxygen diffusibility of approximately 5 μmol kg^{-1} min^{-1} mmHg^{-1} in the smaller pig brain. The evidence also confirms that blood flow rates are down-regulated when the energy demand of neurons falls at low temperatures (Sakoh and Gjedde 2003).

The formulation of Eq. (22.23) is also consistent with evidence from functional MRI experiments which yield changes in oxygen consumption relative to baseline oxygen consumption as a function of changes in blood flow relative to

Flow-Metabolism Coupling During Hypothermia

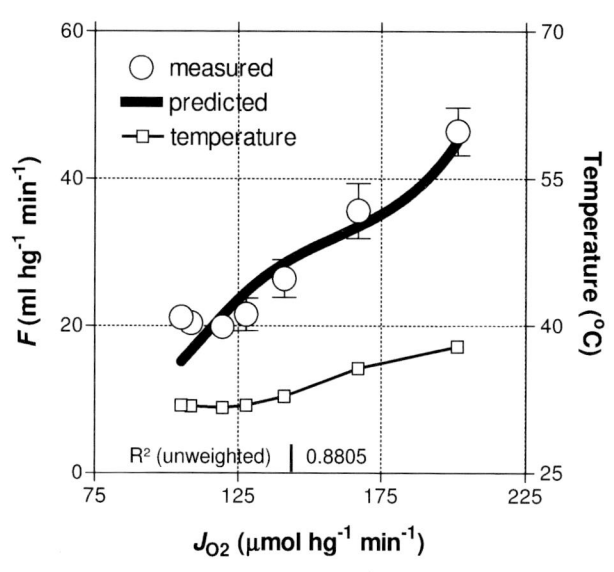

Fig. 22.11 Flow-metabolism coupling during hypothermia, calculated for porcine cortical tissue with oxygen diffusibility of 5 µmol hg^{-1} min^{-1} mm^{-1} Hg. Abscissa: oxygen consumption (µmol hg^{-1} min^{-1}). Ordinate: cortical blood flow (ml hg^{-1} min^{-1}) and cerebral core temperatures (°C). Note that axes have been inverted from previous graphs). (Modified from Sakoh and Gjedde (2003).

the baseline blood flow. This evidence generally shows that the relative change of the oxygen consumption is a linear function of the relative change of blood flow. Equation (22.23) makes the prediction illustrated in Fig. 22.6, in which the slope of the relative oxygen consumption change is 27% of the relative blood flow change for whole human brain. This is similar to the 30% relation reported by Kastrup et al. (2002), shown in Fig. 22.12, but lower than the 50% relation reported for human visual cortex by Hoge et al. (1999), shown in Fig. 22.13. It is important to keep in mind that a linear relation between changes in oxygen consumption and blood flow with a slope different from unity, relative to the respective baselines, is proof of a nonlinear relation between the absolute values of the respective variables.

Motor Activation (M1)

INTERCEPT	0.7003
SLOPE	0.3013
Std. Error	
INTERCEPT	0.03827
SLOPE	0.02706

○ Vafaee et al.
■ Kastrup et al.

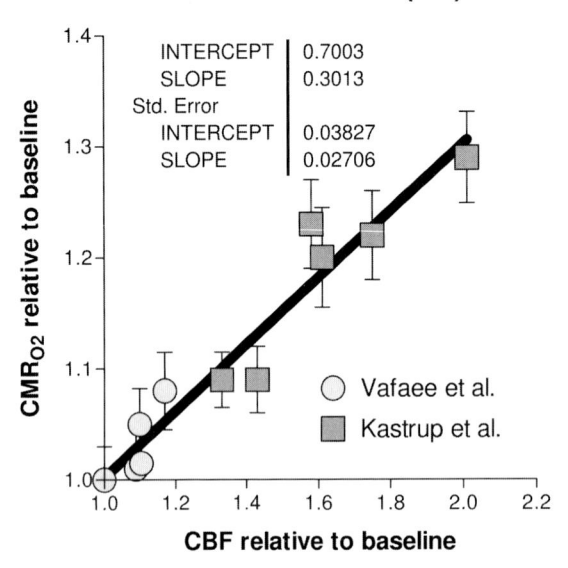

Fig. 22.12 Relative changes of blood flow and oxygen consumption in human motor cortex, measured by functional MRI (Kastrup et al. 2002) and positron emission tomography (Vafaee and Gjedde 2004). Abscissa: change in blood flow, relative to baseline (fractions). Ordinate: change in oxygen consumption in human motor cortex activation by finger motion, relative to baseline (fractions). (Modified from Kastrup et al. 2002 and Vafaee and Gjedde 2004).

22.6.3
Ischemic Limits of Oxygen Diffusibility

Figures 22.2 and 22.3 (and 22.8) show that cerebral blood flow to the human brain limits the oxygen delivery below a threshold of 25 ml kg^{-1} min^{-1}, corresponding to an oxygen extraction of 60%. Below this rate of blood flow, the oxygen supply is not sufficient to satisfy the demand. For standard values of metabolism and diffusibility, listed in Tables 22.1 and 22.2, the threshold is reached when $\bar{E}_{O_2}$ rises to 60%. The threshold extraction of 60% has been confirmed in a number of studies of human brain summarized by Gjedde et al. (2005) as well as in an animal study (Scheufler et al. 2002).

The threshold extraction depends on the magnitude of the diffusibility $\bar{L}$, which in turn can be inferred from the magnitude of the mitochondrial oxygen tension $\bar{P}_{O_2}^{mit}$. Using the well known ischemic limit as an indication that $\bar{P}_{O_2}^{mit}$ is negligible, Eqs. (22.7) and (22.8) give rise to a relationship between $\bar{L}$ and the ischemic threshold shown in Fig. 22.9, according to which the magnitude of $\bar{L}$ is close to 3 µmol hg^{-1} min^{-1} mmHg^{-1} (3.17 µmol hg^{-1} min^{-1} mmHg^{-1} according to Gjedde et al. 2005).

Visual Activation (V₁)

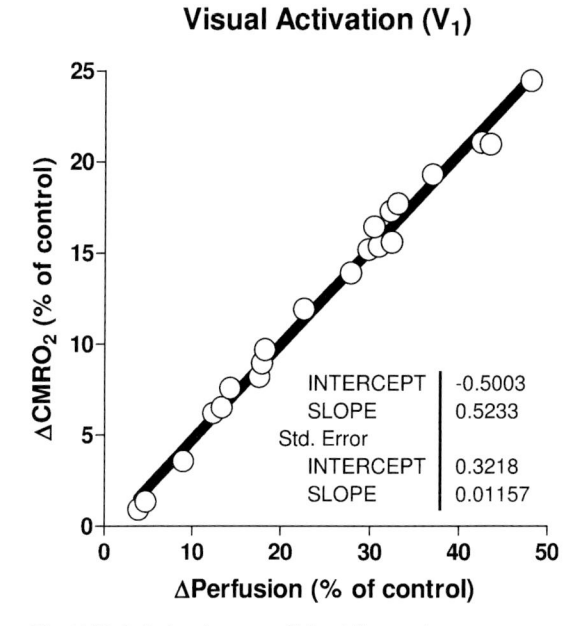

INTERCEPT	-0.5003
SLOPE	0.5233
Std. Error	
INTERCEPT	0.3218
SLOPE	0.01157

Fig. 22.13 Relative changes of blood flow and oxygen consumption during stimulation of human visual cortex, measured by functional MRI. Abscissa: change in blood flow, relative to baseline (%). Ordinate: change in oxygen consumption in human motor cortex activation by photic stimulation of visual cortex, relative to baseline (%). (Modified from Hoge et al. (1999).

References

Aubert A, Costalat R **2002**, A model of the coupling between brain electrical activity, metabolism, and hemodynamics: application to the interpretation of functional neuroimaging, *NeuroImage* 17, 1162–1181.

Buxton RB, Frank LR **1997**, A model for the coupling between cerebral blood flow and oxygen metabolism during neural stimulation, *J Cereb Blood Flow Metab* 17, 64–72.

Crone C **1963**, The permeability of capillaries in various organs as determined by use of the "indicator diffusion" method, *Acta Physiol Scand* 58, 292–305.

Duling BR, Berne RM **1970**, Longitudinal gradients in periarteriolar oxygen tension: a possible mechanism for the participation of oxygen in local regulation of blood flow, *Circ Res* 27, 669–678.

Gjedde A, Ohta S, Kuwabara H, Meyer E **1991**, Is oxygen diffusion limiting for blood-brain transfer of oxygen? in *Brain Work and Mental Activity, Alfred Benzon Symposium 31*, eds. Lassen NA, Ingvar DH, Raichle ME, Friberg L, Munksgaard, Copenhagen, pp. 177–184.

Gjedde A **1997**, The relation between brain function and cerebral blood flow and metabolism, in *Cerebrovascular Disease*, ed. Batjer HH, Lippincott-Raven, Philadelphia, pp. 23–40.

Gjedde A, Marrett S **2001**, Glycolysis in neurons, not astrocytes, delays oxidative metabolism of human visual cortex during sustained checkerboard stimulation in vivo, *J Cereb Blood Flow Metab* 21, 1384–1392.

Gjedde A **2002**, Cerebral blood flow change in arterial hypoxemia is consistent with negligible oxygen tension in brain mitochondria, *NeuroImage* 17, 1876–1881.

Gjedde A, Marrett S, Vafaee M **2002**, Oxidative and nonoxidative metabolism of excited neurons and astrocytes, *J Cereb Blood Flow Metab* 22, 1–14.

Gjedde A, Johannsen P, Cold GE, Ostergaard L **2005**, Cerebral metabolic response to low blood flow: possible role of cytochrome oxidase inhibition, *J Cereb Blood Flow Metab* 25, 1183–1196.

Gnaiger E, Steinlechner-Maran R, Mendez G, Eberl T, Margreiter R **1995**, Control of mitochondrial and cellular respiration by oxygen, *J Bioenerg Biomembr* 27, 583–596.

Gnaiger E, Lassnig B, Kuznetsov A, Rieger G, Margreiter R **1998**, Mitochondrial oxygen affinity, respiratory flux control and excess capacity of cytochrome c oxidase, *J Exp Biol* 201, 1129–1139.

Guilivi C **1998**, Functional implications of nitric oxide produced by mitochondria in mitochondrial metabolism, *Biochem J* 332, 673–679.

Hansen AJ **1984**, Ion and membrane changes in the brain during anoxia, *Behav Brain Res* 14, 93–98.

Hansen AJ **1985**, Effect of anoxia on ion distribution in the brain, *Physiol Rev* 65, 101–148.

Hoge RD, Atkinson J, Gill B, Crelier GR, Marrett S, Pike GB **1999**, Linear coupling between cerebral blood flow and oxygen consumption in activated human cortex, *Proc Natl Acad Sci USA* 96, 9403–9408.

Kassissia IG, Goresky CA, Rose CP, Schwab AJ, Simard A, Huet PM, Bach GG **1995**, Tracer oxygen distribution is barrier-limited in the cerebral microcirculation, *Circ Res* 77, 1201–1211.

Kastrup A, Krüger G, Neumann-Haefelin T, Glover GH, Moseley ME **2002**, Changes of cerebral blood flow, oxygenation, and oxidative metabolism during graded motor activation, *NeuroImage* 15, 74–82.

Krogh A **1919**, The number and distribution of capillaries in muscles with calculations of the oxygen pressure head necessary for supplying the tissue, *J Physiol* 52, 405–415.

Kuwabara H, Ohta S, Brust P, Meyer E, Gjedde A **1992**, Density of perfused capillaries in living human brain during functional activation, *Prog Brain Res* 91, 209–215.

Sakoh M, Gjedde A **2003**, Neuroprotection in hypothermia linked to redistribution of oxygen in brain, *Am J Physiol* 285, H17–H25.

Scheufler KM, Rohrborn HJ, Zentner J **2002**, Does tissue oxygen-tension reliably reflect cerebral oxygen delivery and consumption? *Anesth Analg* 95, 1042–1048.

Vafaee MS, Meyer E, Marrett S, Paus T, Evans AC, Gjedde A **1999**, Frequency-dependent changes in cerebral metabolic rate of oxygen during activation of human visual cortex, *J Cereb Blood Flow Metab* 19, 272–277.

Vafaee MS, Gjedde A **2000**, Model of blood-brain transfer of oxygen explains nonlinear flow-metabolism coupling during stimulation of visual cortex, *J Cereb Blood Flow Metab* 20, 747–754.

Vafaee MS, Gjedde A **2004**, Spatially dissociated flow-metabolism coupling in brain activation, *NeuroImage* 21, 507–515.

Wang CY, Bassingthwaighte J **2001**, Capillary supply regions, *Math Biosci* 173, 103–114.

Weibel ER **1984**, *The Pathway For Oxygen*, Harvard University Press, Cambridge, Mass.

Zauner A, Daugherty WP, Bullock MR, Warner DS **2002**, Brain oxygenation and energy metabolism, part I: biological function and pathophysiology, *Neurosurgery* 51, 289–302.

23
Functional Brain Imaging

Gerald A. Dienel

23.1
Molecular Imaging of Biological Processes in Living Brain

23.1.1
Introduction

Specialized cells in the brain are integrated within a complex, intricate organization of cellular networks and pathways designed to carry out many diverse activities. Modern techniques to visualize human brain have their roots in histochemical and autoradiographic methods used to assess "static structure" (e.g., anatomical localization and relative amounts of cell types, specific proteins, lipids, enzymes, receptors, and biological compounds including metabolites and signaling compounds) and "dynamic functions" (e.g., local rates of blood flow, glucose utilization, and oxygen consumption; DNA, RNA, lipid, amino acid, neurotransmitter, and protein synthesis; and changes in concentrations of levels of metabolites, ions, and signaling molecules) of brain (Fig. 23.1). Imaging of biological processes in vivo is particularly important because the properties and integrative activities of the system of interest can be assessed within the environment of the living brain.

23.1.2
Molecular Imaging

Major goals of neuroscientists who use functional imaging as an experimental tool are to understand behavioral and cognitive functions in terms of cellular activities, interactions of neurons, astrocytes, oligodendroglia, and microglia, and interactions of networks of cells. Broad application of imaging techniques to studies of brain function has been a strong driving force to elucidate the electrophysiological, biochemical, and physiological basis for imaging signals. Blood flow, metabolism, and cellular functions are inseparable aspects of brain activity,

Brain Imaging: Visualizing Structure and Function

Global	Regional	Cellular, Subcellular, Molecular

Whole-brain
Uptake
Clearance
Content
	Anatomy–structure
	Characteristics of
	  brain structures
	Location and levels of
	  molecules
	
	Functional status
	Properties of intrinsic
	molecules change with
	function or physiological
	conditions (e.g., redox
	state, bound oxygen, pH)
	
	Rates of processes
	Tissue levels of tracers
	accumulate in proportion
	to local rates of specific
	processes (flow, transport,
	metabolism, synthesis)
		Cell type & sub-type
		Neurons – neurotransmitter class
		Astrocytes, oligodendrocytes
		Microglia, macrophages
		Blood cells, stem cells
		
		Cells
		Dye filling – volume, territory,
		  connections
		Track movement, migration
		
		Subcellular components
		Pre- and post-synaptic receptors
		Neurotransmitter vesicles
		
		Enzymes, receptors, genes
		Metabolism & cell division
		Neurotransmitter-related functions
		Ligands, suicide substrates
		Reporter genes

Fig. 23.1 Functional imaging techniques obtain specific types of information about the brain and its functions. An integrated approach is necessary to build a framework that facilitates studies spanning the range from global to regional, cellular, subcellular, and molecular analysis.

and these processes are targets of molecular imaging because they underlie and contribute to signals used to construct images of living brain at work (Fig. 23.2). Molecular imaging, as defined by Massoud and Gambhir [1], is "the visual representation, characterization, and quantification of biological processes at the cellular and subcellular levels within intact living organisms." Molecular imaging techniques interface with many disciplines and use a wide variety of probes and technologies designed to detect and measure specific molecules, processes, and functions in cells ranging from gene expression to cognition (Fig. 23.3).

23.1.3
Influence of Blood-Brain Interface on Functional Imaging

Molecular probes to assay brain activities in vivo fall into two major categories: (1) intrinsic compounds that are endogenous components of the blood or brain, and (2) extrinsic compounds that must be inserted into the brain across the blood-brain interface that serves as a protective system for the brain. Together, the blood-brain barrier and the blood-cerebrospinal fluid (CSF) barrier limit the types, amounts, and time-courses of transfer of material from blood to brain by

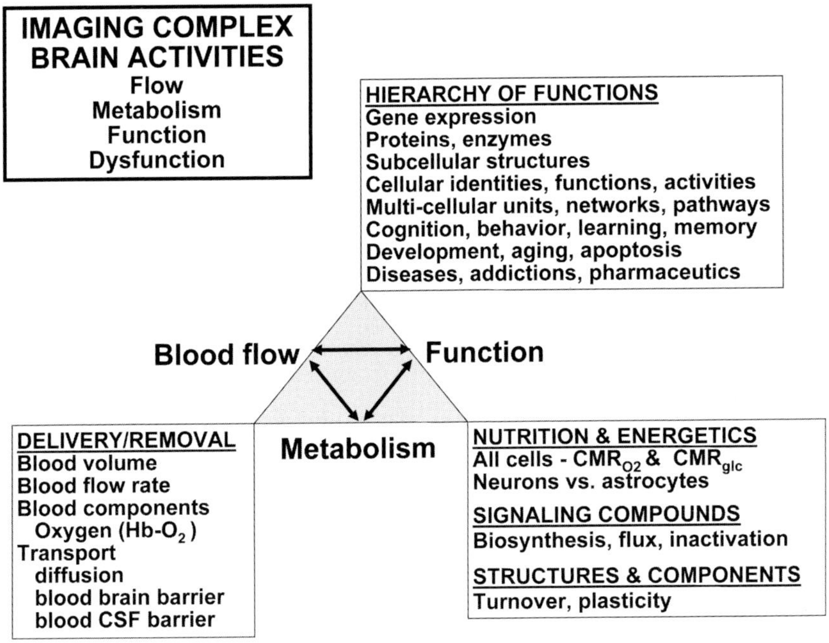

IMAGING COMPLEX BRAIN ACTIVITIES
Flow
Metabolism
Function
Dysfunction

HIERARCHY OF FUNCTIONS
Gene expression
Proteins, enzymes
Subcellular structures
Cellular identities, functions, activities
Multi-cellular units, networks, pathways
Cognition, behavior, learning, memory
Development, aging, apoptosis
Diseases, addictions, pharmaceutics

Blood flow ↔ Function
Metabolism

DELIVERY/REMOVAL
Blood volume
Blood flow rate
Blood components
　Oxygen (Hb-O_2)
Transport
　diffusion
　blood brain barrier
　blood CSF barrier

NUTRITION & ENERGETICS
All cells - CMR_{O_2} & CMR_{glc}
Neurons vs. astrocytes

SIGNALING COMPOUNDS
Biosynthesis, flux, inactivation

STRUCTURES & COMPONENTS
Turnover, plasticity

Fig. 23.2 Interrelationships of cerebral blood flow, metabolism, and brain function. Neural activity requires blood flow and metabolism to provide the fuel for energy-dependent functions. Because local rates of blood flow and metabolism are closely linked to local changes in functional activity they are commonly used as indirect measures of brain function.

various mechanisms. For example, the tight-junction blood-brain barrier system blocks entry hydrophilic substances into brain unless there are specific transporters, thereby excluding neuroactive compounds present in blood, such as glutamate, from entering brain. These transporters control the types and quantities of molecules that enter brain by means of their amounts and substrate affinities (V_{max} and K_m): saturation of a transporter limits uptake into brain when blood levels rise and changes in the number of transporters can help adapt to chronic changes in blood levels of compounds required by brain. Metabolism at the blood-brain interface can also restrict entry into the brain by converting molecules that diffuse through the lipid barrier into other compounds. For example, ammonia readily diffuses into brain but is rapidly converted to glutamine by astrocytes, presumably in their endfeet-surrounding capillaries. Extrusion mechanisms mediated by P-glycoprotein remove specific compounds that gain access to the brain. Although blood-brain barrier systems eliminate use of certain tracers, properties of the blood-brain interface can be used to advantage to design and use extrinsic probes to measure specific processes.

The following sections describe imaging methods, then give examples of biological processes actively being studied, along with some emerging approaches.

Tools for Imaging Brain Functions

Methodologies
Nuclear imaging
Optical imaging
Magnetic resonance imaging
Thermal imaging
Acoustic imaging

Tracers or reporter molecules
Inert gas
Radiolabeled tracers
Fluorescent compounds
Voltage sensitive dyes
Bioluminescence
Endogenous compounds

Molecular, metabolic, physiologic, & cognitive processes; diseases
Gene expression
CBF, $CMRO_2$, CMR_{glc}, $CMR_{acetate}$
Protein synthesis
Phospholipid turnover
Neurotransmitter turnover
Receptors, transporters, channels
Pharmacology
Blood brain barrier transport, integrity
Cell-cell & regional interactions
Cell migration
Interactions of functional pathways
Cognition, learning, plasticity
Pathophysiological conditions
   Localize & characterize
   Monitor treatment efficacy

Fig. 23.3 Brain imaging can employ one or more methodological approaches to evaluate biological processes associated with specific functions. Tools depend on the technological approach and biochemical or physiological process of interest.

Emphasis is placed on in vivo assays of functional activities, but some in vitro studies in brain slices are also described, since these extend analysis of in vivo findings and serve as driving forces for new in vivo studies. Interested readers are directed to the reference materials for more comprehensive coverage of these topics.

23.2
Overview of Brain Imaging Methodologies

Various technologies, including autoradiography, optical imaging of signals derived from intrinsic and extrinsic molecules, single photon emission tomography (SPECT), positron emission tomography (PET), X-ray computed tomography (CT), magnetic resonance imaging (MRI), and functional magnetic resonance imaging (fMRI) are used, both alone and increasingly more commonly in combination, to obtain pictorial representations of brain structure and functional activity under normal and abnormal conditions in experimental animal and human subjects. Many physiological, behavioral, and cognitive activities can only be studied in brains of conscious subjects and brain imaging is used to investigate and map brain functions that require developmental and experiential

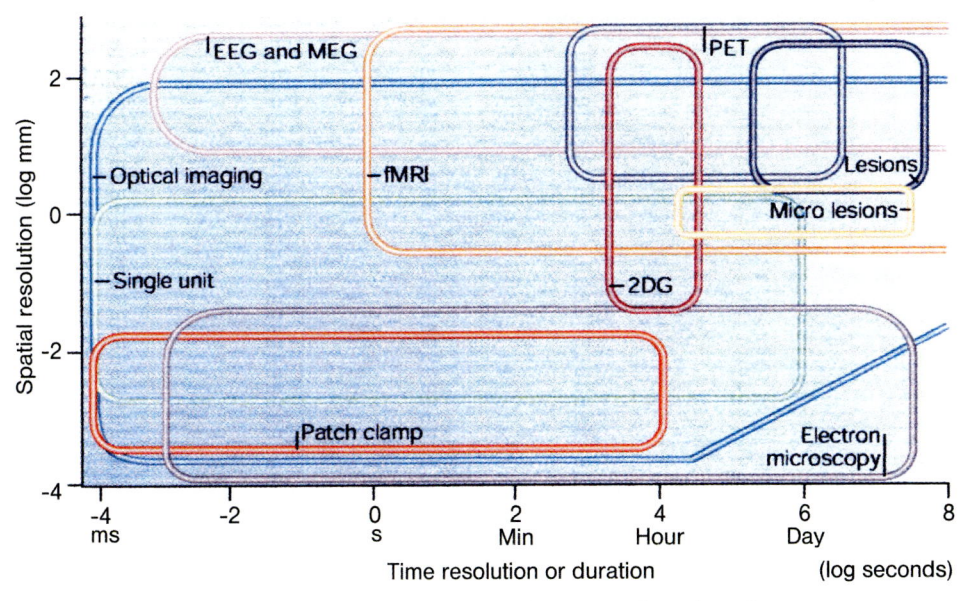

Fig. 23.4 Spatial resolution and temporal discrimination vary with methodological approaches to study brain function. Optical imaging techniques (voltage-sensitive dye imaging, assay of intrinsic signals, or fluorescent dyes) can detect rapid changes with high resolution and repeated measures can be made over long time intervals, depending on experimental design, conditions, and technology. Functional magnetic resonance imaging [fMRI] can detect rapid (<1 s) changes in blood oxygenation and repeated measures. Radiochemical assays for blood flow and oxygen utilization assays are carried out over intervals of about 1–2 min, whereas metabolic imaging with deoxyglucose [DG], fluorodeoxyglucose, and other labeled compounds requires 30–60 min. Most of the signal is acquired within the first 5–10 min after pulse labeling; and the remainder of the experimental interval allows clearance of unmetabolized precursor and a longer time to reduce the impact of estimates of rate constants on calculated rates. Parallel assays to assess electrophysiological activity (patch clamping, electroencephalography [EEG], and magneto-encephalography [MEG]) complement imaging approaches. (Reprinted from [2], with permission from the authors and Nature Publishing Group).

maturation in vivo, as well as evolution of pathophysiological changes. Thus, brain imaging is a very important and widely used technique to simultaneously visualize, localize, and quantify structural and functional properties of the entire brain; imaging is also a valuable diagnostic tool to evaluate debilitating disorders, such as stroke, dementia, and tumors.

Imaging technologies measure local concentrations of molecules and their synthesis, metabolic transformation, binding, and change of state with different spatial-temporal scales that vary over a wide range (Fig. 23.4). For example, voltage-sensitive dye imaging has the highest spatial and temporal resolution to measure cortical functional activity in living brain, registering changes within milliseconds at distances of microns [2]. Activity-dependent changes in optical properties of brain tissue also have millisecond temporal resolution and autofluorescence arising from changes in the redox state of intrinsic molecules, such as NAD(P)H and flavoproteins, can be registered within seconds and millimeters [3–6]. fMRI detects and localizes rapid shifts in the level of oxygenation of hemoglobin in blood within a few seconds with spatial resolution of a few millimeters that is being improved to the tens of milliseconds and sub-millimeter ranges [7]. Nuclear-based imaging can assess a broad array of metabolic processes, metabolite and electrolyte transport, blood-brain integrity, receptor binding, and gene expression over time-scales of tens of seconds to 5–60 min, with spatial resolution in the order of 50 μm for ^{14}C-based autoradiography to 1–2 mm for PET and SPECT [1]. The different technologies also vary in the depth of tissue that can be imaged and sensitivity to detect changes in the molecular probe.

23.2.1
Computed Tomography

Major techniques to visualize brain structure or anatomy are CT and nuclear MRI. Coaxial tomographic systems measure the attenuation of X-ray beams derived from an external source that pass through a target tissue and are registered on detectors. Tissue composition and water content contribute to mass density that is proportional to the attenuation coefficient, and X-ray transmission data are used for image reconstruction by various techniques. No contrast agents are required and the scan time is relatively short (about 20 s) compared to about 5–60 min for SPECT and PET scan times. CT scans are useful for detection of changes in ventricular volume or shape, regional atrophy, tissue density, and, when combined with PET radiotracer imaging, anatomic and diagnostic information improves interpretation of the imaging studies; use of intravenous contrast agents allows CT blood perfusion and arteriography imaging [8–10].

23.2.2
Magnetic Resonance Imaging

Magnetic resonance imaging takes advantage of the magnetic properties of atoms with an odd number of protons and/or neutrons (e.g., ^{1}H, ^{13}C, ^{23}Na, ^{31}P) and the ability of atoms to absorb energy at a very specific (resonance) radio frequency (RF) and later emit energy at the same RF after the excitation pulse ceases. The time-constants of the emitted energy are dependent on interactions with tissue, thereby introducing an influence of biological properties of tissue on the emitted signal that can be analyzed and used to construct images [11, 12]. In tissue, hydrogen in water is the most abundant atom and provides the strongest signals that vary in strength with proton density. In their normal environment, protons spin about their axis and form magnetic fields that have random orientations. When a strong external magnetic field is imposed these nuclei become aligned with the imposed magnetic field (the longitudinal or z-axis). If an RF pulse with a specific frequency is transmitted to the aligned nuclei in the transverse plane (x-y plane), the nuclei absorb energy if their resonance frequency is the same as the frequency of the RF pulse and they shift the direction of their magnetic moments toward the transverse plane. When the radio frequency pulse is turned off, the nuclei return to the original alignment in the strong external magnetic field and give off energy at the resonance frequency in a time-dependent manner. This decay or relaxation time provides a signal that has two major tissue-dependent components, T1 and T2. T1 reflects the time constant for return to magnetic alignment along the longitudinal or z-axis; and the T1 component of the relaxation process is due to the ability of protons to exchange energy with neighboring protons (spin-lattice relaxation time). The T2 signal reflects loss of signal strength due to loss or dephasing of the spin coherence (spin-spin relaxation time) in the x-y plane after the RF pulse ceases. T2* is the loss of spin phase coherence that arises from inhomogeneities in the magnetic field of the external magnet and in the local environment. Tissue components with a short T1 return to equilibrium faster and give a higher signal, whereas those with long T2 give a stronger coherent signal. In general, high signal intensity arises from high proton density, short T1, and long T2; and a low signal is associated with low proton density, long T1 and short T2. RF pulse sequence conditions have been devised to enhance gray-white matter contrast or detection of abnormalities and images can be constructed so that they mainly reflect T1, T2, or T2* signals (T1-, T2-, or T2*-weighted images); and weighting modalities are especially useful for imaging pathophysiological conditions [11]. Three-dimensional MRI images with high spatial resolution are extremely useful for anatomical images [12], particularly when combined with dynamic imaging technologies that provide functional information [13].

23.2.3
Functional MRI

Blood oxygen level-dependent (BOLD) contrast imaging, or fMRI, has been widely used to visualize neural activity in humans and experimental animals since the methodology was devised by Ogawa and colleagues in the early 1990s [14, 15]. The basis for the method is that oxyhemoglobin in blood is diamagnetic (low magnetic susceptibility), whereas deoxyhemoglobin is paramagnetic and significantly influences the T2*, causing signal attenuation or contrast. Thus, the BOLD signal represents the balance between oxy- and deoxyhemoglobin within a sampling volume, and venous blood with high levels of deoxyhemoglobin appear dark on images. Increased oxidative metabolism relative to oxygen delivery causes a negative BOLD response, whereas a higher supply of oxygen compared to demand causes a positive BOLD response because the level

Functional Magnetic Resonance Imaging (fMRI)
Blood Oxygen Level-Dependent (BOLD) Contrast

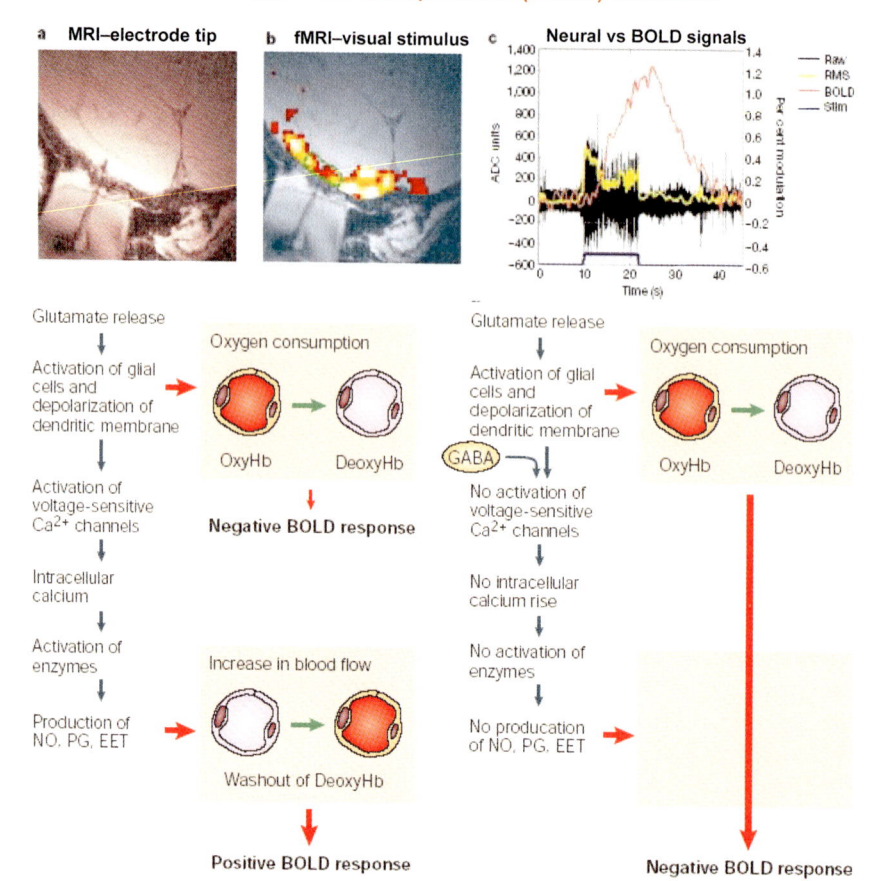

Fig. 23.5 fMRI measures BOLD responses to neural activation. The BOLD contrast MRI registers changes in blood oxygenation and can be influenced by the net balance between oxygen supply and demand arising from changes in metabolic activity of brain cells, blood oxygen level, blood volume, blood flow, and other factors. Oxyhemoglobin is diamagnetic and does not contribute to bold contrast, whereas deoxyhemoglobin is paramagnetic and venous blood containing deoxyhemoglobin contributes BOLD contrast and appears dark in the image. An increase in oxygen consumption that exceeds oxygen delivery will cause a negative BOLD response, whereas an increase in blood flow and vascular volume in excess of oxygen utilization will reduce the amount of deoxyhemoglobin in venous blood and cause a positive BOLD response. Top panels: (a) MRI image showing placement of electrode tip in primary visual cortex of anesthetized monkeys. (b) A localized positive BOLD response (color scale: red>orange>yellow>blue) in striate cortex to rotating checkerboard patterns; activation is measurable around the electrode tip. (c) Hemodynamic (BOLD) response (red line) superimposed on the raw neural signal (black) and the root mean square of the signal (thick yellow line). Neural activity began at the onset of the stimulus (blue line along abscissa) and ceased when the stimulus terminated, whereas the onset of the positive BOLD response lagged by about 2.2 s and peaked

of deoxyhemoglobin in the venous blood is reduced. Functional activity not only influences oxidative metabolism, but also governs regulation of hemodynamic properties, including blood volume, blood flow, and blood oxygen content, all of which can secondarily affect the BOLD signal [14, 15]. Because changes in functional activity are closely linked to hemodynamics and metabolism, fMRI is widely used to map changes in neural activity in human brain, as well in experimental animals (Fig. 23.5). This approach has the advantage that intrinsic signals are measured and repeated studies on the same subject can be carried out within single sessions and in longitudinal studies. The physiological basis of fMRI is not yet fully understood, but emerging evidence indicates that the BOLD signal may or may not show an initial (negative) dip due to rapid oxygen consumption just after onset of activation, whereas the positive BOLD response lags the increase in neural activity by several seconds due to slower hemodynamic responses and it involves a greater tissue volume than the activated cells. The positive signal may reflect neural activity related to input and local processing rather than spiking activity that is thought to be related to output of a functional unit (Fig. 23.5) [16–20].

Exogenous NMR contrast agents extend and complement the use of signals arising from endogenous compounds. For example, a paramagnetic agent, Mn^{2+}, has three major types of applications, assessment of active areas due to uptake of Mn^{2+} into excitable cells via voltage-gated calcium channels, three-dimensional tracing of anatomical pathways in longitudinal studies due to anterograde transport after direct injections, and whole-brain contrast after a peripheral injection of $MnCl_2$, enabling detailed visualization of cytoarchitecture [21]. Also, contrast derived from superparamagnetic iron oxide (SPIO) can be used to

---

Fig. 23.5 (continued)
later. The relationship between neural activity and the fMRI signal remains to be established, but data analysis indicates that the BOLD response appears to be more strongly correlated with local field potentials that reflect input and local integrative processing of signals in a given area than with neuronal spiking activity that is thought to be related to the output of the area. (Reprinted from [16], with permission from the authors and Nature Publishing Group). Bottom figure: A simple model portraying possible effects of synaptic inhibition on activity-dependent blood flow and BOLD contrast signals.
Left panels: Excitatory activity associated with glutamate release activates cellular activity and causes an increase in oxygen consumption, producing deoxyhemoglobin (deoxyHb), thereby leading to a negative BOLD response because utilization exceeds supply. Activation of voltage-sensitive calcium channels and calcium signaling pathways are postulated to then activate enzymes governing neurovascular regulatory mechanisms leading to increased blood flow and vascular volume, causing a subsequent positive BOLD response (see top panels b and c).
Right panels: Activation by glutamate release is associated with GABA-mediated synaptic inhibition that reduces activation of voltage-gated calcium channels, blocking the blood flow response. In this case, metabolic demand would still rise without a commensurate rise in blood flow, perhaps resulting in a more prolonged negative BOLD response. (Reprinted from Lauritzen [20], © 2005, with permission from the author and Nature Publishing Group).

image processes as diverse as apoptosis [22] and macrophage infiltration into damaged tissue [23] (see Section 23.6).

23.2.4
Radionuclide Imaging

Quantitative autoradiography, SPECT, and PET have very important roles in brain imaging studies (Fig. 23.6), and different isotopes are used for these methodological approaches (Table 23.1). Autoradiographic analysis has been used for many decades to visualize and quantify distributions of ^{3}H-, ^{14}C-, and ^{35}S-labeled compounds in tissues [24, 25], as well as movement of labeled elec-

Imaging With Radiolabeled Tracers

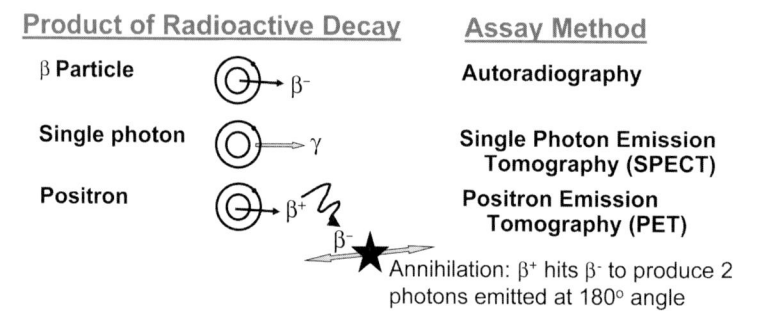

Product of Radioactive Decay	Assay Method
β Particle	Autoradiography
Single photon	Single Photon Emission Tomography (SPECT)
Positron	Positron Emission Tomography (PET)

Annihilation: β⁺ hits β⁻ to produce 2 photons emitted at 180° angle

Tissue Labeling by Different Biological Processes			
Transport	S* + E	ES*	E + S*
Metabolism	S* + E	ES*	E + P*
Irreversible binding	S* + E	ES*	
Ligand-receptor binding	L* + R	L*R	

Fig. 23.6 Radioactive tracers have broad application to molecular imaging studies. Beta (β⁻) emitters, especially the long half-lived isotopes ^{3}H and ^{14}C, are widely used in animal studies because a wide variety of biological compounds can be synthesized and used to study specific pathways and processes by quantitative autoradiographic techniques. This approach has high anatomical resolution (50–100 μm) and can be used as a routine procedure in many laboratories. A disadvantage is that each experimental subject can be used only once. In contrast, external imaging by SPECT or PET permits repeated scans of the same subject, but the technology is much more expensive. SPECT has lower resolution and a more limited number of labeled tracers, compared to PET. Radiolabeled compounds can be used to measure transport, metabolism, and binding of many compounds with different functions and in various metabolic and signaling pathways and neurotransmitter systems. See Table 23.1 for examples of isotopes, half-lives, and processes assayed. S*, radiolabeled substrate; E, enzyme; ES*, enzyme-substrate complex; P*, labeled product; L*, labeled ligand; R, receptor. For more detailed discussion, see [1, 26–28].

Table 23.1 Some examples of radiolabeled tracers and processes assayed with imaging techniques. Values for half lives [117] are rounded off.

Nuclide	Half-life	Representative compounds	Biological processes
PET			
^{82}Rb	1 min	Rb$^+$	Blood-brain barrier integrity, cardiac blood flow, integrity
^{15}O	2 min	H$_2$O; O$_2$; CO	Blood flow; oxidative metabolism; blood volume
^{13}N	10 min	Ammonia	Nitrogen and amino acid metabolism
^{11}C	20 min	Glucose; methionine, thymidine palmitate, acetate; nicotine; SCH 23390, raclopride, WIN 35428; Carfentanil, Ro 15-1788	Glucose utilization; protein synthesis, DNA synthesis; fatty acid metabolism, oxidative metabolism; nicotinic receptors; dopamine D-1 and D-2 receptors, dopamine transporter; opiate receptor, benzodiazepine receptor
^{68}Ga	68 min	Ga-chelates, antisense oligonucleotides	Gene activation, protein binding
^{18}F	110 min	F$^-$; fluorodeoxyglucose (FDG); fluoroDOPA, fluorouracil	Bone scan; glucose utilization; neurotransmitter turnover; nucleic acid turnover
SPECT			
^{133}I	13 h	Radiopharmaceuticals	Blood flow, receptor binding
^{67}Ga	78 h	Ga-chelate	Tumor detection, localization
^{133}Xe	5 days	Xe	Regional blood flow
^{99m}Tc	6 h	Radiopharmaceuticals	Tumors, blood flow
Autoradiography			
^{86}Rb; ^{54}Mn; ^{22}Na; ^{36}Cl	19 days; 312 days; 2.6 years; 3×10^5 years	Rb$^+$; Mn^{++}; Na$^+$; Cl$^-$	Electrolyte transport and homeostasis
^{35}S	87 days	Methionine	Protein synthesis
^{45}Ca	165 days	Ca^{++}	Calcium homeostasis
^{32}P; ^{33}P	14 days; 25 days	PO$_4^-$; ATP	Phosphorylation reactions
^{131}I; ^{125}I	8 days; 60 days	Iodoantipyrine	Blood flow; iodination reactions to label proteins, thyroxin, and other compounds
^{3}H	12 years	Carbohydrates, lipids, nucleic acids, carboxylic acids, etc.	Metabolism, signaling, cell division, pharmacology, transport, enzyme mechanisms
^{14}C	5 730 years	Carbohydrates, lipids, nucleic acids, carboxylic acids, etc.	Blood flow, metabolism, transport, signaling, cell division, receptor pharmacology, enzymes, protein, lipid, DNA, RNA synthesis

trolytes within the body. In brief, the weak beta particles (electrons) emitted during the decay process can be detected by autoradiography using X-ray film, emulsion, or phosphoimager techniques. Autoradiography provides a relatively cheap and easy-to-use procedure for the routine laboratory setting, but it has the disadvantage that external detection of the low-energy beta particles is not possible and tissue from each experimental subject must be processed for analysis; only one time or treatment sample can be obtained for each animal. The external detection of high-energy γ-rays emitted from within the subject is exploited by SPECT and PET. These modalities require expensive equipment and a highly-trained technical staff to carry out the studies but have the advantage that repeated, noninvasive studies can be carried out in the same subject. Single photon γ-emitting isotopes (Table 23.1) can be incorporated into various molecules and detected and quantified by γ-cameras that rotate around the subject so that tomographic images can be produced. PET has an advantage over SPECT in that many more positron-emitting isotopes that have different, short half-lives (Table 23.1) can be used by radiochemists to synthesize a large variety of labeled compounds of biological importance and interest. Also, the annihilation reaction between a positron and electron produces two photons with an energy of 511 keV emitted at about 180° (Fig. 23.6); and these signals are nearly simultaneously detected, then processed to calculate the origin and magnitude of the signal source. The PET detection system, although complex and expensive, can be used for all compounds of interest due to the same 511 keV photons derived from each positron emitter. Extraordinary advances have been made with PET during the past several decades and recent comprehensive reviews and monographs describe progress in this field in detail; many types of radiolabeled tracers are available for evaluating blood components, blood-brain barrier transport, blood-CSF transport, and brain structure and functions [1, 26–28].

23.2.5
Optical Imaging

Neural activity causes electrochemical, biochemical, and physiological changes in tissue that are detectable by altered absorption, light scattering, fluorescence, and other optical parameters (Fig. 23.7), and optical signals can be generated by activity-dependent electrolyte- and metabolism-driven changes in properties of intrinsic or extrinsic molecules [3–6]. For example, light scattering properties of tissue are altered by changes in ion and water flux across cellular membranes. Oxidative metabolism causes deoxygenation of hemoglobin and changes in its absorption or reflection of near-infrared light; oxidation of NADH and NADPH produces nonfluorescence compounds, whereas oxidation of $FADH_2$ yields a fluorescent product [29].

Near-infrared spectroscopy (NIRS), with a spectral window range of 700–1000 nm, was first used for noninvasive in vivo optical imaging through the in-

23.2 Overview of Brain Imaging Methodologies

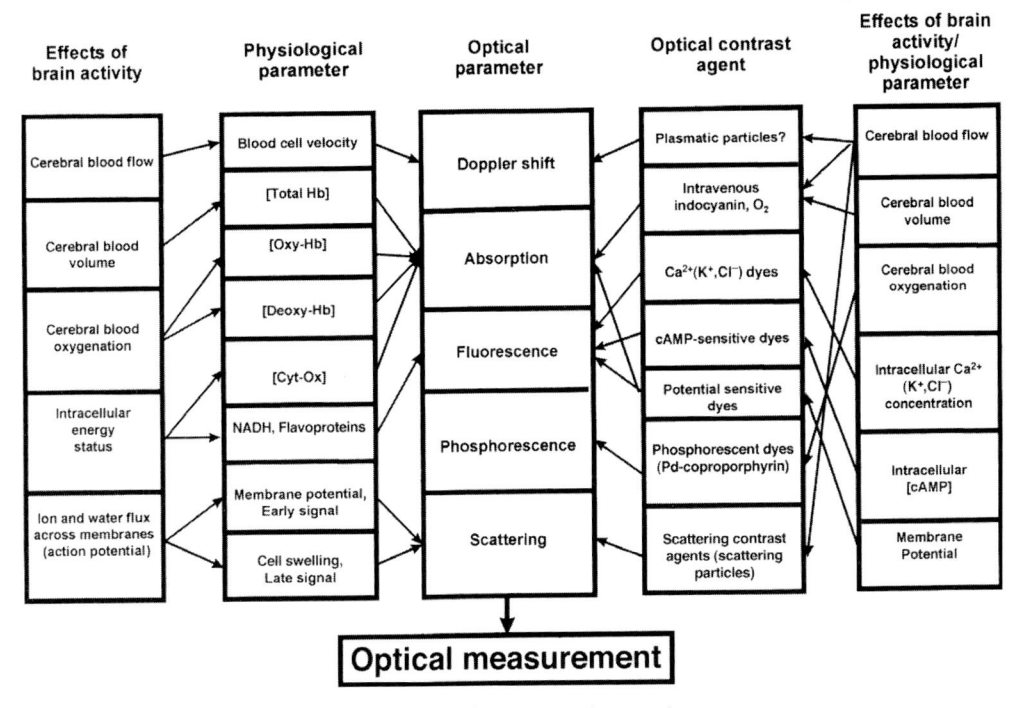

Fig. 23.7 Assays of physiological changes related to brain activity by optical methods. Changes in absorption, fluorescence, and light scattering are the major optical parameters monitored via intrinsic (derived from endogenous compounds) or extrinsic (derived from exogenous molecules) reporter molecules. These optical signals provide information about blood flow, volume, oxygenation, energy status of brain cells, oxidation-reduction status of specific molecules, membrane voltage changes, and other physiological parameters. (Reprinted from [3], with permission from Elsevier).

tact skull by Jöbsis [30]. This approach has less anatomical resolution but good temporal resolution and specificity of the parameters measured. Light absorption is mainly by oxy- and deoxyhemoglobin, water, and lipids, and scattering is due mainly to differences in refractive indices of cellular components [31]. NIRS imaging is sensitive to hemodynamic changes and provides information related to functional activation by noninvasive methods [32]. Real-time imaging of a cerebral thrombosis and vascular leakage after blood-brain barrier damage is achieved using NIR-fluorescent probes [33].

Extrinsic or externally-applied molecules, such as fluorescent dyes, are used to report specific properties of cells related to function-dependent shifts in the free concentrations of major electrolytes and in metabolism (e.g., membrane voltage, calcium level, or pH changes). The uniform loading of exogenous dyes into living brain is difficult because these are lipid-soluble compounds that are not readily delivered by blood; direct application to the brain surface may not result in homogeneous distribution of a dye. Also, the magnitude of many fluorescent signals is small, and photobleaching can limit the usefulness of a dye. Nevertheless, voltage-sensitive and calcium-binding dyes are very useful for optical imaging of fast neural events (milliseconds) [2, 34], whereas a pH-responsive dye, neutral red, reports slower changes (seconds) [35]. A recent advance in optical imaging of endogenous fluorofluors is the construction of genetically encoded reporter protein molecules, thereby circumventing the problem of placing the extrinsic compound in the region of interest. Fluorescent protein-based markers expressed in specific cell types are used to monitor cell movements in brain slices [36], neural development [37], and new voltage-sensitive fluorescent fusion proteins are used to assay changes in membrane potential [38]. Fluorescent proteins are mainly used in cultured cells and brain slices with wide-field, confocal, and two-photon microscopy, but future applications will undoubtedly include in vivo studies.

23.2.6
Thermal and Optico-Acoustic Imaging

Infrared imaging of temperature gradients in brain is an emerging technology that is currently being used to detect temperature differences in brain and between brain and tumor. Tumor-tissue temperature differences can arise from the combined effect of hypoperfusion of tissue surrounding a tumor and increased vascularity of tumors [39]. A high resolution infrared camera detects thermal energy that is emitted from a source in proportion to temperature and the resulting electrical signal is displayed as a thermal image on a video monitor. The sensitivity is on the order of $0.02\,°C$, with a spatial resolution in the 100–400 µm range and scan intervals as low as 10 ms. The brain surface temperature is heterogeneous and the differences between arteries and veins (1.5–2.0 °C), between veins and parenchyma (0.1–1.0 °C), and between vascular occlusion and reperfusion are readily detectable [39]. Conceivably, this technology might eventually be useful to detect thermal changes arising from functional activation or depression and hemodynamic changes.

Another developing technology is ultrasound-mediated imaging that has applications for tumor imaging, but might be useful for functional imaging in the future. Optico-acoustic and photo-acoustic tomography take advantage of the resolution advantage of ultrasound and contrast advantage of optical imaging; biological tissues have poor ultrasound contrast but low attenuation, whereas near-infrared light has better contrast but high scattering and less depth penetration. In brief, the optico-acoustic technology employs very short pulses from a laser light source or radiofrequency source to irradiate tissue, whereas photo-acoustic technology employs ultrasonic modulation of coherent laser light in tissue [40]. With optico-acoustic tomography, absorption of the energy from near-infrared light causes small, focal heating in tissue and this perturbation causes acoustic waves that propagate through the tissue. The acoustic waves are detected by ultrasonic transducers and used to reconstruct the optical absorption pattern. Recent studies illustrate the use of endogenous optical contrast (mainly hemoglobin) and exogenous contrast agents, either a dye or gold nanoparticles; and the resultant images can have sub-millimeter resolution at tissue depths as much as 5–6 cm, with the ability to map blood vessels, internal structures, and antibody-bound nanoparticles targeted to tumor cells in vitro [41–43].

23.2.7
Summary

Technologies for imaging brain in vivo (Figs. 23.4 to 23.7) vary widely in their use of intrinsic and extrinsic signaling molecules, their temporal and spatial resolution, the biological parameters that they measure, and the use of conscious or anesthetized subjects. Some approaches are particularly suited for extremely fast events, others for slower processes. In vitro systems and some in vivo systems have key roles for development of more complex and expensive technologies. Some approaches have specific molecular targets that can be measured directly, whereas others assess neural events indirectly, by means of changes in hemodynamics or metabolism. No single methodology or molecular probe can provide a complete picture of the events taking place in brain; and combinations of technologies and imaging probes improve the analysis of functional changes within identifiable anatomical structures over time.

23.3
Imaging Biological Processes in Living Brain: Watching and Measuring Brain Work

23.3.1
Functional Activity, Brain Work, and Metabolic Imaging

In the broadest sense, functional activity involves all brain processes related to information transfer within cells, from cell to cell, and among networks of cells. The term "brain activation" generally refers to an increase in information trafficking and energy demand over and above that measured in a baseline or "resting" state, with implicit recognition that: (1) there is a considerable amount of conscious and unconscious activity taking place in the absence of the specific stimulatory condition or task being studied, and (2) the specific resting state can vary markedly with experimental condition. For example, in certain types of studies, anesthesia is required to prevent movement artifacts during the recording session, and the resting state in the anesthetized subject is clearly not the same as that in the conscious subject.

Brain work is mainly related to ionic homeostasis in excitable cells and it has been estimated that 50–60%, perhaps even more, of the energy expended by the brain is used to maintain and restore cellular ion gradients [44–46]. Electrical amino acids and some organic acids in blood are also taken up into brain and metabolized, but overall, these are minor energy substrates [47].

Glucose was identified as the major fuel for brain by global methods, i.e., determination of metabolite levels in arterial and venous blood [47]. Quantitative determinations of rates of cerebral blood flow with nitrous oxide [49] permitted the calculation of rates of oxygen and glucose consumption by brain as a whole according to the Fick principle, i.e., CMR = CBF(A-V), where CMR (μmol min^{-1} g^{-1} tissue) denotes cerebral metabolic rate for the substance of interest, CBF is cerebral blood flow (ml min^{-1} g^{-1}), and A and V are the arterial and cerebral venous substrate concentrations (μmol ml^{-1}), respectively. Studies in many laboratories established the overall stoichiometry of utilization of oxygen and glucose, the ratio CMR_{O_2}/CMR_{glc}, in normal adult brain of about 5.5 [47]. Because this value is close to the theoretical maximum of 6 for complete oxidation of glucose, most of the glucose entering brain is oxidized. Thus, in adult brain as a whole, only a small percentage of the carbon derived from glucose is used for biosynthetic reactions or is lost from brain, but functional imaging studies have revealed an apparent predominance of glycolytic compared to oxidative metabolism during activation (see Section 23.4.4).

The brain is a very heterogeneous tissue and specific functions are localized to small nuclei and neuronal networks. Analytical methods that rely on gross tissue dissection for metabolite analysis therefore lose vital information and can obtain misleading results due to averaging of results obtained in adjacent structures that have unrelated functions or different metabolic rates. Highly sensitive, precise microassays developed for the assay of enzymes and metabolites in individual dissected cells in Oliver Lowry's laboratory [50] overcame this problem, but the labor-intensive analyses limited the number of samples from each brain that could be analyzed.

Development of quantitative autoradiographic methods provided sufficient spatial resolution (about 50–100 μm) and permitted the simultaneous determination of local rates of blood flow and glucose metabolism in all brain structures. This approach not only established a close linkage between functional activity (ATP demand), blood flow, and glucose utilization, but also served as the basis for broad application of quantitative autoradiography as a major tool in neuroscience. Because the basic principles and physiological, pharmacological, anatomical, and pathological applications of the autoradiographic and PET methodology have been reviewed in detail by Louis Sokoloff [51, 52] and others [27, 28], only selected examples will be presented below. The autobiographical perspectives of Seymour Kety [53] and Louis Sokoloff [54], both of whom made seminal contributions to the foundation of the functional brain imaging field, are quite interesting.

23.3.2.2 Highly Diffusible Tracers to Measure CBF

The first fully quantitative autoradiographic procedure to measure activity-dependent changes in regional blood flow was developed by Kety and colleagues in 1955 [55]. Tissue levels of [^{131}I]trifluoroiodomethane, an inert radioactive gas, were assayed by autoradiographic analysis of frozen tissue and comparison of optical densities of regions of interest to those of calibrated labeled standards for determination of tracer concentrations per gram tissue; these values yielded local blood flow rates after dividing by the arterial blood time-activity integral and correcting for the tissue-to-blood partition coefficient for the tracer [55, 56]. Sokoloff and colleagues later developed the use of a less volatile tracer, antipyrine [57] and iodoantipyrine [58]. In the early 1960s, Ingvar and Lassen devised an external scanning method using ^{85}Kr, so CBF could be measured in humans [59, 60]. Blood flow assays are short (<1 min) and have useful applications in studies to identify brain structures that respond to brief stimuli.

23.3.2.3 Metabolizable Glucose Analogs to Measure Hexokinase Activity and CMR$_{glc}$

2-Deoxy-D-glucose (DG), a glucose analog lacking the hydroxyl group at carbon 2, was first synthesized in 1920 by Emil Fischer [61], who established some of its physical and chemical properties, such as the extreme acid lability of its glycosidic linkage. In 1954, Sols and Crane [62] assessed the substrate specificity of brain hexokinase and noted that DG isolates the hexokinase reaction, because DG is converted to DG-6-P but not to fructose-6-P. In the late 1950s, Tower used loading doses of unlabeled DG as a competitive inhibitor in assays of glucose utilization and showed the accumulation of its phosphorylated derivative in tissue. In his historical perspective, Tower [63] noted the irony that many researchers who first used DG "failed to see any practical applicability of the effects of this most interesting compound on cerebral metabolism." [^{14}C]Deoxyglucose was elegantly developed by Sokoloff and colleagues as a tracer to determine local rates of glucose utilization in rat brain [51]. Sokoloff recognized that, under steady-state conditions, the rate of each reaction in a multistep pathway is equivalent, so an assay of the rate of any single step yields the flux through the entire pathway. He developed a theoretical model to describe these processes, analyzed the model mathematically to derive an operational equation for calculating the glucose utilization rate from the rate of DG phosphorylation by hexokinase, and developed the practical procedures to carry out the method using quantitative autoradiography [51]. Within 2 years, [^{18}F]fluorodeoxyglucose (FDG), a positron-emitting analog, was in use for PET studies in primates and man [64–66]. Thus, a number of essential but unrelated studies over a long time span contributed to methodology development.

The concept of the DG assay procedure is illustrated in the cartoon in Fig. 23.9. A radiolabeled glucose analog (DG or FDG) is injected intravenously, timed samples of arterial blood are drawn, the quantities of product formed in brain regions of interest are measured, and these values are divided by the integrated specific activity of the precursor pool in brain that is calculated from that measured in blood. Finally, this value for the rate of phosphorylation of the tracer is converted to glucose utilization by correcting for the kinetic differences between glucose and DG or FDG for transport and phosphorylation: approximately two molecules of glucose are phosphorylated to every one of DG. Note

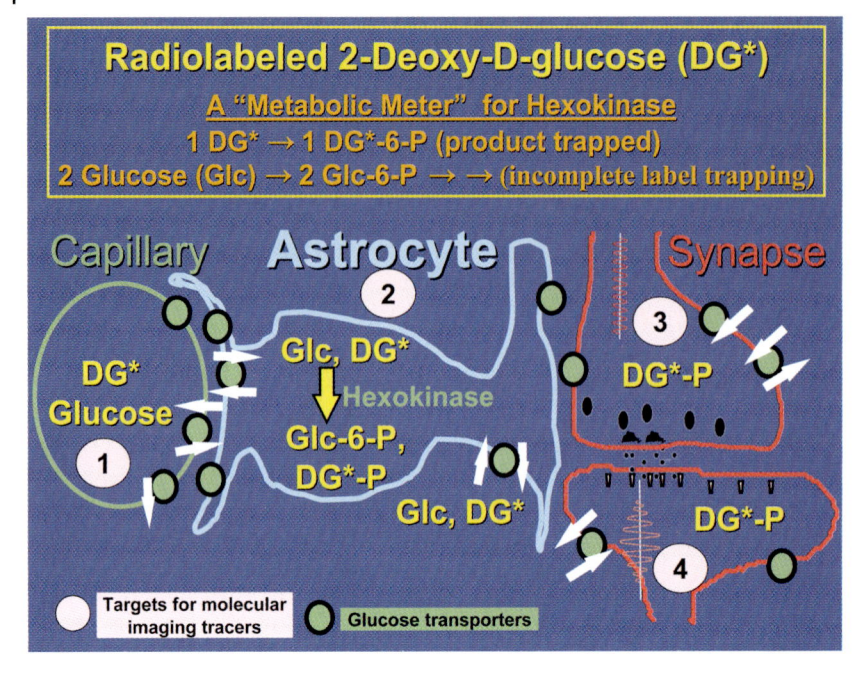

Fig. 23.9 Metabolic trapping of labeled compounds to measure reaction rates. Tracer levels of radiolabeled molecules that are analogs of endogenous compounds are used to quantitatively measure rates of specific reactions in the midst of all the other biochemical reactions taking place in living brain. 2-Deoxy-D-glucose (DG) is a glucose analog that competes with glucose for: (1) bi-directional transport across the blood-brain barrier and brain cell membranes by glucose transporters and (2) phosphorylation by hexokinase. DG-6-phosphate (DG-6-P) is trapped within the cells that metabolize DG for a reasonable duration (about 30–60 min), and in rat brain, about two molecules of glucose are phosphorylated for each molecule of DG [51]. In contrast, products of metabolism of the endogenous substrate, glucose, are diffusible and rapidly lost from activated tissue, causing large underestimates of functional activation. The design and synthesis of many labeled tracers facilitates the investigation of specific neurotransmitter systems in brain of experimental animals and humans. Targets of interest for molecular imaging tracers, indicated by circles, are: (1) blood (flow, volume, and components, including oxygen levels), (2) astrocytes (metabolism, receptors, or transporters), (3) presynaptic neuronal functions (neurotransmitter biosynthesis and levels, transmitter vesicles, transporters, receptors), and (4) postsynaptic neuronal functions (enzymes, receptors). For more details, see [1, 26–28, 94, 95, 98, 114, 115].

Fig. 23.10 Imaging sensory activation and tumor localization.
Left and middle panels, top row: unilateral photic stimulation (15 Hz on/off flash) stimulates the retina and increases neuronal signaling and astrocyte activity in the dorsal superior colliculus of the rat. The local rate of glucose utilization (CMR$_{glc}$) increases, reflected by greater trapping of [^{14}C]deoxyglucose-6-phosphate in the activated cells (arrows) but not in the contralateral tissue that was not stimulated. CMR$_{glc}$ measurements reflect the composite activity of all brain cells in the region of interest, whereas acetate metabolism reports oxidative activity in astrocytes due. Astrocyte activity also

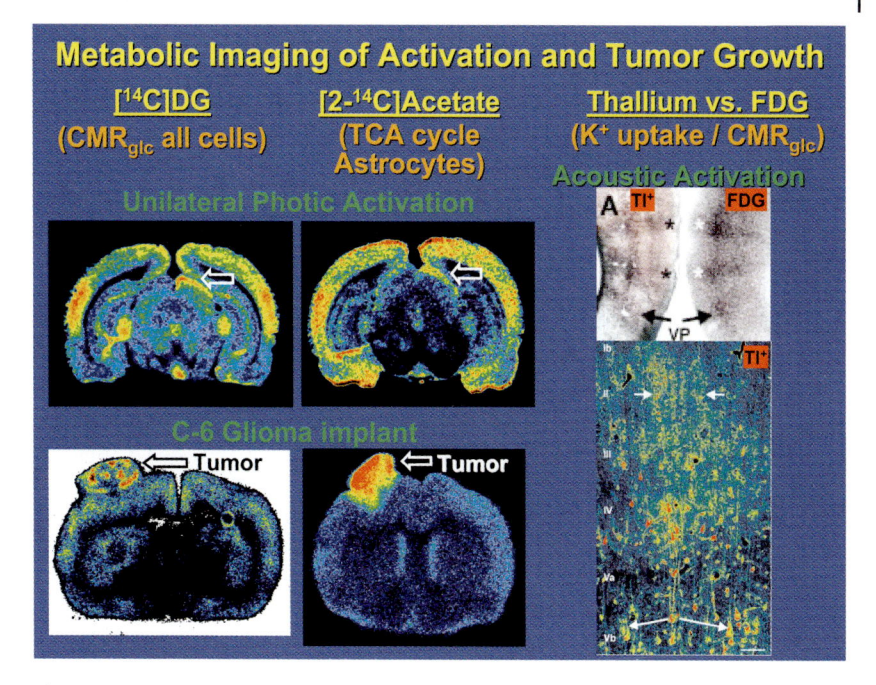

Fig. 23.10 (continued) increases during photic stimulation, and is detected by increased metabolic trapping of labeled products, mainly glutamate and glutamine. (Courtesy of N. F. Cruz and G. A. Dienel).

Left and middle panels, bottom row: although tumors are highly glycolytic, acetate is rapidly incorporated into TCA cycle-derived amino acids; and tumors can readily be detected with a 5–min labeling period with [2-^{14}C]acetate at 2 weeks after implantation of C-6 glioma cells into rat brain. Because labeling of normal brain tissue by acetate is more uniform than that obtained with [^{14}C]DG, the C-6 glioma tumor-to-tissue labeling ratio was higher for acetate compared to DG. (Courtesy of N. F. Cruz and G.A. Dienel; also see [77]). The color scales represent local tissue ^{14}C concentrations that are proportional to metabolic rates. Note that the four autoradiographs have different color scales (not shown), and ^{14}C concentrations are ranked from highest to lowest in the following order: red, yellow, green, purple, blue.

Right panels: acoustic stimulation increases CMR$_{glc}$ (assayed with FDG) and intracellular uptake of thallium (Tl), a marker for K$^+$ uptake, in gerbil auditory cortex. The top panel shows three tonotopic activation columns in auditory cortex (asterisks show the primary field AI, and arrows show the ventroposterior [VP] field) in response to a combined 1-kHz and 8-kHz acoustic stimulus. Tl$^+$ (left) and [^{14}C]FDG uptake (right) were measured in the same animal. The lower panel is a pseudo-color image showing thallium uptake in cortical layers of an 8-kHz tonotopic column at higher magnification (scale bar is 50 μm). Arrows in layer II delineate the borders of increased Tl$^+$ uptake and arrows in layer Vb point from a central cluster of pyramidal neuron to satellite clusters. Tl$^+$ uptake is detectable in different cell types (pyramidal neurons, interneurons, putative astrocytes, and dendrites. (Reprinted from [78], with permission from Elsevier).

that the product trapping of the labeled DG-6-P or FDG-6-P for the duration of the experimental period (30–60 min) is essential for the success of this method.

To summarize, development of the first quantitative procedure to measure local metabolic rates simultaneously in all regions of living brain required a comprehensive knowledge of biochemistry, physiology, mathematics, and radiochemistry. Neuroanatomic correlates of processing of physiological information are readily detected and quantified by the DG method, the effects of pharmacological intervention are easily visualized and determined in all brain structures, and glycolytic tumors can be localized. The [^{14}C]DG autoradiographs in Fig. 23.10 illustrate: (1) the regional heterogeneity of glucose metabolism in brain, (2) specific increases in CMR_{glc} in the dorsal superior colliculus in response to unilateral photic stimulus, and (3) localization of a brain tumor by DG. Note the heterogeneous optical densities of the autoradiographs that reflect regional differences in metabolism: the higher the color-coded optical density, the greater the accumulation of product. In normal rat brain, CMR_{glc} ranges from a low of about 0.3–0.4 $\mu mol^{-1} g^{-1} min^{-1}$ in white matter to about 0.5–1.8 $\mu mol^{-1} g^{-1} min^{-1}$ in various gray matter structures, with the highest values in the auditory structures [51]. In primates, values for CMR_{glc} are about half those in the rat but the rank order is similar [52]. In spinal dorsal root ganglia, electrical stimulation increases CMR_{glc} in the synaptic areas, but not in nerve cell bodies, in proportion to the stimulation frequency; and in the pituitary gland, the magnitude of CMR_{glc} was shown to be related to sodium pump activity [52].

23.3.2.4 Non-Metabolizable Analogs to Assay Transport and Tissue Concentration

3-O-methyl-D-glucose is a non-metabolizable analog that is a substrate for glucose transporters. Radiolabeled methylglucose competes with unlabeled glucose for influx into and efflux from brain; and regional glucose transporter activity and glucose concentrations can be quantitatively measured using this tracer. Methylglucose uptake measures transport without the complication of metabolic trapping of the tracer and steady-state levels of methylglucose concentration are quantitatively related to tissue glucose levels. Methylglucose autoradiographs show that glucose concentrations are relatively uniform throughout brain, indicating that glucose supply matches demand (see [67] and cited references therein). The maximal capacity for glucose transport (T_{max}) into brain from blood is about three times greater than the maximal CMR_{glc} (V_{max}) [68], so supply capacity greatly exceeds demand capacity. Compensatory mechanisms that increase blood flow contribute to the ability of delivery to match and exceed utilization, even with large increases in CMR_{glc}.

23.3.2.5 Cellular Basis of Glucose Utilization

A key issue is assessing the contributions of neurons, astrocytes, and other cell types to overall CMR_{glc} during rest, activation, and disease states. Fully quantitative assays with cellular resolution are not possible due to diffusion and loss of label during tissue processing and staining to identify cell types. [^{3}H]DG studies in the auditory system show no exclusive labeling of one cell type, with high grain density over neurons, fibers, and glia [69]. High-resolution [^{14}C]DG autoradiographic and fluorescence assays with the glucose analog NBDG show similar labeling of neurons and astrocytes under resting conditions [70, 71].

23.3.2.6 Acetate is an "Astrocyte Reporter Molecule"

Astrocytes are a major cell type in brain that have essential roles to support and regulate neuronal activity, and development of methods to measure their functional activity in vivo are important. Early studies that evaluated patterns of labeling of glutamate and glutamine by different precursors led to the identification of two major TCA cycles in brain, described as "large" and "small" glutamate pools; acetate labeled the small pool [72, 73]. Acetate and other short-chain carboxylic acids are present in blood and cross the blood-brain barrier via a monocarboxylic acid transporter (MCT) and diffusion. Preferential transport of acetate into astrocytes compared to neurons by the MCT is the basis for cell-type specificity of acetate metabolism [74] that has been confirmed by NMR studies [75]. Activation of astrocytes in the dorsal superior colliculus in vivo by photic stimulation of retinal neurons is illustrated in Fig. 23.10. During acoustic stimulation, the calculated rate of oxidative metabolism of acetate in rat brain is similar to estimates of oxidative metabolism of glucose by astrocytes [76], suggesting that acetate is an important minor fuel and it provides biologically significant amounts of energy for working astrocytes. Labeled acetate is also an emerging tool for imaging brain tumors [77].

23.3.2.7 Summary

Quantitative autoradiographic assays of cerebral blood flow and metabolism in experimental animals were a major advance in the brain imaging field because they changed data acquisition from mainly "static" information provided by anatomic and histochemical assays to "dynamic" information, consisting of rates of processes that could be simultaneously measured in all regions of conscious subjects under different experimental conditions. A brief history of these methodological approaches was presented to illustrate the complexity of the obstacles facing the initial technological developments that, once surmounted, could be quickly adapted for use in humans. CBF, CMR_{glc}, and CMR_{O_2} are useful measures of overall functioning of brain, but blood flow is not a direct measure of metabolism or of function, and flow and metabolism do not necessarily change proportionately (see Section 23.4.4). Also, the measurement of hexokinase activity does not provide information about the downstream metabolism of glucose, including oxidative metabolism, lactate formation, biosynthetic reactions. Flow-metabolism-function studies paved the way for the development of many sophisticated brain imaging methods and studies, some of which are described in the following sections.

23.4
Molecular Probes Are Used for a Broad Spectrum of Imaging Assays in Living Brain

23.4.1
Potassium Uptake and CMR$_{glc}$ During Functional Activation

Cellular activation is linked to changes in ion fluxes and increases in K$^+$ uptake during functional activation can be visualized with thallium (Tl$^+$), a K$^+$ analog. Thallium uptake into brain after intraperitoneal injection of thallium acetate is detected histochemically by its precipitation as the sulfide salt during perfusion-fixation at the end of the experimental interval [78]. The acoustic stimulation of gerbil with alternating 1 kHz and 8 kHz tones activates specific high and low frequency topographic columns in auditory cortex that are similar in location when visualized with [^{14}C]FDG and Tl$_2$S in the same animal; and at higher magnification, the cellular localization of thallium in neurons of different morphological types and in other cells in cortical layers is readily detected (Fig. 23.10).

23.4.2
Multimodal Assays in Serial Sections of Brain

Brain activation and injury alter many processes that vary in time after the initial event that triggers the biological response, and it is, therefore, essential to be able to quantitatively measure many characteristics of the tissue, as illustrated in Fig. 23.11. One of the advantages of analysis of thin (20 µm) serial coronal sections of brain is that different assays can be carried out in parallel, as long as the tissue preparative method preserves the compounds of interest. Radiolabeling methods are readily combined with fluorescence assays to evaluate the consequences of cortical freezing injury [79] and visualize regional differences in the temporal profiles of immediate and delayed responses (Fig. 23.11). For metabolic studies, the brains must be carefully frozen in situ to preserve labile metabolites that would otherwise be rapidly consumed during the immediate postmortem interval. Alternatively, histochemical and immunohistochemical staining brains can be sampled by routine euthanasia or perfusion-fixation procedures. Local rates of protein synthesis are measured with a radiolabeled precursor amino acid, and if the label on the precursor is not lost via metabolism during the experimental interval, the tissue sections must be washed with acid to remove the unincorporated tracer prior to analysis. Local pH is assayed by fluorescence imaging, whereas ATP, glucose, and lactate levels in freeze-dried sections are assayed by bioluminescence assays after coating the tissue sections with a thin layer of appropriate enzymes and cofactors and exposure to photographic film. Gene expression is assayed by autoradiography after hybridization with ^{35}S-labeled probes for specific mRNAs. Note the severe effects of traumatic brain injury at the site of the lesion, with marked changes in the penumbra re-

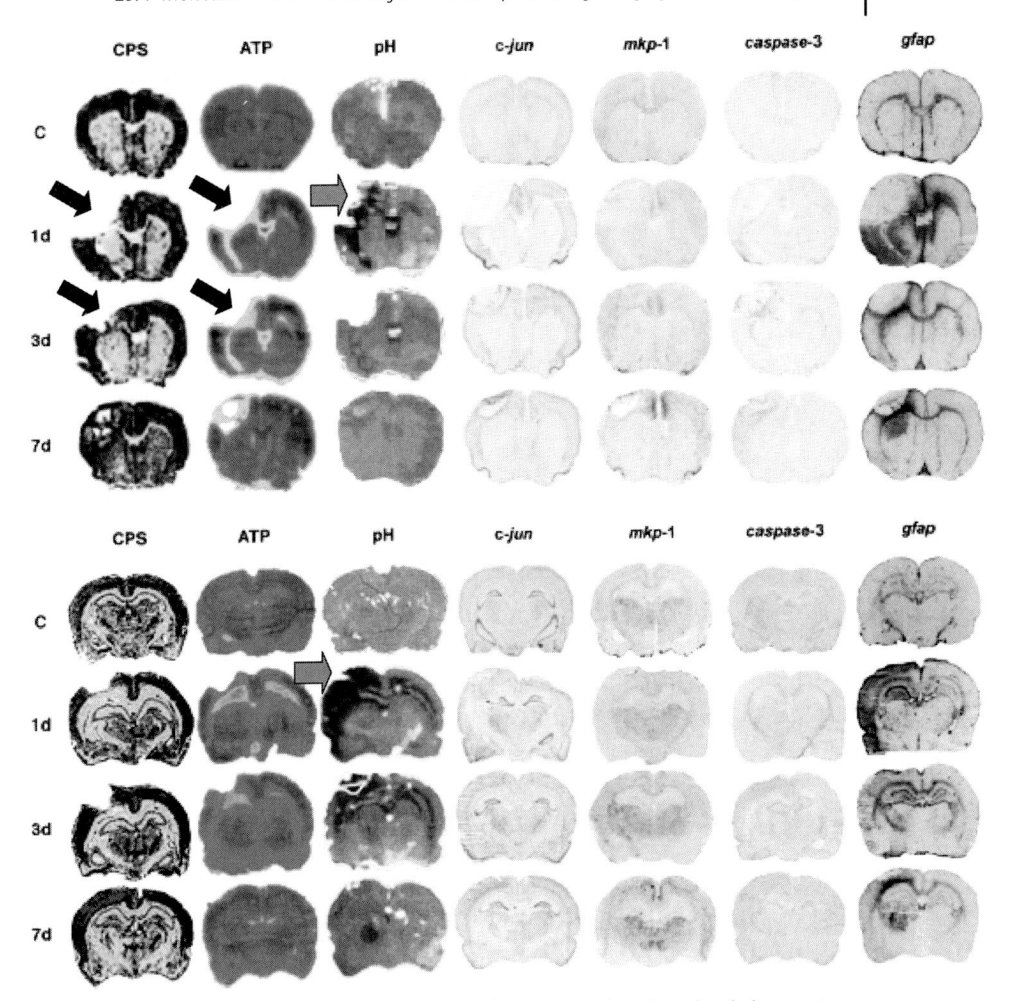

Fig. 23.11 Multimodal assessment of cellular responses after injury. Autoradiographs of coronal sections of rat brain after cortical cold-lesion injury at the level of the striatum (top set) and hippocampus (bottom set) in control (C) and at 1, 3, and 7 days after the lesion. Measured values are: local rates of cerebral protein synthesis (CPS, using [^{3}H]leucine), ATP concentrations (ATP-specific bioluminescence), pH (umbelliferone fluorescence), and c-jun, mkp-1, caspase-3 and GFAP mRNA levels (in situ hybridization with ^{35}S-labeled probes). Note the regionally selective changes in protein synthesis in the cortex and caudate, local changes in ATP level and pH, and heterogeneous, time-dependent responses of gene expression. The cold injury abolished protein synthesis and reduced ATP levels at days 1 and 3 after the cortical lesion (black arrows, left hemisphere; light areas in the autoradiographs), and the surrounding tissue had an increased pH (dark areas in the pH autoradiographs, grey arrows) that extended into more caudal regions (lower panels). Gene expression responses varied considerably. (Reprinted from [79], with permission from Elsevier).

gion, remote changes throughout the brain, and time-varying effects on different aspects of metabolism and gene expression (Fig. 23.11).

23.4.3
Imaging Human Brain Tumors

One of the most prevalent applications of FDG-PET is localization of tumors in cancer patients. Most tumors are highly glycolytic, so their rates of glucose utilization are generally much greater than surrounding tissue. Tumors, such as melanoma metastases, can be readily located by qualitative assays of [^{18}F]FDG uptake in whole-body scans (Fig. 23.12) that can help guide surgical and treatment approaches, as well as evaluate therapeutic responses [80]. Tumors vary in their extent of vascularization, and the effectiveness of tumor radiation therapy is influenced by tissue oxygen level, so it is important to establish the degree of hypoxia in a tumor for adjustment of the oxygen level in the inspired air mixture to enhance treatment efficacy. Hypoxic tissue is detectable with a PET tracer, [^{18}F]FMISO, that is specifically retained in tissue with low oxygen levels [81]. High labeling of a brain tumor with both FDG and FMISO (Fig. 23.12) is consistent with the presence of hypoxic, glycolytic tissue. One advantage of qualitative FDG tumor localization assays is that the blood time-activity integral does not need to be measured; the patient does not need to be placed in the scanner until about 1 h after the tracer injection, when nearly all FDG has been converted to FDG-6-P. However, this waiting time can lead to unanticipated results, and perhaps, misinterpretation of results unless care is taken. For example, in a case study report [82], a patient who was being assessed for tumor recurrence read an engrossing detective novel during the prescan interval, causing functional activation of Broca's area and unanticipated increased FDG uptake in this structure (Fig. 23.12). The examples shown in Fig. 23.12 also demonstrate the importance of anatomical-functional fusion images (PET-CT or PET-MRI) to identify structures of interest. FDG is the predominant tumor tracer, but many other PET tracers have been developed to study, characterize, and localize tumors [28].

23.4.4
Functional Imaging Studies of Sensory and Cognitive Activity Reveal Disproportionate Increases in CBF and CMR$_{glc}$ Compared to CMR$_{O_2}$ During Activation

Physiological stimulus and cognitive studies to identify activated brain regions and evaluate interactions among networks of cells during cognitive tasks in human subjects initially relied on PET flow-metabolism assays, but fMRI is increasingly used because exposure to radioactive material is eliminated and repeated studies with high temporal and spatial resolution are possible. For exam-

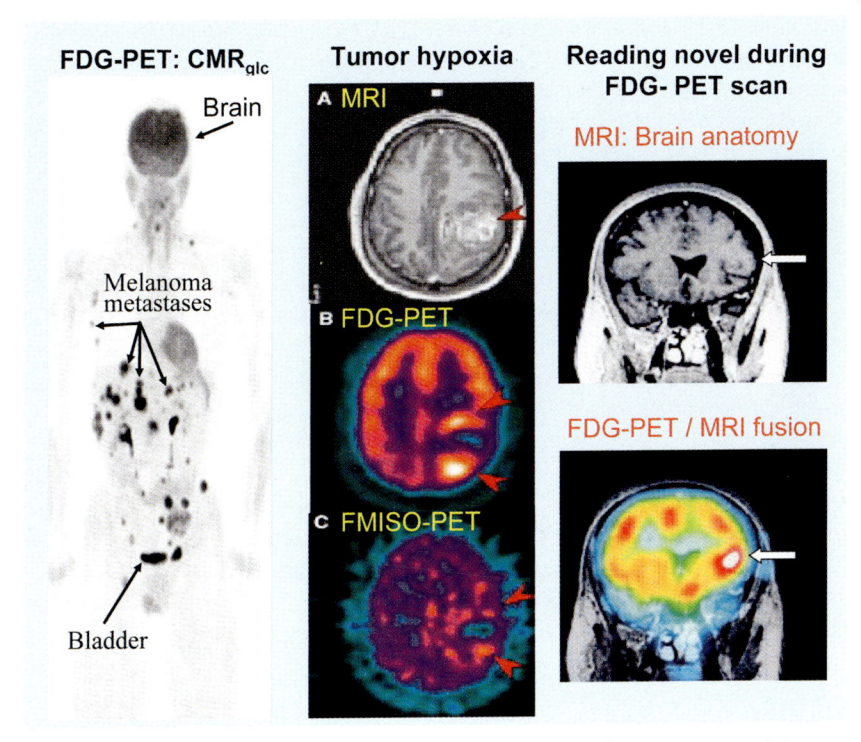

Fig. 23.12 Localization and characterization of tumors in humans.
Left panel: [^{18}F]fluorodeoxyglucose (FDG) localizes metastatic melanoma tumors throughout the body (some are indicated by arrows) that appear as metabolic "hot spots" due to their high glycolytic rates compared to normal tissue. Note that brain has a very high glucose utilization rate compared to other body organs and that the bladder is highly labeled due to clearance of FDG via the kidneys. (Reprinted from [80], with permission from Elsevier).
Center panels: a patient with high-grade glioma indicated as a large mass in the MRI image (A, arrow). FDG-PET shows high uptake in the peripheral regions of the tumor (B, arrows) and low uptake in the center of the mass, consistent with a necrotic core. Retention of [^{18}F]fluoromisonidazole (FMISO) varies with oxygen concentration and is much higher under hypoxic compared to normoxic conditions. The FMISO-PET image shows hypoxic areas (C, arrows) that partially overlap with regions of high FDG uptake. (Reprinted from [81], with permission from Elsevier). Right panels: MRI image (top) and FDG-PET/MRI fusion image (bottom) from a patient who was reading a "page-turner" detective novel in the waiting room during the interval between the FDG injection and the PET scan for assessment of metastatic tumors. The FDG-PET shows focal activation in the Broca's area of cerebral cortex (arrows; high metabolic activity is denoted by color coding of red and white), whereas no anatomical changes were evident in the corresponding MRI images. The interpretation was that reading the exciting novel, not a metastatic tumor, caused local brain activation. (Reprinted from [82], with permission from Lippincott, Williams and Wilkins).

ple, studies by Raichle and colleagues [83, 84] to evaluate functional metabolic and hemodynamic responses to sensory stimulation and to identify neural correlates of consciousness during skill learning (word processing tasks: speaking, hearing, reading, generating words) revealed brain regions with increased and decreased activity, raising interesting questions regarding factors that govern "baseline" activity itself.

One of fascinating findings from activation studies in different laboratories is the apparent dissociation of the stoichiometric relationship between oxygen and glucose utilization in human subjects. Sensory stimulation or mental tasks evoked 30–50% increases in blood flow and CMR_{glc} with little or no change in CMR_{O_2} when assayed by PET [85] or arteriovenous difference [86]. As shown in Fig. 23.13, blood flow and CMR_{glc} rise or fall in parallel during activation or deactivation, respectively, exceeding the much smaller changes in CMR_{O_2} and oxygen extraction fraction (OEF); note that the positive BOLD effect is associated with the small change in CMR_{O_2} and reduced OEF, consistent with a rise in the level of oxyhemoglobin. The CMR_{O_2}/CMR_{glc} mismatch is sometimes called "aerobic glycolysis" because it suggests the possibility of increased production and loss of lactate from the activated tissue, even though there are normal and adequate levels and rates of delivery of oxygen. In an animal model for aerobic glycolysis, the changes in brain contents of lactate and glycogen were not sufficient to explain the glucose consumed in excess of oxygen, and analysis of the metabolic fate of [^{14}C]glucose showed that biosynthetic processes could not fully explain the disproportionate utilization of glucose and that glycogenolysis, which occurs in astrocytes, enhances the magnitude of the oxygen-glucose mismatch [87, 88].

Aerobic glycolysis is a very important phenomenon because it reveals a lack of detailed understanding of the loci and identity of energy-producing reactions required to support functional activation. Also, the failure of CMR_{O_2} to equal CMR_{glc} during activation is strong in vivo evidence against a model attempting to explain the cellular basis for activation-dependent increases in CMR_{glc} by linking neuronal excitatory glutamatergic activity to glycolytic metabolism in astrocytes and parallel oxidative metabolism of lactate in neurons via the so-called "astrocyte-to-neuron lactate shuttle". Some laboratories find that sodium-dependent glutamate uptake stimulates DG metabolism and lactate release from cultured astrocytes (for a review, see [89]), whereas other laboratories find either no effect or an inhibition of CMR_{glc} and lactate release during glutamate exposure; also, glutamate is oxidized by astrocytes and it stimulates oxygen consumption (for reviews, see [90–93]). The basis for these apparently discrepant findings is unknown, but might arise from differences in oxidative capacity in various preparations of cultured astrocytes. Nevertheless, the fact that CMR_{O_2} does not match CMR_{glc} during many activation paradigms in vivo argues against a tight coupling of lactate production, shuttling, and oxidation, since this requires stoichiometric increases in CMR_{O_2} and CMR_{glc}. Thus, if lactate is produced in large amounts during activation, it must be very quickly released from the activated area, since lactate accumulation in brain and oxygen consumption fail to match the rise in glucose metabolism.

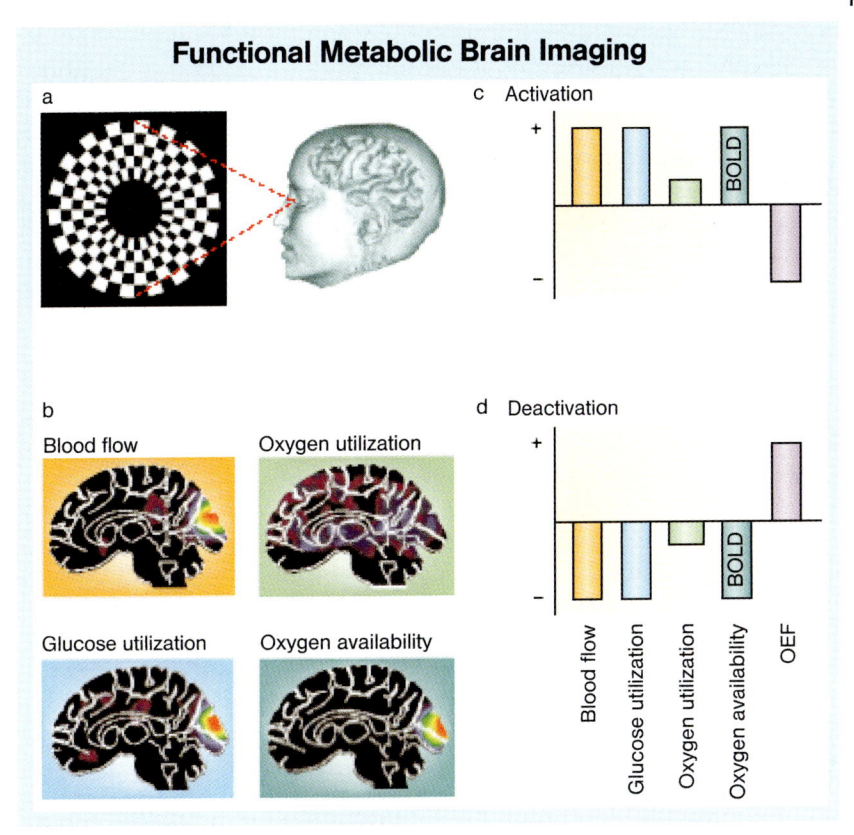

Fig. 23.13 Functional activation during visual stimulation. Relationships among blood flow, BOLD response, glucose and oxygen utilization, and oxygen extraction fraction are shown. Visual stimulation with a checkerboard pattern (a) causes an increase in blood flow (CBF) and glucose utilization (CMR_{glc}) in visual cortex, but only a small, if any, rise in oxygen consumption (CMR_{O_2}) in this structure, leading to greater oxygen availability in blood because delivery exceeds demand (b). Thus, representative metabolic measurements during brain activation using PET typically show increases in flow and metabolism, but with a disproportionate rise in blood flow and CMR_{glc} compared to CMR_{O_2}. This phenomenon is associated with a smaller oxygen extraction fraction (OEF) and excess oxygen availability in blood, causing a positive fMRI BOLD response (c) (also see Fig. 23.5). In contrast, task-induced reductions in brain activity below the baseline condition of the experimental procedure (deactivation) are associated metabolic changes in the opposite direction (d) compared to activation (c). The basis for deactivation is not understood but is thought to arise from lower overall activity rather than simply increased inhibitory activity, which itself is associated with increased CMR_{glc}. (Reprinted from [84], with permission from the authors and Nature Publishing Group).

To summarize, unanticipated findings from human sensory and cognitive activation studies using metabolic markers to provide imaging signals exposed major shifts in energy metabolism that are not understood and not explained by current models for astrocyte-neuron interactions. The excess of oxygen availability is registered by metabolic and fMRI BOLD signals; it serves as a strong impetus to understand the biochemical and physiological basis of functional activation.

23.4.5
Imaging Brain Maturation and Aging, Neurotransmitter Systems, and Effects of Drugs of Abuse

Radioactive probes are now available for many purposes, including assay of the activities of specific enzymes in vivo (e.g., hexokinase [CMR_{glc}], thymidine kinase [cell division], aromatic L-amino acid decarboxylase [dopaminergic system], monoamine oxidase, acetylcholinesterase), biosynthetic processes (protein and lipid synthesis), phospholipase-A_2-initiated signal transduction activities, neurotransmitter systems (dopamine, serotonin, acetylcholine, GABA, glutamate, opiate) and their components (pre- and postsynaptic structures, rates of neurotransmitter synthesis, transporters, receptors, receptor sub-classes), transporters (sugars, amino acids, monocarboxylic acids, multidrug resistance), abnormal formations (Lewy bodies and β-amyloid), tumors (protein, DNA, and lipid synthesis, receptors, hypoxia, and, angiogenesis, apoptosis, antigens), gene expression and gene therapy, pharmacology, addiction to drugs of abuse, and neurological diseases and disorders [1, 26–28, 94–96].

Disruption of programmed development, damage from drugs of abuse or side-effects of pharmaceuticals, and deficits incurred during aging can all impair brain function; brain imaging techniques can help monitor and evaluate these processes. For example, the maternal-fetal placental barrier permits the passage of FDG [96], and functional metabolic assays in both maternal and fetal organs, including brain, can be carried out in utero (Fig. 23.14), permitting assessment of maturation. Transfer of drugs of abuse, such as [^{11}C]cocaine, from the mother to fetus is also measurable [97]. In aging patients, a battery of assays are available to characterize and quantify specific neurotransmitter systems, as illustrated for the cholinergic system in Fig. 23.14. In dementia patients, deficits in overall "status" (registered by CMR_{glc}) are associated with reductions in vesicular acetylcholine transporters and acetylcholinesterase activity and brain atrophy [98].

The dopaminergic system is involved in the neurobiology of addiction to drugs of abuse, and PET studies help elucidate the actions of drugs of abuse, modulation of reward, motivation, and behavior circuits in brain, and effects of therapeutic interventions. Psychostimulant drugs such as cocaine are associated with blockade of the dopamine transporter, causing an increase in dopamine in the synaptic cleft. In contrast nicotine can stimulate dopaminergic cell firing,

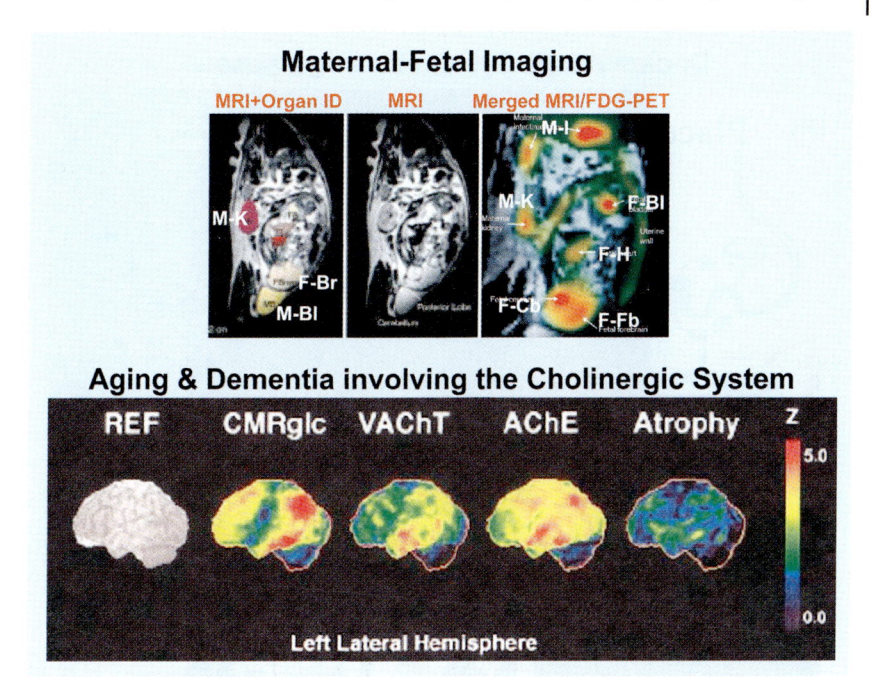

Fig. 23.14 Glucose utilization during fetal development and neurochemical changes in Alzheimer's disease.
Top panels: superimposition of maternal-fetal MRI (left two panels) and FDG-PET images (right panel) shows that FDG injected into the maternal circulation of a pregnant monkey labels maternal organs (M-K, kidney; M-I, intestine; M-Bl, bladder), and FDG readily crosses the placenta to produce high labeling of fetal brain (F-Br), forebrain (F-Fb), cerebellum (F-Cb), heart (F-H), and bladder (F-Bl). (Reprinted from [96], with permission from the Society of Nuclear Medicine).
Bottom panels: images of statistical maps derived from use of MRI and different PET tracers are used to evaluate anatomical and neurochemical changes in brain of Alzheimer's patients, compared to controls of similar age. The highest Z score, denoted by red on the color scale, indicates greatest reduction in glucose utilization (CMR_{glc}) with [^{18}F]FDG-PET, vesicular acetylcholine transporters (VAChT) measured with [^{125}I]iodobenzovesamicol-SPECT, acetylcholinesterase (AChE) activity measured with [^{11}C]methylpiperidyl propionate-PET, and regional atrophy determined by MRI imaging. Most pronounced reductions are in parietotemporal and frontal association cortices, with small or negligible effects in primary sensorimotor cortex and cerebellum. (Reprinted from [98], with permission from Elsevier).

and opiates and alcohol are linked to the disinhibition of dopaminergic cells [99]. Specific PET tracers, such as [^{11}C]raclopride that bind to dopamine receptors can reveal relationships between the perceived drug-induced "high", occupancy of the receptor, and overall functions (assayed with [^{18}F]FDG) of the brain circuits implicated in addition (Fig. 23.15). One of the major mechanisms for inactivation of catecholamine neurotransmitters is oxidation by monoamine oxidase (MAO). Recent studies with a PET tracer that measures the amount of

MAO B (one of the two isoforms of MAO) show a dramatic reduction in the level of this enzyme throughout the body of chronic smokers (Fig. 23.16). This effect is not related to nicotine, indicating that tobacco smoke has widespread effects on many different organs [100]. These examples illustrate the use of PET imaging to investigate specific neurotransmitter systems at the molecular, cellular, and organ level in living subjects.

23.4.6
Imaging Electrolyte Transport Across the Blood-Brain Interface and Shifts in Calcium Homeostasis Under Pathophysiological Conditions

Transport of electrolytes across the blood-brain interface is very restricted, compared to the rapid, high-capacity transport of glucose. The highest rate of transfer electrolytes from blood to brain is across the choroid plexus; it is less extensive across arachnoid-pial membranes, and quite low across endothelial cells. Autoradiographic analysis of blood-to-brain transfer of a representative cation, $^{45}Ca^{2+}$, shows highest labeling of periventricular tissue at 1 h after a pulse intravenous injection of tracer; by 5 h the $^{45}Ca^{2+}$ distributes widely within brain, followed by clearance at 24 h (Fig. 23.17) [101]. Studies with other radiolabeled electrolytes, including Na^+, K^+, Mg^{2+}, Mn^{2+}, Cl^-, and PO_4^{2-}, show similar labeling patterns (see [102] and references therein).

Brain regions vulnerable to transient ischemia exhibit abnormal calcium homeostasis that becomes manifest along with the evolution of ischemic cell change. Increased uptake and trapping of calcium and greater entry of blood-borne calcium into tissue compartments not labeled in normal tissue is readily detected and quantified by ^{45}Ca autoradiography after transient ischemia (Fig. 23.17). Accumulation of blood-borne ^{45}Ca in the ischemic-damaged choroid plexus is evident at 2–5 h after onset of reperfusion, in the caudate within 5 h, and at 24–72 h in the CA1 hippocampus, consistent with delayed onset of ischemic damage in the hippocampus compared to caudate.

Local secondary calcium responses that occur after a primary excitotoxic challenge of hippocampal neurons in brain slices with kainic acid are readily detected by optical imaging using a calcium-binding fluorescent dye [103]. These calcium responses originate in dendrites, and over a period of minutes, spread to involve cell bodies, then affect the entire cell and eventually lead to death (Fig. 23.17). These results illustrate an advantage of combining in vivo and in vitro studies because temporal profiles of regional disruption of calcium homeostasis can be identified, and detailed analysis of the dynamics and mechanisms of calcium overloading can be assessed.

Control of hydrogen ion concentration is essential for normal cell function. For example, regulated transfer of protons across mitochondrial membranes are used to generate energy, and proton-coupled transport of metabolites, such as facilitated cotransport of H^+ plus lactate by the MCT, can clear metabolic by-products and help control intracellular pH during functional activation or tissue

Ion transport Across BBB, Distribution, Challenge-Response

^{45}Ca Autoradiography

Normal adult rat **Transient ischemia (30 min)**

Ca^{2+} Indicator Dye

Kainate exposure (hippocampal neurons)

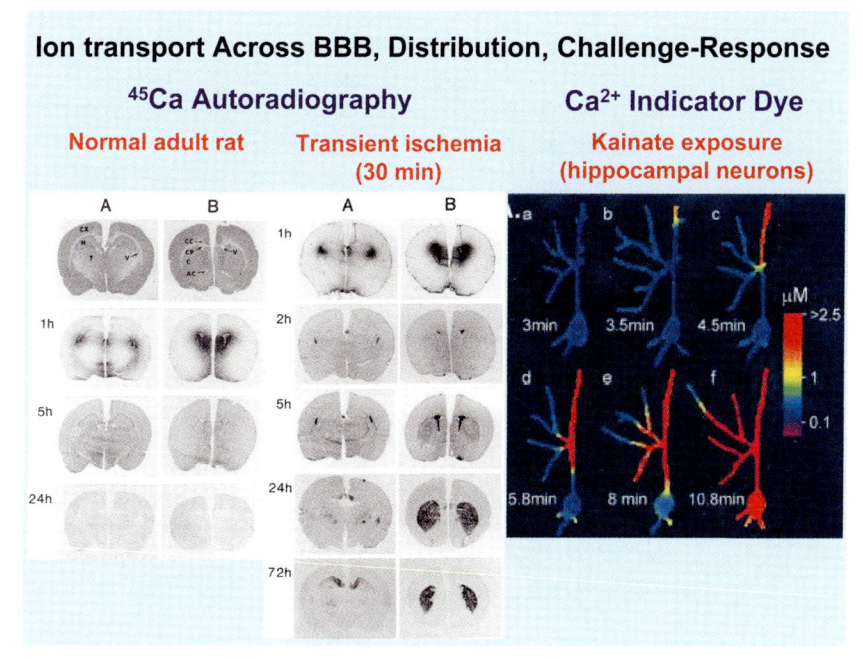

Fig. 23.17 Calcium homeostasis in normal and ischemic-damaged brain and during excitotoxic challenge.
Left two columns: top row is hematoxylin and eosin-stained coronal sections of rat brain at the level of the anterior hippocampus (A) and caudate (B). Cx = cerebral cortex; H = hippocampus; T = thalamus; V = lateral ventricle that contains choroid plexus (CP); cc = corpus callosum; c = caudate; ac = anterior commisure. ^{45}Ca autoradiographs prepared from normal rats at the indicated intervals after a pulse intravenous injection of ^{45}Ca show entry of calcium into brain mainly via the choroid plexus (1 h), spreading within brain to produce more uniform labeling of the brain at 5 h, and clearance from brain at 24 h. Middle two columns: rats subjected to 30 min of complete forebrain ischemia by four-vessel occlusion incur ischemic damage that becomes maximal in the caudate within 24 h and is delayed in the hippocampus, becoming evident at 24 h and maximal at 72 h. When ^{45}Ca is injected immediately after onset of reperfusion there is intense labeling of the choroid plexus, which is damaged by ischemia (1–5 h; A, B), increased levels of ^{45}Ca in the ischemic-damaged CA1 hippocampus at 24 h and increasing at 72 h (A); labeling of the dendritic regions is most prominent. In contrast, accumulation of ^{45}Ca in the ischemic-damaged caudate is evident within 5 h and high levels persist at 24–72 h. Note that ^{45}Ca is gradually cleared from undamaged regions in the rats subjected to ischemia (e.g., cerebral cortex and thalamus; A, B). (Reprinted from [101], with permission from Blackwell Publishing).
Right panels: the fluorescent calcium indicator dye reveals local Ca^{2+} responses in hippocampal dendrites in mouse brain slices exposed to kainic acid. Within 3 min after washout of the kainic acid, dendrites recovered to resting Ca^{2+} levels (a), but shortly thereafter, secondary Ca^{2+} responses were evident in dendrites (b), followed by propagation into secondary dendrites (c–d), before reaching (e) and involving the soma (f). Cascades leading to elevated levels of Ca^{2+} can involve the entire cell and lead to neuronal death. (Reprinted from [103], with permission from The Society for Neuroscience).

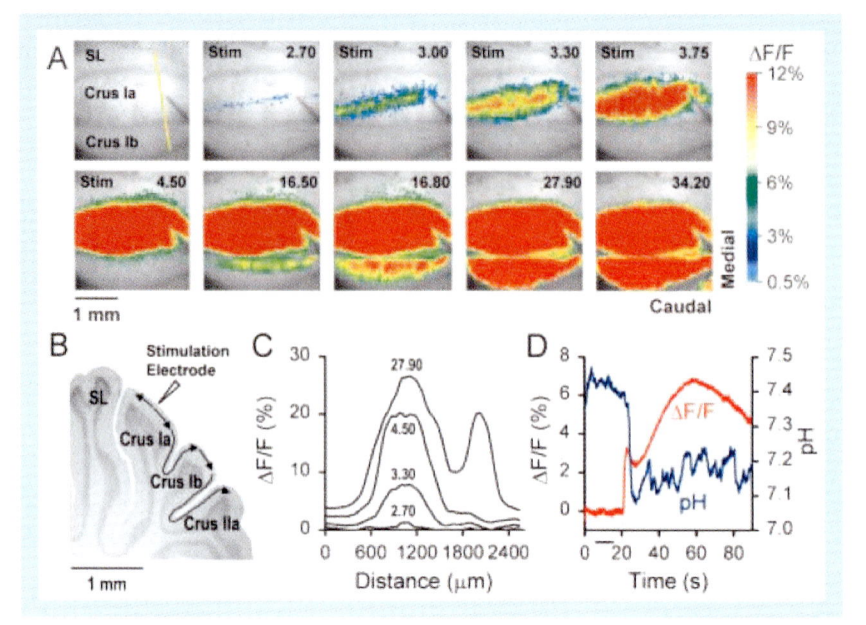

Fig. 23.18 Rapid waves of stimulus-induced spreading acidification. (A) Neutral red, a pH-sensitive dye was administered before the stimulus session. Color-coded images show spreading acidification in rat cerebellar cortex induced by a 10-s pulse train superimposed on the surface image of the brain. The first image (top left) is a background image showing the electrode position; the yellow line indicates the plane of the parasagittal section shown in (B). The series of images shows the progressive spreading of acidification across the tissue; the numbers in the upper right corner of each successive image indicates the time (s) after stimulus onset. The distance of spreading at four time-intervals is shown in (C); and (D) shows that the rise in $\Delta F/F$ corresponds to a fall of about 0.3 pH units, as measured with a pH-sensitive electrode. (Reprinted from [35], with permission from Sage Publications).

injury that increases glycolytic metabolism. pH-sensitive fluorescent dyes have been used to image pH changes in tissue slices from funnel-frozen brain (Fig. 23.11) and living brain (Fig. 23.18). Stimulus-induced changes in neutral red fluorescence reveal a rapidly spreading wave of acidification associated with depression of pre- and postsynaptic neuronal activity. The acid-shift is about 0.2 pH units, and it spreads across the cerebellar surface at about 500 μm s^{-1}, peaking at twice that speed; these rates are about ten times greater than those of spreading cortical depression or calcium waves. The phenomenon of spreading acidification depends on intact circuitry and can be modulated by various interventions, but the mechanisms of acid production and neuronal depression remain to be clarified [35]. Given the general tendency for greater utilization of glucose compared to oxygen during activation (Section 23.4.4), it is tempting to speculate that spreading acidification waves might involve, at least in part, lactate formation and transport.

23.4.7
Summary

Molecular imaging probes are used to assess a broad spectrum of biological processes and functions ranging from enzyme activities to transmitter systems and complex cognitive processing. These techniques have revealed unexpected and poorly understood phenomena (aerobic glycolysis, BOLD contrast, and spreading acidification) that are now stimulating more studies designed to better understand the cellular interactions and regulatory processes involved in relationships among blood flow, metabolism, and function.

23.5
Optical Imaging of Functional Activity by Means of Extrinsic and Intrinsic Fluorescent Compounds

The use of voltage-sensitive dyes to register and report very fast changes in membrane potential has made enormous progress in the past decade, and the approach has great promise for future studies of membrane-related events linked to cortical dynamics and sensory perception [2]. Optical imaging has the advantage of high spatial resolution (50 µm), and assays can now be carried out in conscious monkeys during behavioral tasks, with reproducible results in studies extending over periods approaching a year (Fig. 23.19). The voltage-sensitive dyes bind to the exterior membranes and have millisecond time resolution; the amplitude of the signals is directly proportional to membrane voltage and area, corresponding to electrical activities of cells [2]. These properties provide advantages over assays based on hemodynamic changes that are indirect measures of cellular activity plus blood flow responses. This technology has applications in multimodal studies, and will be useful in analysis of single cell and multicellular activities during perception, behavior, learning, memory, plasticity, and other complex, higher brain functions [2].

The analysis of intrinsic signals derived from autofluorescence of NADH plus NADPH and oxidized flavoproteins is an emerging field that shows considerable promise in helping to understand the cellular and subcellular roles of glycolytic and oxidative metabolism during functional activation [4–6]. For example, sensory stimulation of the anesthetized rat caused an increase in flavoprotein fluorescence (Fig. 23.19) that had the same distribution as cortical field potentials; and the oxidative metabolic change was followed by an increase in blood flow that could be attenuated by nitric oxide synthase inhibitor, whereas the fluorescence signal was not altered. The combined use of flavoprotein autofluorescence to evaluate oxidative metabolism and laser speckle imaging to assess local changes in blood flow showed a rapid increase in oxidative metabolism in rat whisker barrel cortex, with a lag of about 100 ms after onset of vibrissa stimulation in anesthetized rats. In contrast, the blood flow response was slower, rising at 600 ms and peaking at about 1.4 s after stimulus onset and involving about

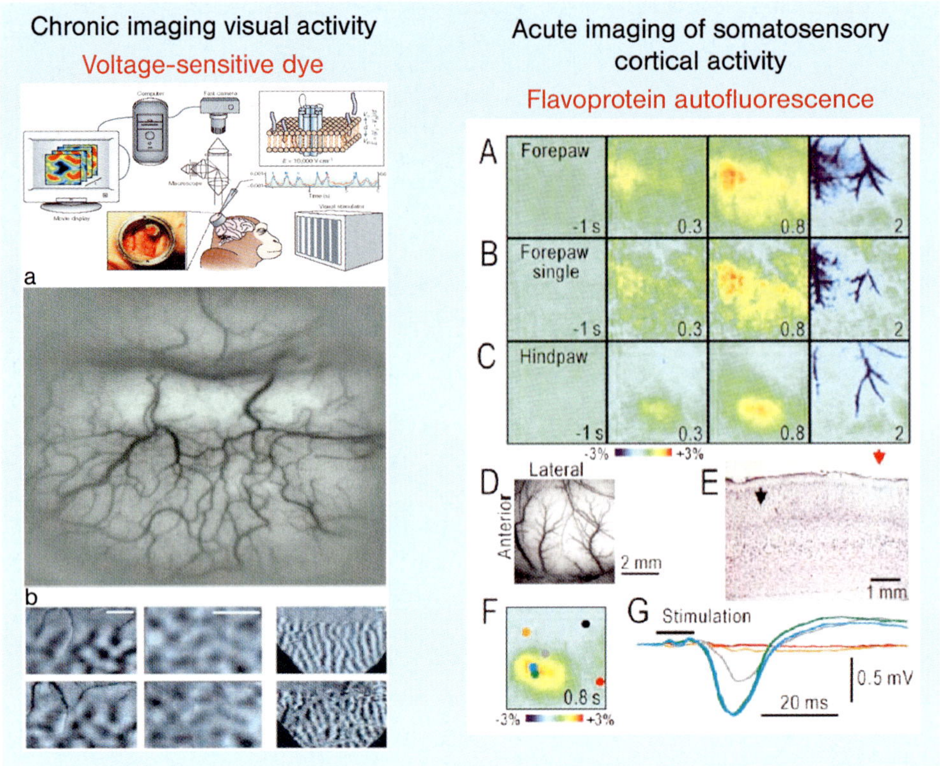

Fig. 23.19 Optical imaging of changes in membrane potential and cellular redox state during functional activation.
Left panels: top figure illustrates an in vivo system for chronic optical imaging of cortical dynamics in conscious monkey during visual stimulation, using voltage sensitive dyes. The exposed cortical tissue is covered with a sealed chamber (inset, lower left) and the cortex is stained with a suitable dye and illuminated with an external light source so that serial fluorescent images can be captured with high temporal (milliseconds) and spatial (~200 μm) resolution during visual stimulation, digitized, analyzed, and displayed as functional maps. Electrical recording along with the optical imaging shows congruence of the intracellular activity (blue) and voltage-sensitive dye population activity (red; inset, middle right). The lower figure illustrates the area of exposed cortex in the sealed chamber (a) and the functional voltage-sensitive dye imaging maps (b) of ocular dominance (left and right) and orientation (middle) recorded from the same monkey in recording sessions about a year apart (compare top and bottom panels). (Reprinted from [2], with permission from the authors, Nature Publishing Group, and Springer).
Right panels: autofluorescence of mitochondrial flavoproteins changes in response to vibratory stimulation of the skin of the rat forepaw, revealing changes in redox state: responses to (A) forepaw stimulation, (B) a single forepaw stimulus trial, (C) hindpaw stimulation. The brain surface image is shown in (D). The cresyl violet-stained parasagittal section of cortex (E) indicates the boundary between motor and somatosensory cortices (black arrow) and an electrolytic lesion (red arrow) made at the center of the autofluorescence responses in (C). Autofluorescence responses to 50 Hz, 1 s hindpaw stimulation and field potentials (G) were elicited by a single 10-ms stimulus and recorded at the colored spots in F: colors identify corresponding spots and traces. (Reprinted from [4], with permission from the authors and Blackwell Publishing).

2.5 times more tissue than metabolic activation [104]. In cerebellar cortex of anesthetized mice, surface stimulation evoked a flavoprotein-derived fluorescence signal that was linearly related to the stimulus amplitude, frequency, and duration. Also, the signal could be reduced by impairment of mitochondrial respiration with cyanide without altering the pre- and postsynaptic components of the electrophysiological response [6]. In hippocampal slices, the stimulus-induced rise in NAD(P)H fluorescence was an "inverted match" of the FAD fluorescence; and it was attributed mainly to postsynaptic neuronal activity because it was abolished by ionotropic glutamate receptor block [5]. Taken together, studies of autofluorescence signals derived from intrinsic molecules should help understand the cellular basis of shifts in glycolytic and mitochondrial metabolism as neurons and astrocytes respond to various stimuli.

23.6
Tracking Dynamic Movement of Cellular Processes and Cell Types

Synaptic remodeling during development and cellular responses to brain injury are associated with the movement of subcellular structures and migration of specific cell types within brain. For example, the mobility of subcellular components and dynamic movement of synaptic spines are essential for plasticity and reorganization. Exposure of cultured astrocytes to the excitatory neurotransmitter glutamate causes filopodial extension and elongation within 30 s [105]; spontaneous movement of astrocytic fine processes in the territory of active synapses is visualized in brain slices of transgenic mice that express high levels of green fluorescent protein. Astrocytes exhibit the gliding type of movement and process extension that takes place over periods of minutes (Fig. 23.20). The significance of astrocytic process mobility remains to be established, but it raises the fascinating possibility that interactions among brain cells might involve temporary changes in proximity to better control synaptic activity and signaling strength, perhaps by regulating glutamate uptake into astrocytes [36, 105].

Macrophages and microglia "patrol" the brain and contribute to inflammatory responses after brain infection and injury. These cell types express peripheral-type benzodiazepine-binding sites that selectively bind the ligand PK 11195; macrophages and microglia responding to ischemic damage can be localized and quantified in animals and human patients with ^{3}H- and ^{11}C-labeled PK 11195 [106, 107]. [^{11}C]PK 11195 has also been used to identify brain regions involved in multiple system atrophy [108]. Noninvasive assay of cellular migration within brain can also be carried out by using ultrasmall superparamagnetic iron oxide (USPIO) as an MRI contrast agent [109, 110], and movement of USPIO-labeled macrophages and stem cells into the territory of ischemic damage can be readily visualized and tracked over periods of days (Fig. 23.21). Thus, new strategies involving radiolabeled ligands that are specific for cell types or prelabeling cells with NMR contrast agents enable the in vivo analysis of cellular movements.

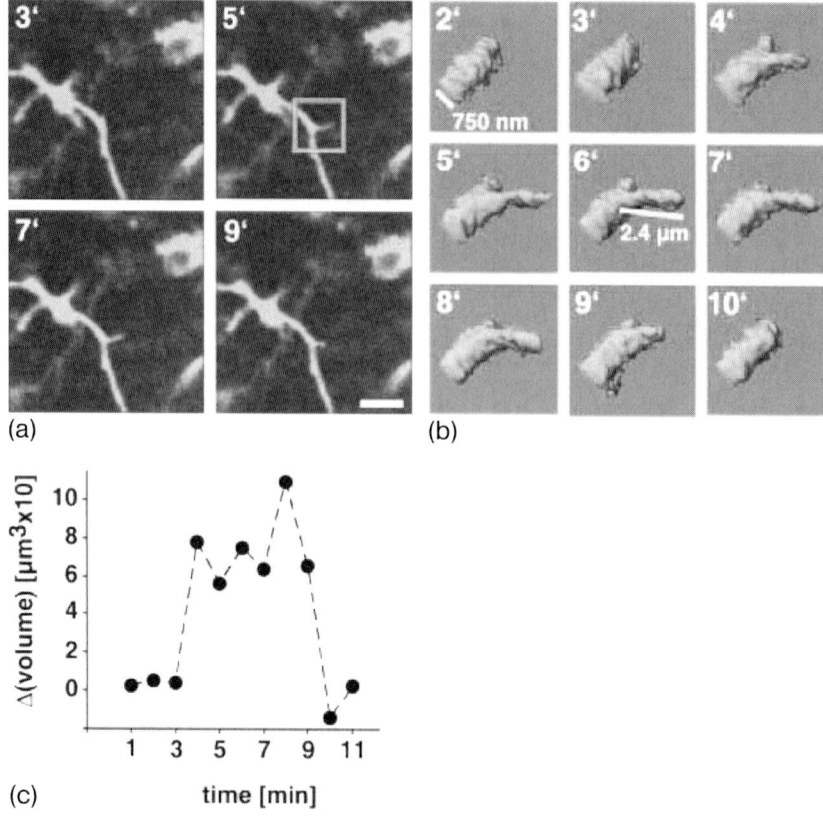

Fig. 23.20 Dynamic mobility of astrocytic filopodia near active synaptic terminals. Transient extension of filopodia (a) in an acute brain slice preparation from a transgenic mouse that expresses green fluorescent protein in astrocytes is visualized in time-lapse images taken in a single plane at 2-min intervals. (b) 3-D reconstruction of time-lapse images after deconvolution. (c) Temporal profile of filopodial movement, expressed as volume changes determined from the reconstructed 3-D images. The astrocyte processes are in close proximity to active synapses identified by uptake of FM1-43, a fluorescent dye that is taken up during recycling of synaptic vesicles. Two types of astrocyte process movement were identified, lamellipodia-like membrane gliding movements and filopodial process extensions. (Reprinted from [36], with permission from Blackwell Publishing).

23.7
Evaluation of Exogenous Genes, Cells, and Therapeutic Efficacy

Movement disorders associated with Parkinson's disease arise from the slow, progressive degeneration of dopaminergic neurons in the substantia nigra. Nigral cell death causes depletion of dopamine levels in the striatum and gradual acquisition of motor symptoms, tremor, disruption of gait, and other complications that become evident when dopamine levels are reduced to about 20–40%

NMR Imaging Movement of Specific Cell Types in Brain
USPIO-Labeled Macrophages

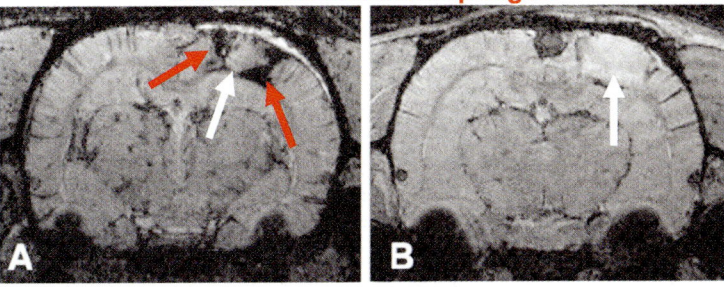

USPIO-Labeled Stem Cells

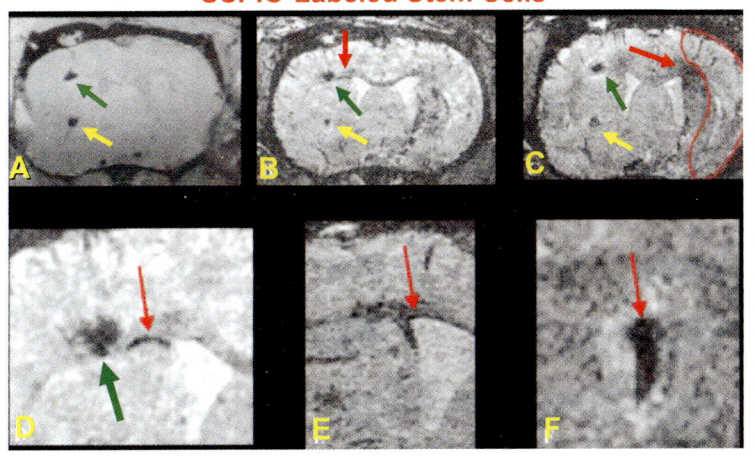

Fig. 23.21 Tracking cellular movements within brain after ischemic injury.
Top panels: macrophages in damaged brain are visualized by NMR by means of ultrasmall superparamagnetic iron oxide (USPIO)-containing particles that are engulfed by the macrophages. At 5 days after photothrombolytic infarction, rats were injected with USPIO (A) or saline (B) and imaged 24 h later by MRI, to reveal a rim of hypointensity (red arrows, A) around the zone of infarction characterized by signal loss on T2*-weighted images. The middle region of the infarct remained hyperintense, as did the entire infarcted region in the saline-injected rat (white arrows, A, B). The USPIO contrast image coincided with those obtained by histochemical detected iron and ED1 immunohistochemical assay of macrophages (not shown). (Reprinted and modified from [109], with permission from John Wiley and Sons).
Bottom panels: embryonic stem cells labeled with USPIO using a lipofection technique were injected into two sites (ventral cortical layer near the border of the corpus callosum [green arrow] and in the caudate [yellow arrow]) 2 weeks after induction of transient focal ischemia in the contralateral hemisphere by 60-min occlusion of the middle cerebral artery. MRI imaging shows loss of contrast (dark areas indicated by arrows) at the injected sites on the day of the injection (A) and 6 days (B) or 8 days (C) later. Migration of stem cells is shown (red arrows) along the corpus callosum (B, and at higher magnification in D, arrow) and the ventricular wall (E) and to the choroid plexus (E). The necrotic zone is outlined in the right hemisphere in (C); a dark region in the dorsal part of the damaged territory reflects the arrival of the USPIO-labeled stem cells. (Reprinted and modified from [110], with permission from the United States National Academy of Sciences).

of normal. Because dopamine cannot cross the blood-brain barrier, neurotransmitter replacement therapy involves the treatment of Parkinson's patients with the precursor, L-dopa. Dopa crosses the blood-brain interface and is decarboxylated in brain by aromatic L-amino acid decarboxylase (AADC) to form dopamine that can then interact with postsynaptic receptors in the striatum and ameliorate dysfunction. Unfortunately, this treatment paradigm has side-effects, and with time, its efficacy diminishes due, in part, to progressive loss of the decarboxylase as the presynaptic neurons die.

Alternative therapeutic approaches, including gene replacement via viral vectors or cell injection, have been tested in animal models for Parkinson's disease produced by treatment with MPTP. This toxin selectively damages dopaminergic neurons and causes motor disturbances. The efficacy of a viral vector system to incorporate AADC into the striatum of an MPTP-treated monkey is shown in Fig. 23.22. Functioning of the AADC gene after its incorporation into brain tissue was visualized by PET using [^{18}F]fluoro-L-m-tyrosine (FMT), a substrate for AADC: the enzymatic product is a tyramine derivative that accumulates in dopaminergic terminals and serves as a measure of dopamine synthesis capacity [111]. A different approach to help restore striatal function of MPTP-treated monkeys is implantation of human retinal pigment epithelial cells that have the capability to produce L-dopa [112]. PET imaging with [^{18}F]fluorodopa showed that cell-implant therapy enhanced AADC activity and increased the rate of formation of fluorodopamine that is retained in presynaptic terminals; also, receptor binding of [^{11}C]raclopride fell after treatment, consistent with increased competition of the PET tracer with higher endogenous dopamine levels (Fig. 23.22). Thus, pre- and postsynaptic activity of the dopaminergic system can be assessed in vivo and correlated with behavioral improvement after therapeutic intervention.

In vivo assessment of tumor cell proliferation and response to gene therapy treatment employs genetic engineering and various imaging strategies involving optical, NMR, and PET modalities [113]. For example, insertion of the luciferase gene into mice allows the monitoring of tumor growth in brain in vivo (Fig. 23.22). The quantity of light emitted has been shown to be proportion to tumor size, and longitudinal imaging of tumor response to treatment via different signaling pathways can be followed in individual animals, thereby circumventing problems associated with inherent variability arising from initial differences in tumor size and depth [114]. Another approach to monitoring gene therapy is to incorporate an engineered reporter gene into the system so that when the gene is expressed and translated, the reporter protein can be detected by

---

Fig. 23.22 Imaging activities of exogenous genes and cells in brain.
(A) PET imaging of enzyme replacement therapy by means of gene restoration or cell implantation. The top panels show imaging of a monkey previously treated by unilateral infusion via the right internal carotid artery of MPTP (1-methyl-4-phenyl-1,2,3,6-tetrahydropyridine), a dopaminergic toxin that causes clinical signs of Parkinson's disease. Partial loss of dopamine-producing cells in Parkinson's patients is treated by providing

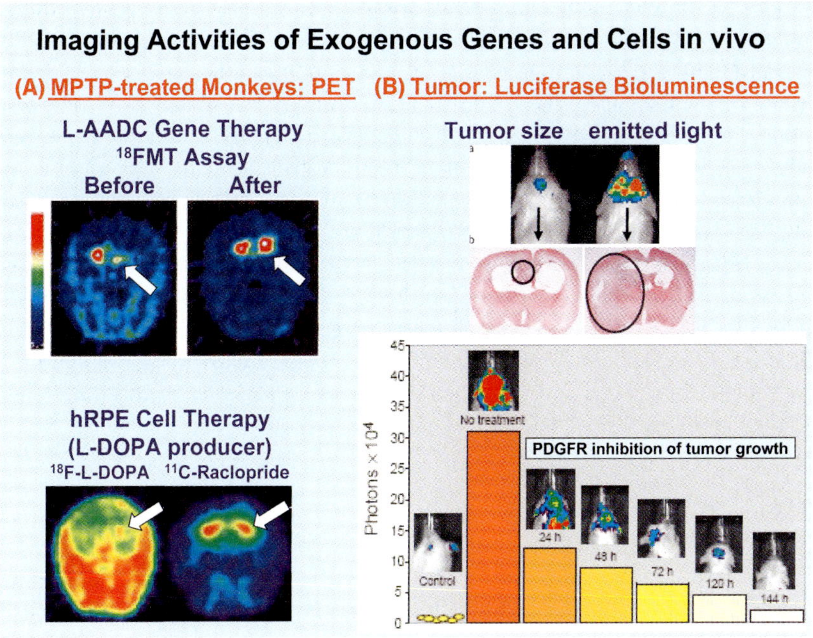

23.7 Evaluation of Exogenous Genes, Cells, and Therapeutic Efficacy

Imaging Activities of Exogenous Genes and Cells in vivo

(A) MPTP-treated Monkeys: PET   **(B) Tumor: Luciferase Bioluminescence**

L-AADC Gene Therapy
^{18}FMT Assay
Before   After

hRPE Cell Therapy
(L-DOPA producer)
^{18}F-L-DOPA ^{11}C-Raclopride

Tumor size   emitted light

PDGFR inhibition of tumor growth

Fig. 23.22 (continued) the precursor, L-DOPA, that can cross the blood-brain barrier; one approach is to restore aromatic L-amino acid decarboxylase (AADC) activity so that the enzyme can synthesize dopamine from peripherally administered precursor. Before gene therapy with a viral vector containing the AADC gene (left panel), the monkey was imaged with an AADC tracer, 6-[^{18}F]fluoro-L-m-tyrosine (FMT), that showed low FMT uptake in the MPTP-treated (right) striatum (arrow). After gene therapy (right panel), AADC activity increased, causing a large rise in FMT uptake in the treated (right) striatum (arrow). (Reprinted from [111], with permission from Academic Press).

The bottom two panels show an alternative approach, i.e., direct injection of human retinal pigment epithelial cells (hRPE) attached to gelatin microcarriers into the putamen of bilaterally MPTP-treated monkeys; hRPE cells produce L-DOPA, the precursor for dopamine that would be produced by residual dopaminergic neurons. Imaging of the effects of these cells with L-[^{18}F]fluoroDOPA (tracer for decarboxylation to fluorodopamine and storage in presynaptic dopamine terminals) shows increased uptake in the hRPE-injected putamen (arrow, left panel) and decreased binding of [^{11}C]raclopride, consistent with higher endogenous production of DOPA that would bind to the dopamine receptors, inhibiting binding of [^{11}C]raclopride in the hRPE-treated putamen (arrow) compared to the contralateral structure (also see Fig. 23.15). (Reprinted from [112], with permission from Academic Press).

(B) Bioluminescence imaging of tumor growth and regression during treatment. Top panels show that that platelet-derived growth factor (PDGF)-driven oligodendrogliomas induced in transgenic mice expressing the luciferase gene emit more light (top two panels, a) when tumor size is larger, shown by histological staining (b). The graph in the lower panel shows light emitted from control mice and from an untreated tumor-bearing mouse, followed by blockade of the PDGF receptor causing tumor growth cycle arrest and reduction of tumor size and bioluminescence in serial assays carried out over a 6-day period. (Reprinted from [114], with permission from the authors and Nature Publishing Group).

PET or NMR imaging. Imaging depends upon: (1) binding to a suitable radiolabeled or paramagnetic reporter probe, (2) transport, or (3) metabolism of the probe by the reporter protein so that the probe or its labeled product is trapped in the cells where the gene is expressed. For example, transfection of a viral thymidine kinase gene into tumors is an experimental therapy because a prodrug (e.g., ganciclovir) can be administered systemically, taken up, and trapped by phosphorylation by the viral thymidine kinase, followed by further phosphorylation by cellular enzymes and incorporation into tumor DNA where it acts as a chain terminator to kill the tumor cells. Expression of the enzyme can be detected by PET assays of ^{18}F-labeled reporter probes that are taken up and phosphorylated [115, 116]. Clever applications of genetic engineering coupled with brain imaging techniques using unique reporter systems and pharmaceuticals will enable both direct and indirect assessment of disease states and novel therapeutic advances.

23.8
Summary and Perspectives

The physical structure and biochemical composition of brain are designed to support and maintain a large and very complex range of functional activities by different types of cells in brain. To visualize and quantify biological processes in living brain, neuroscientists must take into account the properties of the blood-brain interface and the regional complexity and heterogeneity of the brain. The basic principles and methods first devised for the assay of local rates of blood flow and glucose utilization have been widely applied, refined, and extended. During the past 40 years, novel new technologies, extraordinarily useful probes, and insightful experimental strategies have been developed for molecular imaging to localize and quantify biological processes ranging from gene expression, specific enzymatic reactions, signaling activities and pathways, metabolite and electrolyte transport, concentrations and redox state of key components of blood and brain, cellular movement, and complex higher functions of brain in living organisms. Different imaging modalities have advantages and disadvantages in terms of temporal and spatial resolution, single or repeated assays in longitudinal studies, cost and complexity, and types of applications. The fusion of anatomical and functional images derived from extrinsic and intrinsic reporter molecules and multimodal approaches are essential for the analysis of dynamic aspects of brain function. Challenges for the future include understanding the biochemical and physiological basis for many of the key imaging signals and neurological and mental diseases, and developing effective treatments for brain disorders and tumors.

Acknowledgments

Supported by NIH grants NS36728 and NS38230.

References

1 T. F. Massoud, S. S. Gambhir **2003**, *Genes Dev.* 17, 545–580.
2 A. Grinvald, R. Hildesheim **2004**, *Nat. Rev. Neurosci.* 5, 874–885.
3 A. Villringer, B. Chance **1997**, *Trends Neurosci.* 20, 435–442.
4 K. Shibuki, R. Hishida, H. Murakami, M. Kudoh, T. Kawaguchi, M. Watanabe, S. Watanabe, T. Kouuchi, R. Tanaka **2003**, *J. Physiol.* 549, 919–927.
5 C. W. Shuttleworth, A. M. Brennan, J. A. Connor **2003**, *J. Neurosci.* 23, 3196–3208.
6 K. C. Reinert, R. L. Dunbar, W. Gao, G. Chen, T. J. Ebner **2004**, *J. Neurophysiol.* 92, 199–211.
7 S. G. Kim, S. Ogawa **2002**, *Curr. Opin. Neurobiol.* 12, 607–615.
8 S. F. Halpin **2004**, *Br. J. Radiol.* 77, S20–S26.
9 A. M. Alessio, P. E. Kinahan, P. M. Cheng, H. Vesselle, J. S. Karp **2004**, *Radiol. Clin. N. Am.* 42, 1017–1032.
10 G. K. von Schulthess **2004**, *Mol. Imaging Biol.* 6, 183–187.
11 R. M. Dijkhuizen, K. Nicolay **2003**, *J. Cereb. Blood Flow Metab.* 23, 1383–1402.
12 H. Benveniste, S. Blackband **2002**, *Prog. Neurobiol.* 67, 393–420.
13 T. Z. Wong, T. G. Turkington, T. C. Hawk, R. E. Coleman **2004**, *Cancer J.* 10, 234–242.
14 S. Ogawa, T. M. Lee, A. R. Kay, D. W. Tank **1990**, *Proc. Natl. Acad. Sci. USA* 87, 9868–9872.
15 S. Ogawa, D. W. Tank, R. Menon, J. M. Ellermann, S. G. Kim, H. Merkle, K. Ugurbil **1992**, *Proc. Natl. Acad. Sci. USA* 89, 5951–5955.
16 N. K. Logothetis, J. Pauls, M. Augath, T. Trinath, A. Oeltermann **2001**, *Nature* 412, 150–157.
17 M. Lauritzen **2001**, *J. Cereb. Blood Flow Metab.* 21, 1367–1383.
18 N. K. Logothetis, B. A. Wandell **2004**, *Annu. Rev. Physiol.* 66, 735–769.
19 M. Lauritzen, L. Gold **2003**, *J. Neurosci.* 23, 3972–3980.
20 M. Lauritzen **2005**, *Nat. Rev. Neurosci.* 6, 77–85.
21 A. C. Silva, J. H. Lee, I. Aoki, A. P. Koretsky **2004**, *NMR Biomed.* 17, 532–543.
22 J. H. Hakamäki, K. M. Brindle **2003**, *Trends Pharmacol. Sci.* 24, 146–149.
23 M. Bendszus, G. Stoll **2003**, *J. Neurosci.* 23, 10892–10896.
24 A. W. Rogers **1979**, *Techniques of Autoradiography*, 3rd edn., Elsevier/North-Holland Biomedical Press, Amsterdam.
25 W. E. Stumpf, H. F. Solomon, eds. **1995**, *Autoradiography and Correlative Imaging*, Academic Press, San Diego.
26 S. H. Britz-Cunningham, S. J. Adelstein **2003**, *J. Nucl. Med.* 44, 1945–1961.
27 M. E. Phelps, J. C. Mazziotta, H. R. Schelbert, eds. **1986**, *Positron Emission Tomography and Autoradiography: Principles and Applications for the Brain and Heart*, Raven Press, New York.
28 M. E. Phelps **2004**, *PET: Molecular Imaging and its Biological Applications*, Springer, New York.
29 B. Chance, P. Cohen, F. Jöbsis, B. Schoener **1962**, *Science* 137, 499–508.

30 F. Jöbsis **1977**, *Science* 198, 1264–1267.
31 X. Intes, B. Chance **2005**, *Radiol. Clin. N. Am.* 43, 221–234.
32 H. Obrig, A. Villringer **2003**, *J. Cereb. Blood Flow Metab.* 23, 1–18.
33 D. E. Kim, F. A. Jaffer, R. Weissleder, C. H. Tung, D. Schellingerhout **2005**, *J. Cereb. Blood Flow Metab.* 25, 226–233.
34 F. Helmchen, J. Waters **2002**, *Eur. J. Pharmacol.* 447, 119–129.
35 T. J. Ebner, G. Chen **2003**, *Neuroscientist* 9, 37–45.
36 J. Hirrlinger, S. Hulsmann, F. Kirchhoff **2004**, *Eur. J. Neurosci.* 20, 2235–2239.
37 C. M. Niell, S. J. Smith **2004**, *Annu. Rev. Physiol.* 66, 771–798.
38 T. Knöpfel, K. Tomita, R. Shimazaki, R. Sakai **2003**, *Methods* 30, 42–48.
39 A. M. Gorbach, J. D. Heiss, L. Kopylev, E. H. Oldfield **2004**, *J. Neurosurg.* 101, 960–969.
40 L. V. Wang **2003**, *Dis. Markers* 19, 123–138.
41 X. Wang, Y. Pang, G. Ku, G. Stoica, L. V. Wang **2003**, *Opt. Lett.* 28, 1739–1741.
42 X. Wang, G. Ku, M. A. Wegiel, D. J. Bornhop, G. Stoica, L. V. Wang **2004**, *Opt. Lett.* 29, 730–732.
43 J. A. Copland, M. Eghtedari, V. L. Popov, N. Kotov, N. Mamedova, M. Motamedi, A. A. Oraevsky **2004**, *Mol. Imaging Biol.* 6, 341–349.
44 M. Erecińska, I. A. Silver **1989**, *J. Cereb. Blood Flow Metab.* 9, 2–19.
45 D. Attwell, S. B. Laughlin **2001**, *J. Cereb. Blood Flow Metab.* 21, 1133–1145.
46 D. Attwell, C. Iadecola **2002**, *Trends Neurosci.* 25, 621–625.
47 B. K. Siesjö **1978**, *Brain Energy Metabolism*, John Wiley & Sons, New York.
48 A. Nehlig **1996**, *Dev. Neurosci.* 18, 426–433.
49 S. S. Kety, C. F. Schmidt **1948**, *J. Clin. Invest.* 27, 476–483.
50 O. H. Lowry **1990**, *Annu. Rev. Biochem.* 59, 1–27.
51 L. Sokoloff, M. Reivich, C. Kennedy, M. H. Des Rosiers, C. S. Patlak, K. D. Pettigrew, O. Sakurada, M. Shinohara **1977**, *J. Neurochem.* 28, 897–916.
52 L. Sokoloff **1986**, in *Positron Emission Tomography and Autoradiography: Principles and Applications for the Brain and Heart*, eds. M. Phelps, J. Mazziotta, H. Schelbert, Raven Press, New York, pp. 1–71.
53 S. S. Kety **1996**, in *The History of Neuroscience in Autobiography*, Vol. 1, ed. L. R. Squire, Society for Neuroscience, Washington, D.C., pp. 382–413.
54 L. Sokoloff **1996**, in *The History of Neuroscience in Autobiography*, vol. 1, ed. L. R. Squire, Society for Neuroscience, Washington, D.C., 454–497.
55 W. M. Landau, W. H. Freygang Jr, L. P. Roland, L. Sokoloff, S. S. Kety **1955**, *Trans. Am. Neurol. Assoc.* 1956, 125–129.
56 W. H. Freygang Jr, L. Sokoloff **1958**, *Adv. Biol. Med. Phys.* 6, 263–279.
57 M. Reivich, J. Jehle, L. Sokoloff, S. S. Kety **1969**, *J. Appl. Physiol.* 27, 296–300.
58 O. Sakurada, C. Kennedy, J. Jehle, J. D. Brown, G. L. Carbin, L. Sokoloff **1978**, *Am. J. Physiol.* 234, H59–H66.
59 N. A. Lassen, D. H. Ingvar **1961**, *Experientia* 17, 42–43.
60 N. A. Lassen, K. Hoedt-Rasmussen, S. C. Sorensen, E. Skinhoj, S. Cronquist, B. Bodforss, D. H. Ingvar **1963**, *Neurology* 13, 719–727.

61 E. Fischer, M. Bergmann, H. Schotte **1954**, *Ber. Dtsch Chem. Gesellsch.* 53, 509–547.
62 A. Sols, R. K. Crane **1954**, *J. Biol. Chem.* 210, 581–595.
63 D. B. Tower **1980**, in *Cerebral Metabolism and Neural Function*, eds. J. V. Passonneau, R. A. Hawkins, W. D. Lust, F. A. Welsh, Williams & Wilkins, Baltimore, pp. 1–8.
64 M. Reivich, D. Kuhl, A. Wolf, J. Greenberg, M. Phelps, T. Ido, V. Casella, J. Fowler, B. Gallagher, E. Hoffman, A. Alavi, L. Sokoloff **1977**, *Acta Neurol. Scand. Suppl.* 64, 190–191.
65 M. Reivich, D. Kuhl, A. Wolf, J. Greenberg, M. Phelps, T. Ido, V. Casella, J. Fowler, E. Hoffman, A. Alavi, P. Som, L. Sokoloff **1979**, *Circ. Res.* 44, 127–137.
66 M. E. Phelps, S. C. Huang, E. J. Hoffman, C. Selin, L. Sokoloff, D. E. Kuhl **1979**, *Ann. Neurol.* 6, 371–388.
67 G. A. Dienel, N. F. Cruz, K. Adachi, L. Sokoloff, J. E. Holden **1997**, *Am. J. Physiol.* 273, E839–E849.
68 J. E. Holden, K. Mori, G. A. Dienel, N. F. Cruz, T. Nelson, L. Sokoloff **1991**, *J. Cereb. Blood Flow Metab.* 11, 171–182.
69 A. F. Ryan, F. R. Sharp **1982**, *Brain Res.* 252, 177–180.
70 Y. Itoh, T. Abe, R. Takaoka, N. Tanahashi **2004**, *J. Cereb. Blood Flow Metab.* 24, 993–1003.
71 A. Nehlig, E. Wittendorp-Rechenmann, C. D. Lam **2004**, *J. Cereb Blood Flow Metab.* 24, 1004–1014.
72 R. Balázs, J. E. Cremer, eds. **1972**, *Metabolic Compartmentation in the Brain*, John Wiley & Sons, New York.
73 S. Berl, D. D. Clarke **1969**, Structural neurochemistry, in *Handbook of Neurochemistry*, vol. 2, ed. A. Lajtha, Plenum Press, New York, pp. 447–472.
74 R. A. Waniewski, D. L. Martin **1998**, *J. Neurosci.* 18, 5225–5233.
75 B. Hassel, U. Sonnewald, F. Fonnum **1995**, *J. Neurochem.* 64, 2773–2782.
76 N. F. Cruz, A. Lasater, H. R. Zielke, G. A. Dienel **2005**, *J. Neurochem.* 92, 934–947.
77 G. A. Dienel, D. Popp, P. D. Drew, K. Ball, A. Krisht, N. F. Cruz **2001**, *J. Nucl. Med.* 42, 1243–1250.
78 J. Goldschmidt, W. Zuschratter, H. Scheich **2004**, *NeuroImage* 23, 638–647.
79 D. M. Hermann, K. A. Hossmann, G. Mies **2004**, *Neuroscience* 123, 371–379.
80 K. P. Friedman, R. L. Wahl **2004**, *Semin. Nucl. Med.* 34, 242–253.
81 S. S. Foo, D. F. Abbott, N. Lawrentschuk, A. M. Scott **2004**, *Mol. Imaging Biol.* 6, 291–305.
82 P. W. Wiest, R. R. Lee, M. F. Hartshorne **2001**, *Clin. Nucl. Med.* 26, 964–965.
83 M. E. Raichle **1998**, *Philos. Trans. R. Soc. Lond. B Biol. Sci.* 353, 1889–1901.
84 D. A. Gusnard, M. E. Raichle **2001**, *Nat. Rev. Neurosci.* 2, 685–694.
85 P. T. Fox, M. E. Raichle, M. A. Mintun, C. Dence **1988**, *Science* 241, 462–464.
86 P. L. Madsen, S. G. Hasselbalch, L. P. Hagemann, K. S. Olsen, J. Bulow, S. Holm, G. Wildschiodtz, O. B. Paulson, N. A. Lassen **1995**, *J. Cereb. Blood Flow Metab.* 15, 485–491.

87 P. L. Madsen, N. F. Cruz, L. Sokoloff, G. A. Dienel **1999**, *J. Cereb. Blood Flow Metab.* 19, 393–400.
88 G. A. Dienel, R. Y. Wang, N. F. Cruz **2002**, *J. Cereb. Blood Flow Metab.* 22, 1490–1502.
89 L. Pellerin, P. J. Magistretti **2004**, *Neuroscientist* 10, 53–62.
90 G. A. Dienel, N. F. Cruz **2004**, *Neurochem. Int.* 45, 321–351.
91 C.-P. Chih, E. L. Roberts Jr. **2003**, *J. Cereb. Blood Flow Metab.* 23, 1263–1281.
92 G. A. Dienel, N. F. Cruz **2003**, *Neurochem. Int.* 43, 339–354.
93 L. Hertz **2004**, *J. Cereb. Blood Flow Metab.* 24, 1241–1248.
94 C. Y. Shiue, M. J. Welch **2004**, *Radiol. Clin. North Am.* 42, 1033–1053,
95 S. I. Rapoport **2003**, *J. Pediatr.* 143, S26–S34.
96 H. Benveniste, J. S. Fowler, W. D. Rooney, D. H. Moller, W. W. Backus, D. A. Warner, P. Carter, P. King, B. Scharf, D. A. Alexoff, Y. Ma, P. Vaska, D. Schlyer, N. D. Volkow **2003**, *J. Nucl. Med.* 44, 1522–1530.
97 H. Benveniste, J. S. Fowler, W. Rooney, Y. S. Ding, A. L. Baumann, D. H. Moller, C. Du, W. Backus, J. Logan, P. Carter, J. D. Coplan, A. Biegon, L. Rosenblum, B. Scharf, J. S. Gatley, N. D. Volkow **2005**, *J. Nucl. Med.* 46, 312–320.
98 S. Minoshima, K. A. Frey, D. J. Cross, D. E. Kuhl **2004**, *Semin. Nucl. Med.* 34, 70–82.
99 N. D. Volkow, J. S. Fowler, G. J. Wang **2004**, *Neuropharmacology* 47 [Suppl. 1], 3–13.
100 J. S. Fowler, J. Logan, G. J. Wang, N. D. Volkow, F. Telang, W. Zhu, D. Franceschi, N. Pappas, R. Ferrieri, C. Shea, V. Garza, Y. Xu, D. Schlyer, S. J. Gatley, Y. S. Ding, D. Alexoff, D. Warner, N. Netusil, P. Carter, M. Jayne, P. King, P. Vaska **2003**, *Proc. Natl Acad. Sci. USA* 100, 11600–11605.
101 G. Dienel **1984**, *J. Neurochem.* 43, 913–925.
102 G.-A. Dienel, W. A. Pulsinelli **1986**, *Ann. Neurol.* 19, 465–472.
103 C. W. Shuttleworth, J. A. Connor **2001**, *J. Neurosci.* 21, 4225–4236.
104 B. Weber, C. Burger, M. T. Wyss, G. K. von Schulthess, F. Scheffold, A. Buck **2004**, *Eur. J. Neurosci.* 20, 2664–2670.
105 A. H. Cornell-Bell, G. T. Prem, S. J. Smith **1990**, *Glia* 3, 322–334.
106 R. Myers, L. G. Manjil, B. M. Cullen, G. W. Price, R. S. Frackowiak, J. E. Cremer **1991**, *J. Cereb. Blood Flow Metab.* 11, 314–322.
107 A. Gerhard, J. Schwarz, R. Myers, R. Wise, R. B. Banati **2005**, *NeuroImage* 24, 591–595.
108 A. Gerhard, R. B. Banati, G. B. Goerres, A. Cagnin, R. Myers, R. N. Gunn, F. Turkheimer, C. D. Good, C. J. Mathias, N. Quinn, J. Schwarz, D. J. Brooks **2003**, *Neurology* 61, 686–689.
109 A. Saleh, D. Wiedermann, M. Schroeter, C. Jonkmanns, S. Jander, M. Hoehn **2004**, *NMR Biomed.* 17, 163–169.
110 M. Hoehn, E. Kustermann, J. Blunk, D. Wiedermann, T. Trapp, S. Wecker, M. Focking, H. Arnold, J. Hescheler, B. K. Fleischmann, W. Schwindt, C. Buhrle **2002**, *Proc. Natl. Acad. Sci. USA* 99, 16267–16272.

111 K. S. Bankiewicz, J. L. Eberling, M. Kohutnicka, W. Jagust, P. Pivirotto, J. Bringas, J. Cunningham, T. F. Budinger, J. Harvey-White **2000**, *Exp. Neurol.* 164, 2–14.
112 D. J. Doudet, M. L. Cornfeldt, C. R. Honey, A. W. Schweikert, R. C. Allen **2004**, *Exp. Neurol.* 189, 361–368.
113 S. Gross, D. Piwnica-Worms **2005**, *Cancer Cell.* 7, 5–15.
114 L. Uhrbom, E. Nerio, E. C. Holland **2004**, *Nat. Med.* 10, 1257–1260.
115 R. G. Blasberg, J. G. Tjuvajev **2003**, *J. Clin. Invest.* 111, 1620–1629.
116 H. R. Herschman **2004**, *Crit. Rev. Oncol. Hematol.* 51, 191–204.
117 Merck & Co. **1983**, *Merck Index*, 10th edn, Merck & Co., Rahway, N.J.

Part VII
Disease-Related Response

24
Inflammatory Response of the Blood-Brain Interface

Pedro M. Faustmann and Claus G. Haase

24.1
Introduction

The central nervous system (CNS) is considered an immunoprivileged site where the blood-brain barrier (BBB) controls cellular and serum protein entry into the brain. Under pathological conditions, such as inflammation, trauma, and neurodegeneration, a large number of white blood cells (WBCs) immigrate into the CNS. Additionally, changes of protein concentrations in the cerebrospinal fluid (CSF) indicate a disruption of the BBB. The mechanisms allowing WBCs to pass the BBB, and which protein in the CSF is disease-related represent central questions in neurology, neuropathology, and neuroimmunology. Obviously, they are important for our understanding of the pathogenesis of inflammatory brain diseases.

24.2
Diagnostic Features of Cerebrospinal Fluid

Normal CSF is crystal clear. However, as few as 200 WBCs mm^{-3} or 400 red blood cells (RBCs) mm^{-3} cause CSF to appear turbid. Xanthochromia (yellow-colored) is most often caused by lysis of RBCs, resulting in hemoglobin leakage and breakdown to oxyhemoglobin, methemoglobin, and bilirubin. CSF protein levels of at least 150 mg dl^{-1} (1.5 g l^{-1}), as seen under infectious and inflammatory conditions, or as a result of a traumatic tap that contains more than 100 000 RBCs mm^{-3}, also results in xanthochromia. Newborn CSF is often xanthochromic because of the frequent elevation of bilirubin and protein levels.

Blood-Brain Interfaces: From Ontogeny to Artificial Barriers.
Edited by R. Dermietzel, D.C. Spray, M. Nedergaard
Copyright © 2006 WILEY-VCH Verlag GmbH & Co. KGaA, Weinheim
ISBN: 3-527-31088-6

24.2.1
Cell Count and Cell Differentiation

Normal CSF may contain up to 5 WBCs mm^{-3} in adults and 20 WBCs mm^{-3} in newborns. About 90% of patients with bacterial meningitis have a WBC count higher than 1000 cells mm^{-3} while 99% have more than 100 cells mm^{-3}. Less than 100 WBCs mm^{-3} is more common in patients with viral meningitis. Elevated WBC counts (>20, <100) may also occur after seizure, in intracerebral hemorrhage, brain tumors, and under a variety of inflammatory conditions. Table 24.1 lists common CSF parameters under various types of meningitis.

The WBC count seen in normal adult CSF is comprised of approximately 70% lymphocytes and 30% monocytes. The majority of patients with polyradiculitis will have <10 monocytes mm^{-3}. Up to 50 monocytes mm^{-3} are seen in about 25% of patients with multiple sclerosis. The cell composition of CSF does not allow to differentiate between bacterial and nonbacterial meningitis. For instance, lymphocytosis is frequently seen in viral, fungal, and tuberculous infections of the CNS, although a predominance of polymorphonuclear leukocytes (PMN) may be present in the early stages of these infections. CSF in bacterial meningitis is typically dominated by the presence of PMNs. However, more than 10% of cases of bacterial meningitis shows a lymphocytic predominance, especially at early stages of the clinical course and when the number of WBCs are less than 1000 cells mm^{-3}. Eosinophilic meningitis is characterized by >10 eosinophils mm^{-3} or, when a total CSF cell count is made, >10% eosinophils, a feature that is suspective for a parasitic infection. Other possible causes for eosinophilic CSF may include viral, fungal, or rickettsial meningitis; ventriculoperitoneal shunts with or without coexisting infection; malignancy; and adverse drug reactions.

Table 24.1 Cerebrospinal fluid findings in meningitis.

Test	Bacterial	Viral	Fungal	Tubercular
White blood cell count	≥1000 per mm^3	<100 per mm^3	Variable	Variable
Cell differential	Predominance of PMNs	Predominance of lymphocytes	Predominance of lymphocytes	Predominance of lymphocytes
Protein	Mild to marked elevation	Normal to elevated	Elevated	Elevated
CSF:serum glucose ratio	Normal to marked decrease	Usually normal	Low	Low

24.2.2
Protein Level

CSF protein concentration is one of the most sensitive indicators for a pathology in the CNS. Newborns show up to 150 mg dl^{-1} (1.5 g l^{-1}) of protein. The adult range is 18–58 mg dl^{-1} (0.18–0.58 g l^{-1}). The adult level is achieved between six and 12 months of age. Elevated CSF protein levels are seen in infections, intracranial hemorrhages, multiple sclerosis, Guillain Barré syndrome (polyradiculitis), malignancies, some endocrine abnormalities, certain medication, and a variety of inflammatory conditions (Table 24.2). Protein concentration is falsely elevated by the presence of RBCs in a traumatic tap situation.

24.2.3
Glucose Level

CSF glucose renders about two-thirds of serum glucose. This ratio decreases with increasing serum glucose levels. CSF glucose levels generally do not exceed 300 mg dl^{-1} (16.7 mmol l^{-1}) regardless of serum levels (see Chapter 4). CNS infections can cause lowered CSF glucose levels, although glucose levels are usually normal under viral infections. Normal glucose levels do not rule out infection. Up to 50% of patients who have bacterial meningitis have normal CSF glucose levels.

Chemical meningitis, inflammatory conditions, subarachnoid hemorrhage, and hypoglycaemia may also be responsible for hypoglycorrhachia (low glucose level in CSF) [1–6].

24.3
Acute Bacterial Meningitis

More than 50 years after the advent of antibiotic therapies, the mortality and rates of morbidity associated with bacterial meningitis remain high. Meningitis due to *Streptococcus pneumoniae* has the highest fatality rate of about 20–30%,

Table 24.2 Average and range of cerebrospinal fluid protein.

Condition	Average: mg dl^{-1} (g l^{-1})	Average: mg dl^{-1} (g l^{-1})
Bacterial meningitis	418 (4.18)	21–2220 (0.21–22.2)
Brain tumor	115 (1.15)	15–1920 (0.15–19.2)
Brain abscess	69 (0.69)	16–288 (0.16–2.88)
Aseptic meningitis	77 (0.77)	11–400 (0.11–4.0)
Cerebral hemorrhage	270 (2.7)	19–2110 (0.19–21.1)
Neurosyphilis	68 (0.68)	15–4200 (0.15–42.0)

and up to one-third of survivors are left with significant disabilities or handicaps [7]. The unfavorable clinical outcome of bacterial meningitis is often due to intracranial and cerebral complications comprising cerebrovascular insults, vasogenic and cytotoxic brain oedema, hydrocephalus, and increased intracranial pressure. An overshooting immune response of the host is thought to be responsible for brain damage, which may result in neurological sequelae including hearing loss, symptomatic epilepsy, focal neurological deficits, and mental retardation. Experimental studies using animal models and cell culture systems have substantially increased our knowledge of the pathogenesis and pathophysiological mechanisms during bacterial meningitis. Acute breakdown of the blood-brain barrier, accumulation of PMN in the CSF, and intrathecal production of cytokines are key features in disease development which have proven fundamental for brain oedema, cerebral vasculitis, brain ischemia, and ultimately for neuronal cell death [8–11].

The clinical diagnosis of bacterial meningitis is established by identification of the bacterial microorganism in the CSF and by a high number of CSF cells (>1000 leukocytes μl^{-1}) consisting of more than 60% PMNs [1].

24.3.1
Bacteria and the Blood-Brain Barrier

Meningitis caused by *Haemophilus influenzae* type b in children has been eliminated due to routine immunization of infants against this pathogen. This fact has favored the preponderance of *S. pneumoniae* and *Neisseria meningitidis* as the most prominent pathogen of meningitis in young children; and meningitis by *H. influenzae* type b is now a disease predominantly occurring in adults. Among neonates, *S. agalactiae*, *Escherichia coli*, and *Listeria monocytogenes* are the leading pathogens of bacterial meningitis.

Bacteria enter the CNS via the vasculature or through focal infections located close to the CNS. Bacterial binding to the cerebral microvascular endothelium is considered an important initial step in the pathogenesis of meningitis. In vitro models of bacterial meningitis have been used to study the interaction of bacteria and bacterial components with cultured cerebral microvascular endothelial cells, but the exact interactions are still incompletely understood. We will discuss a few of them in the following [12].

Invasion of *S. pneumoniae* into an endothelial cell layer is dependent on the presence of pneumococcal choline-binding protein A, and invasion is partially inhibited by antagonists to the platelet-activating factor (PAF) receptor on the endothelial cell membranes [13]. Pneumococcal surface protein A (PspA) is regarded as an important virulence factor and has been found to interfere with complement activation [14]. Additionally, it was shown that pneumococci expressing pneumolysin were able to breach the endothelial cells, whereas mutant pneumococci deficient in pneumolysin were unable to penetrate the cell barrier [15]. Pneumolysin is a 53-kDa protein, lytic to all cells with cholesterol-containing membranes. Pneumolysin has been shown a key virulence factor which seems to be responsible for the high morbidity and mortality of pneumococci infections [16]. Whereas pneumolysin inhibits neutrophilic and monocyte function, it also induces the production of proinflammatory mediators, such as tumor necrosis factor alpha (TNF-α), interleukin-1 (IL-1), and IL-6 [17].

For *N. meningitidis*, surface structures of type IV pili (a bacterial appendage) have been identified as being essential for meningeal invasion [18]. The pili interact with CD46, a human cell-surface protein involved in the regulation of complement activation [19]. Additionally, fibronectin has been identified as an uptake-promoting serum factor, which binds to human brain-derived endothelial cells via integrin receptors [20].

E. coli enter the endothelial cell of the BBB by interaction of various bacterial proteins with endothelial receptors [21].

Within the CSF, bacteria multiply, lyse spontaneously and release proinflammatory and toxic compounds by autolysis and secretion. Lipopolysaccharide (LPS) is derived from bacterial cell walls and affects the passage of other proteins across the blood-brain barrier through the release of cytokines and disruption of the BBB [22, 23].

24.3.2
Leukocyte Migration into the CNS

Under normal conditions, the concentrations of leukocytes, complement factors, and antibodies in the CSF are low. PMNs are attracted into the subarachnoid space by proinflammatory mediators which are initially produced by meningeal and perivascular macrophages [24] and later by immigrated PMNs and local microglia. Generally, the high influx of PMNs into the subarachnoid space is associated with the degree of neurological damage in patients with bacterial meningitis. However, the effects of PMNs in the CSF are dependent on the bacterial titer in the CSF and the clinical time-course of the meningitis. An initial high CSF-PMN count prior to antibiotic therapy is associated with a greater chance of survival, suggesting suppression of bacterial multiplication by the PMNs. A continuous CSF-PMN invasion during antibiosis is associated with a greater chance of death, indicating detrimental effects of CSF pleocytosis in the context of effective therapy [25]. Several experimental studies suggest two strategies to decrease PMNs immigration into the CSF: (a) inhibition of adhesion of PMNs to the luminal endothelial surface, and (b) inhibition of attraction (chemotaxis) of PMNs to the subarachnoid space. PMN adhesion to the endothelial surface is initially mediated by the interaction of endothelial selectins with CD15 on the leukocyte surface inducing the phenomenon of leukocytic rolling. Selectin-mediated rolling of PMNs is essential for the transendothelial migration of PMNs, as was shown in selectin-deficient mice [26]. Inhibition of PMNs rolling by a selectin-blocker, the polysaccharide fucoidin, attenuates the inflammatory response in experimental pneumococcal meningitis [27, 28]. Interestingly, another polysaccharide, glucuronoxylomannan (GXM), a capsular component of the fungus *Cryptococcus neoformans*, delays translocation of PMNs across the BBB in an animal model of acute bacterial meningitis [29].

A similar inhibitory effect on PMN adhesion is observed when antibodies directed against the intercellular adhesion molecule (ICAM)-1 [30] are used in experimental meningitis [31]. Transgenic deletion of ICAM-1 led to a more than 50% reduction in the number of infiltrating PMNs in an experimental model of inflammation using systemic lipopolysaccharide (LPS) injection [32]. Antibodies against the junctional adhesion molecule (JAM), a cerebral endothelial tight junction component, inhibit PMN extravasation only in a model of cytokine-induced meningitis in mice [33], but not in a model of infectious meningitis. Moreover, anti-JAM antibodies induce a complement-mediated damage of the endothelium [34].

The process of PMN attraction to the subarachnoid space is regulated by chemotactic substances such as complement factor C5a [35–37], platelet-activating factor, and chemotactic cytokines called chemokines. Chemokines are produced at sites of inflammation and subsequently presented at the luminal side of the endothelium of the postcapillary venules, sometimes in cooperation with surface heparin sulfates [38]. Increased concentrations of interleukin-8 (IL-8), monocyte chemotactic protein-1 (MCP-1), and macrophage inflammatory proteins MIP-1α and MIP-1β are found in the CSF of patients with bacterial meningitis [39, 40].

In experimental models of meningitis, PMN infiltration of the subarachnoid space is only slightly or not at all diminished by intracisternal administration of anti-IL-8 antibodies, but is dramatically reduced by systemic neutralization of IL-8 via intravenously administered anti-IL-8 antibodies. These data show that IL-8 plays an important role in the recruitment of PMNs during meningitis and that the physiological activity of IL-8 during this process appears to be on the blood site of the microvascular endothelium rather than on the parenchymal site. Thus, inhibition of IL-8 provides a possible therapeutic target for adjunct treatment of bacterial meningitis [41, 42]. The idea of an innate immune response in the brain during systemic inflammation is strongly supported by MCP-1 expression within the endothelium of the cerebral vessels, the circumventricular organs, the choroid plexuses, and in a few macrophages and microglial cells in response to systemically administered proinflammatory cytokines [43]. Intraperitoneal administration of anti-MCP-1 antibodies results in significant reduction of macrophages in the CSF and administration of anti-MIP-2 or anti-MIP-1α significantly decreases PMN infiltration during experimental bacterial meningitis [44]. At this point, it is appropriate to ask for the structural avenues which invading leukocytes follow during the process of immigration. Postcapillary venules represent the segment of the microvasculature most vulnerable to inflammatory processes. In experimental studies using transmission and scanning electron microscopy, it was found that PMNs migrate through the endothelium transcellularly. This initial process of extravasation includes: rolling and sticking of PMNs at the endothelium, adherence at the parajunctional endothelial domains, subsequent direct penetration of the

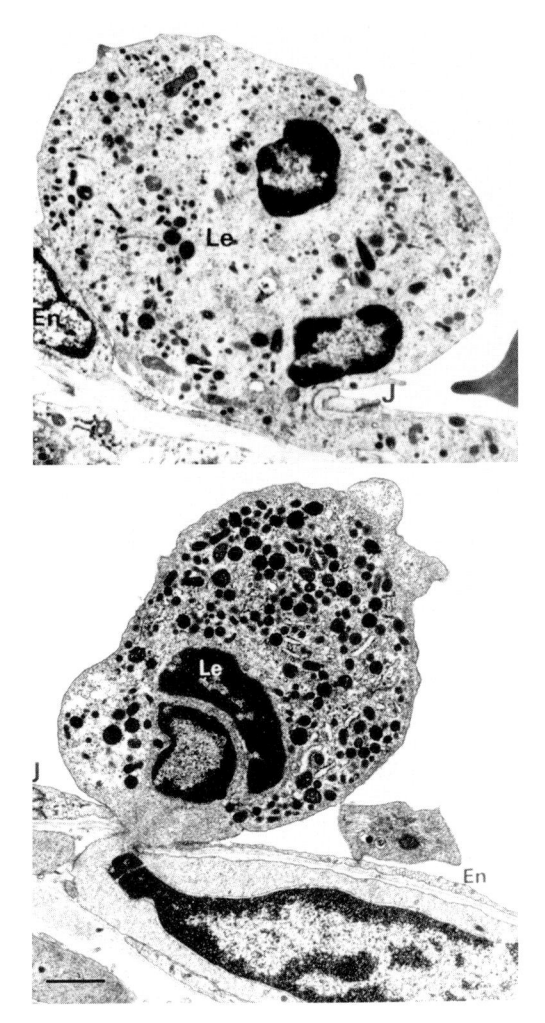

Fig. 24.1 Transmission electron microscopic studies show a direct extravasation of PMNs (Le) through the endothelium (En) in the initial phase of the inflammatory process (<2 h). Note that the interendothelial junction (J) is not affected by the migrating leukocyte (Le). Bar=1.5 µm.

endothelium or activation of the endothelial cells and engulfment of the PMNs by the endothelium, transendothelial passage, and subsequently penetration of the PMNs through the subendothelial basement membrane [45–47]. Morphological evidence for the involvement of interendothelial tight junctions in the initial process of PMNs extravasation through the inflamed BBB could not be obtained in any case (Fig. 24.1). A similar transendothelial extravasation is found for mononuclear cells under acute inflammation (Fig. 24.2) as well as under experimental autoimmune encephalomyelitis [48–50].

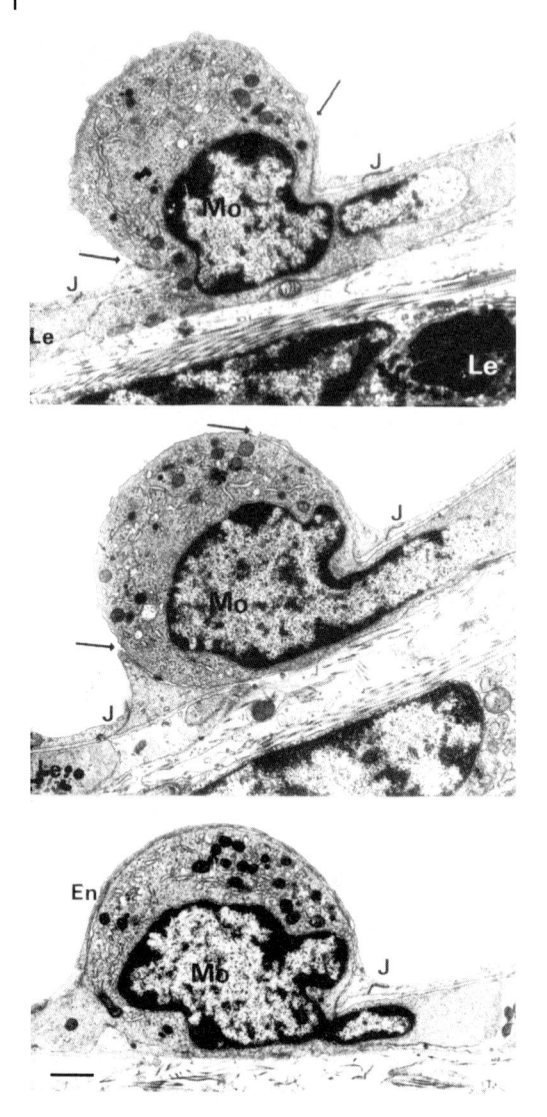

Fig. 24.2 Set of transmission electron microscopic figures showing a sequence of serial sections of one mononuclear cell (Mo) in the initial phase of inflammation. Note that the emigrating cell is engulfed (arrows) by the endothelium (En). The interendothelial junctions (J) are not affected by the emigrating mononuclear cell. Bar=1.0 μm.

24.3.3
Cytokines

Cytokines are key regulators of the immune response. Proinflammatory cytokines include TNF-α, IL-1β, -2, -6, -8, and interferon-gamma (INF-γ). These cytokines are released early in the inflammatory process by endothelial and perivascular cells like macrophages and microglia [51, 52]. Cytokines influence the integrity of the BBB [53]. Increased CSF concentrations of TNF-α, IL-1β, IL-6, and IL-8 are established markers to discriminate acute bacterial meningitis from aseptic (viral) meningitis [54, 55]. In view of acute meningitis it is of interest that S. suis is able to stimulate the production of IL-6, IL-8, and MCP-1 by human brain microvascular endothelial cells but not human umbilical vein endothelial cells, in a time- and concentration-dependent manner [56].

Experimental interventions aimed at suppression of the effect of proinflammatory cytokines during bacterial meningitis may thus target different levels:

1. Direct blocking of the cytokine. TNF-α and IL-1β antibodies reduce meningeal inflammation in experimental meningitis when instilled into the CSF at the beginning of the infection [57]. Endogenous inhibitors of cytokines, such as recombinant IL-1 receptor antagonist or soluble TNF receptor, are effective against cytokine-induced meningitis but are not effective against LPS meningitis [58].

2. Blocking of procytokine activation by proteolytic enzyme inhibition. Studies in a model of neonatal rat experimental bacterial meningitis show significantly down-regulated CSF levels of TNF-α by combined inhibition of matrix metalloproteinases and TNF-α-converting enzyme. In addition, the incidence of seizures, mortality, and neuronal injury is significantly decreased [59]. Similarly, treatment of rabbits and rats with experimental pneumococcal meningitis with the caspase-1 inhibitor z-VAD-fmk (benzyloxycarbonyl-Val-Ala-Asp-fluoromethyl ketone) reduces CSF-PMN counts and IL-1β activation [60].

3. Blocking of proinflammatory cytokines by anti-inflammatory cytokines. IL-10 and transforming growth factor beta (TGF-β) are considered prototypic antiinflammatory cytokines and potently inhibit the production of proinflammatory cytokines in vitro and in vivo [61, 62]. In experimental studies, therapeutic administration of IL-10 or TGF-β is consistently effective in modulating meningeal inflammation [63, 64]. However, in IL-10 gene-deficient mice, the absence of IL-10 is associated with higher proinflammatory cytokine and chemokine concentrations and a more pronounced PMN infiltrate, but antibacterial defense or survival of pneumococcal meningitis is not influenced [65].

4. Regulation of cytokine production via interference with transcription factors, e.g. NF-κB. Various anti-inflammatory drugs including corticosteroids and anti-inflammatory cytokines such as IL-10 inhibit the NF-κB pathway. Pharmacological manipulation of NF-κB activation is investigated in experimental meningitis. Two agents have been tested, a NF-κB inhibitor, which interferes with IκB proteolysis, and another, which inhibits IκB phosphorylation. Both agents improved clinical status and reduced the BBB permeability and PMN extravasation into the CSF [66].

24.4
Inflammatory Response in Acute Trauma

Traumatic injuries of the CNS initiate a complex cascade of events including an acute inflammatory response mediated by proinflammatory cytokines (see Chapter 11). IL-10 levels in the CSF and serum of patients with severe traumatic brain injury are related to IL-6, TNF-α, TGF-β1 and BBB function [67]. TNF-α levels have been associated with the neurological deficits in rodents subjected to experimental traumatic brain injury [68, 69]. In vivo experiments based on the neutralization of TNF-α after traumatic brain injury using common immunosuppressive agents or specific inhibitors showed a better neurological outcome and decreased brain oedema and less BBB dysfunction [68]. But recent studies on the role of cytokines after acute brain injury have yielded conflicting results with regards to their action contributing to repair mechanisms or exacerbating the pathophysiology of trauma [70]. Interestingly, posttraumatic mortality is significantly increased in TNF/lymphotoxin-deficient mice compared with wild-type animals, suggesting a protective effect of these cytokines. Moreover, neither the degree of BBB dysfunction nor the number of infiltrating PMNs in the injured hemisphere is different between wild-type and cytokine-deficient mice [71]. Neuropeptides like substance P and calcitonin gene-related peptide (CGRP) increase microvascular permeability, leading to oedema formation and neurogenic inflammation. Recent studies applied a capsaicin-induced neuropeptide depletion to examine the role of neurogenic inflammation following traumatic brain injury. Neuropeptide depletion yielded a significant reduction in posttraumatic oedema formation, BBB permeability, and both motor and cognitive deficits [72].

Traumatic injury is significantly greater in the spinal cord than in the cerebral cortex, as indicated by the number of PMNs recruited to the lesion site, which is significantly higher in the spinal cord as compared to the brain. Also, the area of BBB breakdown with concurrently occurring long-lasting vascular damage is substantially larger in the spinal cord [73].

A novel therapeutic peptide derived from apolipoprotein E (apoE) reduces brain inflammation and improves outcome after closed head injury [74]. Interestingly, apoE4, which was initially identified as a susceptibility gene for the development of Alzheimer's disease, seems also to be associated with poor outcome after acute brain injury. One mechanism by which apoE may influence neurological outcome is by down-regulating the neuroinflammatory response. ApoE has been demonstrated to reduce glial activation and the CNS inflammatory response in vitro and in vivo in an isoform-specific action [75, 76], with the apoE4 isoform being less effective than apoE3 at down-regulating proinflammatory cytokines [75, 77].

24.5
Inflammatory Response in Alzheimer's Disease

Alzheimer's disease (AD) is a chronic neurodegenerative condition that affects approximately 10% of individuals over 65 years. Five percent of this group of patients suffers from severe dementia. Memory decline as well as the inability to process new information are the most prominent clinical signs. Imaging techniques revealed disturbances in cerebral blood flow and glucose metabolism (see Chapter 25). A deficiency in the neurotransmitter acetylcholine has been suggested to be involved in AD, since cortical acetylcholine synthesis is markedly diminished in Alzheimer's patients. Other neurotransmitters which may be involved in the course of the disease are dopamine, γ-aminobutyric acid, vasoactive intestinal peptide, and glutamate.

The pathology of AD is characterized by the extracellular deposition of a particular form of β-amyloid protein, derived from the larger precursor protein (APP). This amyloid peptide induces the release of the so-called β-peptide or the A4 peptide (A4P), which becomes stacked in the form of a β-pleated sheet structure with a high degree of intermolecular hydrogen bonding. Usually, the amyloid peptide precursor with an apparent molecular mass of 112 kDa is found in the brain. During AD, the A4P is processed abnormally into a 43-amino acid amyloiditic peptide that arises from near the C-terminus of the A4P. The pathology of an AD brain reveals granulovacuolar degeneration of neurons within the region of the hippocampus. Furthermore, neurofibrillary tangles are detected which are intraneuronal accumulations of dense non-membrane-bound fibrillary material forming paired helices. Other lesions are composed of amyloid, predominantly associated with microglial cells [78, 79].

Inflammatory reactions, surrounding the cerebral microvasculature, are frequently observed during AD. The number of perivascular macrophages increases and hypertrophy of astrocytes and microglia is observed in brain sections of AD patients [79]. IL-1 is found in the CNS of patients suffering from AD and its expression is thought to be an early event in the onset of AD. IL-1 affects the synthesis of β-amyloid precursor protein and the subsequent deposition of β-amyloid [80]. Increased levels of IL-6 and MCP-1 in the plasma and CSF of AD patients may also be good candidates as biomarkers for monitoring the inflammatory process in AD [81]. A stringent sign that the brain microvasculature is involved in the generation of AD pathophysiology is the accumulation of ICAM-1 in endothelial cells as well as in senile plaques [82], indicating that infiltration of lymphocytes into the brain tissue accompanies the inflammatory reactions during AD. CSF studies showed an increased level of soluble ICAM-1, but not soluble E-selectin (s-ELAM-1) in AD, suggesting a more neural than endothelial s-ICAM origin in patients with AD [83].

Albumin concentrations are also enhanced in the CSF of Alzheimer's patients at the early onset of AD, apparently resulting from an increased permeability in the BBB [84]. Two-dimensional gel electrophoresis showed the presence of a protein, haptoglobulin, with a molecular mass of 13.5 kDa in the CSF of AD patients. The presence of this protein was suggested to be the result of an increased penetration of haptoglobin across an impaired BBB [85]. AD patients have a diminished density of mitochondria and feature incomplete interendothelial junctions in the cerebral endothelial cells, suggesting leakiness of the vessels [86]. The diminished cerebral metabolic rate observed during AD can be partly explained by a decreased density of glucose transporters (GLUT-1) in the brain capillaries [87, 88]. Changes in intracellular signal transduction in cerebral microvessels of AD patients, in comparison with age-matched controls, have also been observed. Protein kinase C activity is strongly diminished in the brain microvessels, which leads to altered protein phosphorylation. Protein kinase C activity during AD seems to be important in the inhibition of the processing of β-amyloid precursor protein into soluble Aβ protein. Intracellular signaling in the cerebrovasculature may be one of the pathophysiological targets in AD; and the balance of various secondary messenger systems may be disturbed [89]. In AD transgenic mice, the therapeutic effects of the PKC activator bryostatin-1 included improvement of behavioral outcomes [90].

Perivascular inflammatory reactions during AD are likely to cause additional changes in the function of the BBB. Recent studies showed that the BBB is disrupted in brains of AD patients. In addition, cultures of endothelial cells derived from brain capillaries of AD patients revealed the production of the β-amyloid precursor protein. Apolipoprotein E4, which is considered a risk factor for the development of AD, is also produced by cerebral endothelial cells [91]. Brain microvessels from AD patients express high levels of inflammatory proteins, suggesting that the brain microcirculation could be a source of neurotoxic factors in AD [92].

Taking together, all indicated studies provide substantial evidence that the function of the BBB is impaired in AD patients. Disruption of the BBB can lead to the entry of neurotoxic environmental factors into the brain or circulating amyloid. One hypothesis is that BBB opening may be the initial insult that causes AD, although more research needs to be conducted to substantiate this hypothesis.

References

1 H.W. Pfister, T.O. Bleck **1996**, Bacterial infections, in *Neurological Disorders – Course and Treatment*, ed. T. Brandt, L.R. Caplan, J. Dichgans, H.C. Diener, C. Kennard, Acadamic Press, San Djiego, p. 381.
2 K.L. Roos **1997**, *Central Nervous System Infectious Diseases and Therapy*, Dekker, New York.
3 A.A. Pruitt **1998**, *Neurol. Clin.* 16, 419.
4 L. Wubbel, G.H. McCracken Jr. **1998**, *Pediatr. Rev.* 19, 78.
5 L.E. Davis **1999**, *Neurol. Clin.* 17, 761.
6 J.R. Zunt, C.M. Marra **1999**, *Neurol. Clin.* 17, 675.
7 D. Van de Beck, J. de Gans, L. Spanjaard, M. Weisfelt, J.B. Reitsma, M. Vermeulen **2004**, *N. Engl. J. Med.* 351, 1849.

8 K. L. Roos **1997**, Bacterial meningitis, in *Central Nervous System Infectious Diseases and Therapy*, ed. K. L. Roos, Dekker, New York, p. 99.
9 H. W. Pfister, W. M. Scheld **1997**, *Curr. Opin. Neurol.* 10, 254.
10 U. Koedel, H. W. Pfister **1999**, *Brain Pathol.* 9, 57.
11 R. Nau, W. Brück **2002**, *Trends Neurosci.* 25, 38.
12 G. C. Townsend, W. M. Scheld **1995**, *Trends Microbiol.* 3, 441.
13 A. Ring, J. N. Weiser, E. I. Tuomanen **1998**, *J. Clin. Invest.* 102, 347.
14 B. Ren, A. J. Szalai, S. K. Hollingshead, D. E. Briles **2004**, *Infect. Immun.* 72, 114.
15 G. Zysk, B. K. Schneider-Wald, J. H. Hwang, L. Bejo, K. S. Kim, T. J. Mitchell, R. Hakenbeck, H. P. Heinz **2001**, *Infect. Immun.* 69, 845.
16 R. A. Hirst, A. Kadioglu, C. O'Callaghan, P. W. Andrew **2004**, *Clin. Exp. Immunol.* 138, 195.
17 S. Houldsworth, P. W. Andrew, T. J. Mitchell **1994**, *Infect. Immun.* 62, 1501.
18 X. Nassif, S. Bourdoulous, E. Eugene, P. O. Couraud **2002**, *Trends Microbiol.* 10, 227.
19 L. Johansson, A. Rytkonen, P. Bergman, B. Albiger, H. Kallstrom, T. Hokfelt, B. Agerberth, R. Cattaneo, A. B. Jonsson **2003**, *Science* 301, 373.
20 A. Unkmeir, K. Latsch, G. Dietrich, E. Wintermeyer, B. Schinke, S. Schwender, K. S. Kim, M. Eigenthaler, M. Frosch **2002**, *Mol. Microbiol.* 46, 933.
21 Y. Xie, K. J. Kim, K. S. Kim **2004**, *FEMS Immunol. Med. Microbiol.* 42, 271.
22 N. Nonaka, S. Shioda, W. A. Banks **2005**, *Exp. Neurol.* 191, 137.
23 N. S. Ivey, E. N. Martin Jr., W. M. Scheld, B. R. Nathan **2005**, *J. Neurosci. Methods* 142, 91.
24 M. M. J. Polfiet, P. J. G. Zwijnenburg, A. M. van Furth, T. van der Poll, eds. A. Döpp, C. Renardel de Lavalette, E. M. L. van Kesteren-Hendrikx, N. van Rooijen, C. D. Dijkstra, T. K. van den Berg **2001**, *J. Immunol.* 167, 4644.
25 M. van der Flier, S. P. M. Geelen, J. L. L. Kimpen, I. M. Hoepelman, E. I. Tuomanen **2003**, *Microbiol. Rev.* 16, 415.
26 T. Tang, P. S. Frenette, R. O. Hynes, D. D. Wagner, T. N. Mayadas **1996**, *J. Clin. Invest.* 97, 2485.
27 K. Angstwurm, J. R. Weber, A. Segert, W. Bürger, M. Weith, D. Freyer, K. M. Einhäupl, U. Dirnagl **1995**, *Neurosci. Lett.* 191, 1.
28 C. Ostergaard, R. V. Yieng-Kow, T. Benfield, N. Frimodt-Moller, F. Espersen, J. D. Lundgren **2000**, *Infect. Immun.* 68, 3153.
29 M. M. Lipovsky, L. Tsenova, F. E. J. Coenjaerts, G. Kaplan, R. Cherniak, A. I. M. Hoepelman **2000**, *J. Neuroimmunol.* 111, 10.
30 J. B. Dietrich **2002**, *J. Neuroimmunol.* 128, 58.
31 J. R. Weber, K. Angstwurm, W. Bürger, K. M. Einhäupl, U. Dirnagl **1995**, *J. Neuroimmunol.* 63, 63.
32 M. Bohatschek, A. Werner, G. Raivich **2001**, *Exp. Neurol.* 172, 137.
33 A. del Maschio, A. de Luigi, I. Martin-Padura, M. Brockhaus, T. Bartfai, P. Fruscella, L. Adorini, G. V. Martino, R. Furlan, M. G. De Simoni, E. Dejana **1999**, *J. Exp. Med.* 190, 1351.

34 F. Lechner, U. Sahrbacher, T. Suter, K. Frei, M. Brockhaus, U. Koedel, A. Fontana **2000**, *J. Infect. Dis.* 182, 978.
35 J.D. Ernst, K.T. Hartiala, I.M. Goldstein, M.A. Sande **1984**, *Infect. Immun.* 46, 81.
36 R.E. Ganz, P.M. Faustmann **1994**, *Int. J. Neurosci.* 76, 177.
37 P.M. Faustmann, D. Krause, R. Dux, R. Dermietzel **1995**, *Acta Neuropathol.* 89, 239.
38 J. Middleton, S. Neil, J. Wintle, I. Clark-Lewis, H. Moore, C. Lam, M. Auer, E. Hub, A. Rot **1997**, *Cell* 91, 385.
39 C. Ostergaard, T.L. Benfield, F. Sellebjerg, G. Kronborg, N. Lohse, J.D. Lundgren **1996**, *Eur. J. Clin. Microbiol. Infect. Dis.* 15, 166.
40 K.S. Spanaus, D. Nadal, H.W. Pfister, J. Seebach, U. Widmer, K. Frei, S. Gloor, A. Fontana **1997**, *J. Immunol.* 158, 1956.
41 R.A. Dumont, B.D. Car, N.N. Voitenok, U. Junker, B. Moder, O. Zak, T. O'Reilly **2000**, *Infect. Immun.* 68, 5756.
42 C. Ostergaard, R.V. Yieng-Kow, C.G. Larsen, N. Mukaida, K. Matsushima, T. Benfield, N. Frimodt-Moller, F. Espersen, A. Kharazmi, J.D. Lundgren **2000**, *Clin. Exp. Immunol.* 122, 207.
43 I. Thibeault, N. Laflamme, S. Rivest **2001**, *J. Comp. Neurol.* 434, 461.
44 A. Diab, H. Abdalla, H.L. Li, F.D. Shi, J. Zhu, B. Höjberg, L. Lindquist, B. Wretlind, M. Bakhiet, H. Link **1999**, *Infect. Immun.* 67, 2590.
45 P.M. Faustmann, R. Dermietzel **1985**, *Cell Tissue Res.* 242, 399.
46 P.M. Faustmann, S. Teutrine, R. Dermietzel **1987**, Functional ultrastructure of the blood-brain barrier under acute inflammatory reactions, in *Stroke and Microcirculation*, eds. J. Cervos-Navarro, R. Ferszt, Raven Press, New York, p. 285.
47 R. Dermietzel, P.M. Faustmann, D. Krause **1999**, Inflammatory reaction of the blood-brain barrier, in *Brain Barrier Systems*, eds. O. Paulson, G. Moos Knudsen, T. Moos, Munksgaard, Copenhagen, p. 403.
48 H.M. Wisniewski, A.S. Lossinsky **1991**, *Brain Pathol.* 1, 89.
49 B. Engelhardt, H. Wolburg **2004**, *Eur. J. Immunol.* 34, 2955.
50 H. Wolburg, K. Wolburg-Buchholz, B. Engelhardt **2005**, *Acta Neuropathol.* 109, 181.
51 H.E. de Vries, J. Kuiper, A.G. de Boer, T.J.C. van Berkel, D.D. Breimer **1997**, *Pharmacol. Rev.* 49, 143.
52 A.A. Webb, G.D. Muir **2000**, *J. Vet. Intern. Med.* 14, 399.
53 H.E. de Vries, M.C. Blom-Roosemalen, M. van Oosten, A.G. de Boer, T.J. van Berkel, D.D. Breimer, J. Kuiper **1996**, *J. Neuroimmunol.* 64, 37.
54 R.B. Tang, B.H. Lee, R.L. Chung, S.J. Chen, T.T. Wong **2001**, *Childs Nerv. Syst.* 17, 453.
55 T.O. Kleine, P. Zwerenz, P. Zofel, K. Shiratori **2003**, *Brain Res. Bull.* 61, 287.
56 N. Vadeboncoeur, M. Segura, D. Al-Numani, G. Vanier, M. Gottschalk **2003**, *FEMS Immunol. Med. Microbiol.* 35, 49.
57 M.G. Tauber, B. Moser **1999**, *Clin. Infect. Dis.* 28, 1.

58 M. M. Paris, I. R. Friedland, S. Ehrett, S. M. Hickey, K. D. Olsen, E. Hansen, E. J. Thonar, G. H. McCracken Jr. **1995**, *J. Infect. Dis.* 171, 161.
59 S. L. Leib, J. M. Clements, R. L. Lindberg, C. Heimgartner, J. M. Loeffler, L. A. Pfister, M. G. Tauber, D. Leppert **2001**, *Brain* 124, 1734.
60 U. Koedel, F. Winkler, B. Angele, A. Fontana, R. A. Flavell, H. W. Pfister **2002**, *Ann. Neurol.* 51, 319.
61 M. Swada, A. Suzumura, H. Hosoya, T. Marunouchi, T. Nagatsu **1999**, *J. Neurochem.* 72, 1466.
62 F. Molina-Holgado, R. Grencis, N. J. Rothwell **2001**, *Glia* 33, 97.
63 U. Koedel, A. Bernatowicz, K. Frei, A. Fontana, H. W. Pfister **1996**, *J. Immunol.* 157, 5185.
64 H. W. Pfister, K. Frei, B. Ottnad, U. Koedel, A. Tomasz, A. Fontana **1992**, *J. Exp. Med.* 176, 265.
65 P. J. G. Zwijnenburg, T. van der Poll, S. Florquin, J. J. Roord, A. M. van Furth **2003**, *Infect. Immun.* 71, 2276.
66 U. Koedel, I. Bayerlein, R. Paul, B. Sporer, H. W. Pfister **2000**, *J. Infect. Dis.* 182, 1437.
67 E. Csuka, M. C. Morganti-Kossmann, P. M. Lenzlinger, H. Joller, O. Trentz, T. Kossmann **1999**, *J. Neuroimmunol.* 101, 211.
68 E. Shohami, I. Ginis, J. M. Hallenbeck **1999**, *Cytokine Growth Factor Rev.* 10, 119.
69 S. M. Knoblach, L. Fan, A. I. Faden **1999**, *J. Neuroimmunol.* 95, 115.
70 M. C. Morgant-Kossmann, M. Rancan, P. Stahel, T. Kossmann **2002**, *Curr. Opin. Crit. Care* 8, 101.
71 P. F. Stahel, E. Shohami, F. M. Younis, K. Kariya, V. I. Otto, P. M. Lenzlinger, M. B. Grosjean, H. P. Eugster, O. Trentz, T. Kossmann, M. C. Morganti-Kossmann **2000**, *J. Cereb. Blood Flow Metab.* 20, 369.
72 A. J. Nimmo, I. Cernak, D. L. Heath, X. Hu, C. J. Bennett, R. Vink **2004**, *Neuropeptides* 38, 40.
73 L. Schnell, S. Fearn, H. Klassen, M. E. Schwab, V. H. Perry **1999**, *Eur. J. Neurosci.* 11, 3648.
74 J. R. Lynch, H. Wang, B. Mace, S. Leinenweber, D. S. Warner, E. R. Bennett, M. P. Vitek, S. McKenna, D. T. Laskowitz **2005**, *Exp. Neurol.* 192, 109.
75 D. T. Laskowitz, A. D. Thekdi, S. D. Thekdi, S. K. Han, J. K. Myers, S. V. Pizzo, E. R. Bennett **2001**, *Exp. Neurol.* 167, 74.
76 J. R. Lynch, D. Morgan, J. Mance, W. D. Matthew, D. T. Laskowitz **2001**, *J. Neuroimmunol.* 114, 107.
77 J. R. Lynch, W. Tang, H. Wang, M. P. Vitek, E. R. Bennett, P. M. Sullivan, D. S. Warner, D. T. Laskowitz **2003**, *J. Biol. Chem.* 278, 48529.
78 C. A. Marotta, R. E. Majocha, B. Tate **1992**, *J. Mol. Neurosci.* 3, 111.
79 A. Minagar, P. Shapshak, R. Fujimura, R. Ownby, M. Heyes, C. Eisdorfer **2002**, *J. Neurol. Sci.* 202, 13.
80 W. S. T. Griffin, J. G. Sheng, M. C. Royston, S. M. Gentleman, J. E. McKenzie, D. I. Graham, G. W. Roberts, R. E. Mrak **1998**, *Brain Pathol.* 8, 65.

81 Y. X. Sun, L. Minthon, A. Wallmark, S. Warkentin, K. Blennow, S. Janciauskiene **2003**, *Dement. Geriatr. Cogn. Disord.* 16, 136.
82 M. M. Verbeek, I. Otte-Holler, P. Wesseling, D. J. Ruiter, R. M. de Waal **1996**, *Acta Neuropathol.* 91, 608.
83 M. Rentzos, M. Michalopoulou, C. Nikolaou, C. Cambouri, A. Rombos, A. Dimitrakopoulos, E. Kapaki, D. Vassilopoulos **2004**, *J. Geriatr. Psychiatry Neurol.* 17, 225.
84 P. Mecocci, L. Parnetti, G. P. Reboldi, C. Santucci, A. Gaiti, C. Ferri, I. Gernini, M. Romagnoli, D. Cadini, U. Senin **1991**, *Acta Neurol. Scand.* 84, 210.
85 K. M. Mattila, T. Pirtilä, K. Blennow, A. Wallin, M. Vitanen, H. Frey **1994**, *Acta Neurol. Scand.* 89, 192.
86 P. A. Stewart, K. Hayakawa, M. A. Akers, H. V. Vinters **1992**, *Lab. Invest.* 67, 734.
87 S. I. Harik **1992**, *Can. J. Physiol. Pharmacol.* 20[suppl], S113.
88 I. A. Simpson, K. R. Chundu, T. Davies-Hill, W. G. Honer, P. Davies **1994**, *Ann. Neurol.* 35, 546.
89 P. Grammas, P. Moore, T. Botchelet, O. Hanson-Painton, D. R. Cooper, M. J. Ball, A. Roher **1995**, *Neurobiol. Aging* 4, 563.
90 R. Etcheberrigaray, M. Tan, I. Dewachter, C. Kuiperi, I. van der Auwera, S. Wera, L. Qiao, B. Bank, T. J. Nelson, A. P. Kozikowski, F. van Leuven, D. L. Alkon **2004**, *Proc. Natl Acad. Sci. USA* 101, 11141.
91 J. M. Wells, A. Amaratunga, D. C. McKenna, C. R. Abraham, R. E. Fine **1995**, *Int. J. Exp. Clin. Invest.* 2, 229.
92 P. Grammas, T. Ottman, U. Reichmann-Philipp, J. Larabee, P. H. Weigel **2004**, *J. Alzheimer Dis.* 6, 275.

25
Stroke and the Blood-Brain Interface

Marilyn J. Cipolla

25.1
Introduction

Ischemic injury of the brain is one of the most common pathophysiologic processes affecting more than 750,000 people per year in the United States in the form of stroke [1]. Brain edema with a subsequent rise in intracranial pressure is the most dangerous complication of stroke [2], often occurring during the reperfusion period after recanalization of an occluded artery (i.e., reperfusion injury) [3]. Increased cerebrovascular permeability is considered the most important factor for development of cerebral edema and is determined by the cerebral endothelial cells (EC) that form the blood-brain barrier (BBB) [4, 5]. The morphologic features of cerebral ECs that prevent the extravasation of large and small solutes are the presence of tight junctions that reduce paracellular transport and the low rate of pinocytotic vesicle formation that limits transcellular transport [5–7]. Recently, it has become evident that these processes are regulated by factors that influence both EC structure and function, including hemodynamic factors such as flow and pressure [8–10], the production of autocoids and second messengers such as nitric oxide (NO), superoxide anion (O_2^-), and protein kinase C (PKC) [11–13], and the organizational state of the EC actin cytoskeleton [14–16]. Many of these same factors that regulate permeability under normal conditions are known to be altered during ischemia and/or reperfusion, causing enhanced vascular permeability and edema formation. The purpose of this chapter is to present our current understanding of alterations in cerebral EC structure and function due to ischemia and reperfusion (I/R) that promote BBB disruption and cerebral edema formation.

Blood-Brain Interfaces: From Ontogeny to Artificial Barriers.
Edited by R. Dermietzel, D.C. Spray, M. Nedergaard
Copyright © 2006 WILEY-VCH Verlag GmbH & Co. KGaA, Weinheim
ISBN: 3-527-31088-6

25.2
Brain Edema Formation During Stroke

The development of brain edema is one of the most detrimental consequences of stroke and is a significant determinant of stroke outcome [17, 18]. Studies have demonstrated that edema, increased intracranial pressure, and brain herniation are the major determinants of not just patient outcome, but survival as well [19]. Secondary, progressive ischemia due to vascular compression and elevated intracranial pressure are also thought to contribute to the pathogenesis of stroke, demonstrating the significant impact of brain edema [20]. Furthermore, while it is difficult to separate evolving edema due to ischemia from the damage that occurs during the reperfusion phase, there have been numerous cases of severe edema thought to be due to postischemic reperfusion [21–23], indicating that damage is not restricted to the ischemic periods, but occurs during reperfusion as well (i.e., reperfusion injury).

25.2.1
Cytotoxic Versus Vasogenic Edema

Klatzo first characterized brain edema as cytotoxic versus vasogenic, depending on whether or not the BBB is disrupted [24]. In the simplest terms, cytotoxic edema occurs when brain cells swell at the expense of the extracellular space while the BBB remains intact. In vasogenic edema, cerebrovascular permeability is increased due to BBB disruption that allows an influx of plasma constituents and expansion of the extracellular space. During stroke, features of both cytotoxic edema and vasogenic edema occur simultaneously (Fig. 25.1, see pp. 622/623). Ischemia causes cells to lose their normal ionic gradients due to diminished ATP and loss of energy to drive ionic pumps [25]. Since sodium entry exceeds potassium loss, a net increase in intracellular ions occurs [26–28]. Consequently, water enters the cell due to abnormal osmotic gradients and cytotoxic edema ensues [25–30]. Theoretically, because the exchange of water and ions is between the cell and the extracellular space (i.e., no BBB disruption), there is no net increase in brain water content and therefore this is not true cerebral edema. However, the ionic gradient will eventually be sufficient to affect the cerebral endothelium, causing a net increase in water from the plasma [29]. This enhanced hydraulic conductivity increases the bulk flow of ions into brain tissue from the plasma due to hydrostatic pressure. A close correlation exists between the net increase in brain cations and edema formation. In fact, the rate of sodium flux across the endothelium may control the rate of edema formation [26, 27, 29, 30]. During continuous ischemia, the BBB becomes disrupted and ions enter the brain more rapidly, causing significant brain edema and the passage of potentially toxic substances into brain tissue [31].

Under normal conditions, the barrier properties of the cerebral endothelium that forms the BBB has low hydraulic conductivity that prevents bulk flow of water, ions and proteins into the brain from hydrostatic forces such as blood pressure [32]. When the BBB is disrupted (e.g., during ischemia), hydrostatic forces become significant enough that the rate of protein entry into the brain is directly related to the pressure gradient between the blood and the brain. However, while the passage of protein from the plasma into the brain is a measure of BBB disruption, it does not significantly contribute to edema formation. Albumin and other proteins passing into the brain have been used as a measure of the BBB disruption that leads to edema formation. However, the concentration of a large protein is several orders of magnitude smaller than that of ions. Therefore, the increase in osmolality that occurs when albumin enters the brain is small compared to that of ions [32]. Albumin and protein entry into the brain may be an indicator of BBB disruption, but its contribution to edema formation is small compared to that of ions.

It is important to note that cell swelling associated with cytotoxic edema may not be entirely detrimental. Under certain conditions, the cell volume increase may be compensatory or even protective. Precapillary astrocyte endfeet are the first cellular elements to swell during ischemia [33], a process thought to normalize the composition of the extracellular environment for normal neuronal activity [34]. Glial cells can also inactivate neurotransmitters [35], take up excess potassium ions produced during neuronal activity [36], and scavenge reactive oxygen species [37].

25.3
Role of Astrocytes in Mediating Edema During Ischemia

While it is the structural properties of the cerebral endothelium that make up the BBB (e.g., complex tight junctions, low rate of pinocytosis, lack of fenestrations), surrounding cells within the brain parenchyma are known to contribute to and induce BBB properties, most notably astrocytes. An early view of the BBB was that astrocytic endfeet, which are anatomically in close apposition to the cerebral endothelium and brain parenchymal vessels, provided a physical barrier and thus contributed to BBB properties. However, studies by Reese and Karnovsky showed that the site of barrier function was at the level of the cerebral endothelium and not the astrocytes [38, 39]. A role for astrocytes in inducing the BBB properties of the cerebral endothelium, as opposed to being a physical barrier, emerged around the same time [40]. Subsequent in vitro cell culture studies confirmed the important role of glial cells in inducing BBB phenotype, including upregulation of tight junction proteins [41, 42]. There is also considerable evidence that the cerebral endothelium influences and signals astrocytes. For example, the water channel aquaporin 4 (AQP4) is primarily expressed only in astrocytic endfeet surrounding parenchymal vessels, but not in astrocytes just interacting with neurons [43]. In support of this concept, AQP4 has been shown to upregulate in astrocytic endfeet when cocultured with endothelium [44] (see Section 25.3.1). Therefore, the maintenance of BBB properties and function likely depends on cross-talk between the endothe-

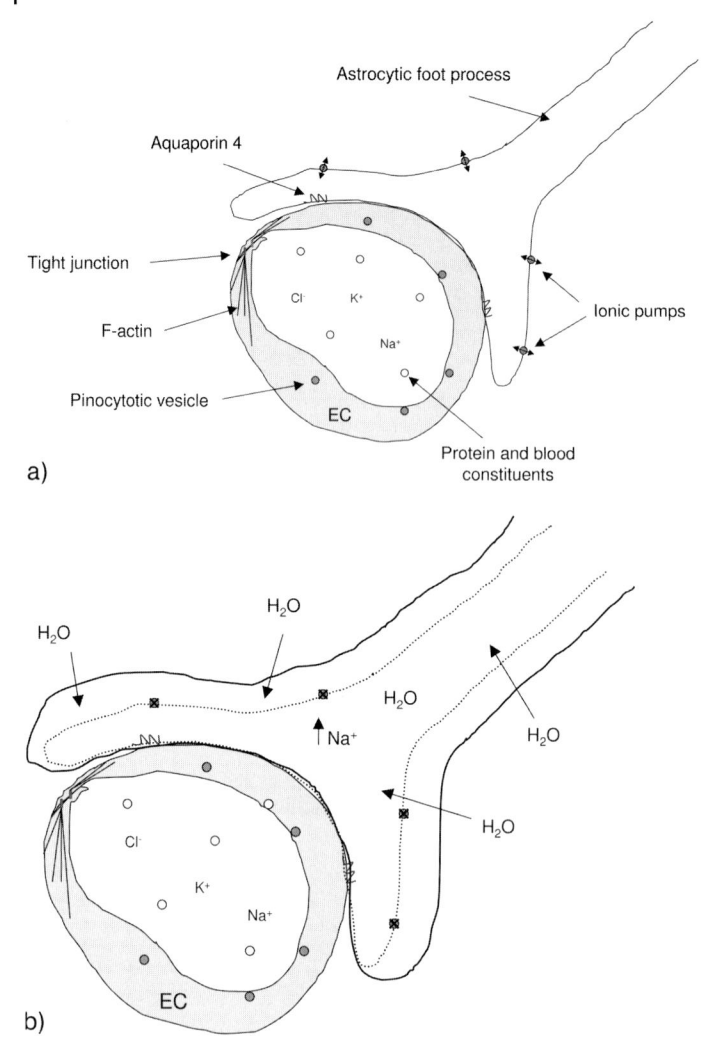

Fig. 25.1 Development of cerebral edema formation during ischemia.
(a) Normal blood-brain barrier properties. Astrocytic foot processes closely appose the cerebral endothelium and contain aquaporin water channels and ionic pumps (Na-K-ATPase) that control the intracellular ionic environment and cell swelling. The cerebral endothelium, due to its structure, has low hydraulic conductivity that limits bulk flow of water from hydrostatic forces. Tight junction proteins are linked to the actin cytoskeleton and allow the regulation of paracellular flux by mediators that affect the contractile state of the cell and/or actin cytoskeletal dynamics. Pinocytotic vesicles are present in low number and contribute to normal brain homeostasis.
(b) Initial stages of ischemia. Ischemia causes cells to lose their normal ionic gradients due to the loss of energy to drive ionic pumps (Na-K-ATPase). Because sodium influx exceeds potassium loss, there is a net increase in intracellular sodium that causes water to enter the cell due to osmotic forces. The astrocytes swell (cytotoxic edema), but since cell swelling is at the expense of the extracellular space, this is not true edema formation.

c)

d)

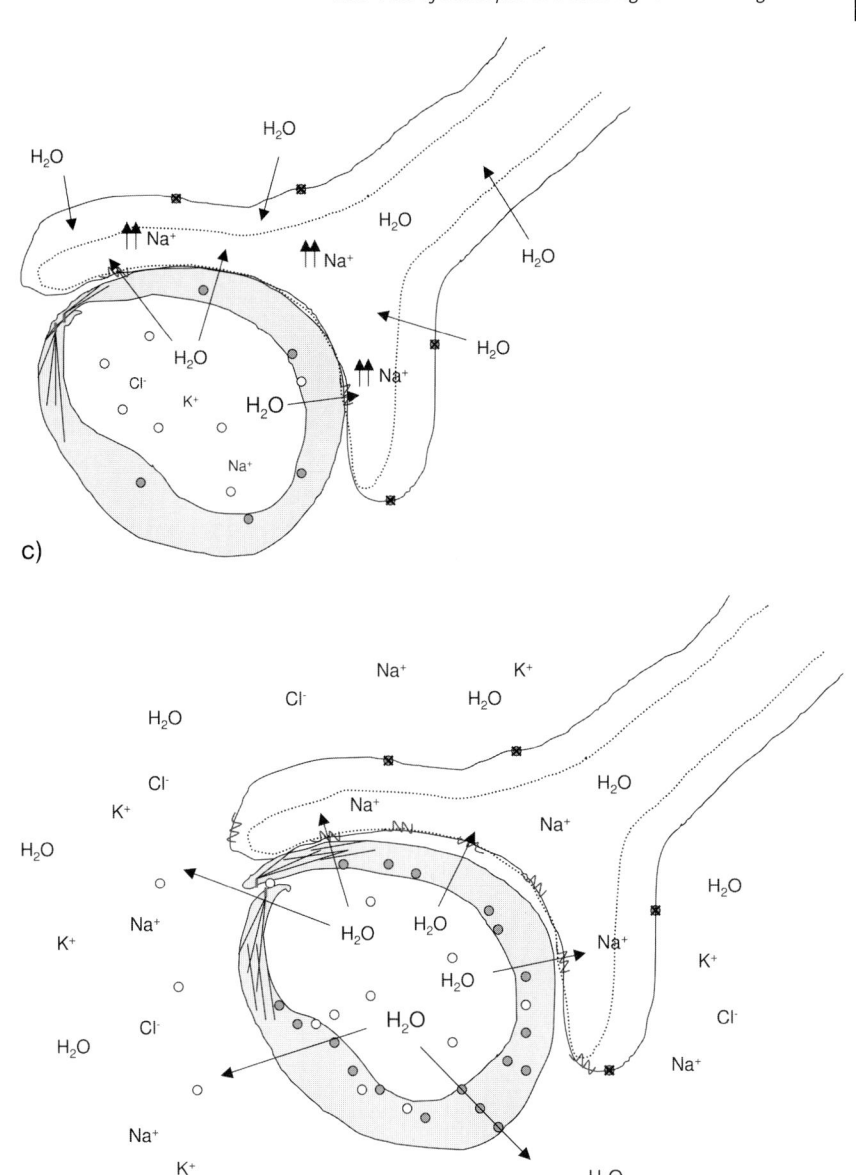

(c) Later stages of ischemia. Eventually the ionic gradient becomes sufficient to pull water across the cerebral endothelium from the plasma, increasing hydraulic conductivity. This is true cerebral edema formation, because of the net increase in brain water content.

(d) Prolonged ischemia. During continuous ischemia, the blood-brain barrier becomes disrupted and ions enter more rapidly, causing significant cerebral edema. At this stage, tight junctions are disrupted and there is an increase in vesicular transport that also increases extravasation of proteins and other toxic substances from the blood into the brain parenchyma. Aquaporin water channels become upregulated in an attempt to normalize intracellular water content.

lium and astrocytes. Because of the interdependence of endothelium and glia, disruption of any of these signals during ischemia can result in disruption of function.

Pial vessels also have BBB properties, even though they are not directly in contact with astrocytes [45, 46]. The BBB extends throughout the brain with the exception of the circumventricular organs. The fact that pial vessels show BBB properties suggests that soluble factors from the glia limitans or subarachnoid CSF may be the source [46]. Butt et al. measured transendothelial electrical resistance (TEER) in pial vessels in situ and found that they have an electrical resistance of around $1800\ \Omega\ cm^{-2}$, a number that is somewhat less than the microcirculation, but considerably greater than cell culture models of BBB, even when cocultured with astrocytes.

During ischemia, astrocytes are known to be protective of both neurons and endothelium and maintain barrier properties. For example, Fischer et al. [47] showed that hypoxia-induced hyperpermeability of cultured porcine brain-derived microvascular endothelial cells was significantly attenuated when either cocultured with astrocytes or in glial-conditioned media. This effect was shown to be due to prevention of hypoxia-induced restructuring of the tight junction protein zona occludens-1 (ZO-1). Other studies have shown that astrocytes confer a protective role on the BBB during hypoxia and reoxygenation. Cerebral microvascular endothelial cells cocultured with astrocytes have reduced decrease in barrier properties, due to hypoxia/reoxygenation compared to endothelial cells alone [48]. The presence of astrocytes has also been shown to prevent upregulation of hypoxia-induced vascular endothelial growth factor (VEGF) expression in endothelial cells, a known contributor to increased vascular permeability during stroke [49].

Another role of astrocytes that may be important in BBB function during ischemia is providing a secondary barrier to endothelial disruption by the ability to swell and maintain ionic homeostasis. This cytotoxic edema, likely mediated by aquaporins [50, 51], can influence ionic composition of the extracellular space and metabolize unwanted protein and neurotransmitter that passes from the blood into the brain [33, 34]. The protective effect of astrocytes is limited though. Eventually, with the BBB disrupted, hydrostatic pressure becomes a significant influence, and together with this cytotoxic edema, intracranial pressure increases and causes neurologic complications. However, this protective role of astrocytes makes them a potential therapeutic target for ischemia-induced edema formation.

25.3.1
Aquaporins and Cerebral Edema During Ischemia

An important factor in determining cytotoxic swelling of astrocytes is the aquaporins (AQPs). The AQPs are a family of channel-forming transmembrane proteins that facilitate the movement of water, glycerol, and other solutes across the plasma membrane of cells [52–55]. AQP4, the predominant aquaporin in the brain, has been localized in the astrocytic endfeet surrounding blood vessels, but has also recently been found in the vascular endothelium [56–59]. One role of astrocytic AQP4 may be in the actual regulation of water transport across the BBB, mainly because of the highly polarized localization of AQP4 in the perivascular endfeet of astrocytes (see Chapter 9) [54, 55, 60]. This concept is supported by numerous studies that have shown increased AQP4 expression during conditions that cause vasogenic brain edema, including brain tumors, focal ischemia, and brain injury [60–64].

Whether the increase in AQP4 expression during ischemia is beneficial or detrimental is still unclear. AQP4 in astrocytes is thought to contribute to BBB properties by taking up excess water brought into the brain by disruption of the BBB. However, excessive AQP4 may be detrimental and promote increased intracranial pressure, due to excessive cytotoxic edema formation. For example, knockout mice in which deletion of α-syntrophin, a membrane protein anchoring AQP4 into the perivascular astrocytic endfeet, resulted in the loss of AQP4 at these membranes [59]. Following middle cerebral artery (MCA) occlusion and reperfusion, α-syntrophin knockout mice showed decreased levels of hemispheric enlargement and brain edema compared to wild-type mice, suggesting that the presence of perivascular AQP4 enhanced brain edema formation [59]. Another study using AQP4 knockout mice found that, following vasogenic edema and BBB disruption from MCA occlusion and focal cerebral ischemia, $AQP4^{+/+}$ mice had a higher mortality rate and greater neurological complications than $AQP^{-/-}$ mice. In addition, the $AQP^{-/-}$ mice had significantly less brain edema, as measured by the increase in the size of the cerebral hemisphere, suggesting AQP4 promotes edema formation [63]. In a model of permanent focal brain ischemia, the pattern of AQP4 expression followed both the formation and resolution of edema formation, suggesting that AQP4 has a role in post-ischemic edema formation and possibly resolution of edema [62].

Investigating specific changes in AQP4 during ischemia and reperfusion may not provide a complete picture of the pathologic consequences of edema formation. An important consideration of the impact of edema formation is edema resolution. Because the cerebral endothelium does not contain fenestrations and the brain lacks a lymphatic system, edema resolution is through the choroid plexus epithelium and cerebral spinal fluid (CSF) that turns over every 4 h [65]. AQP1 is localized to the choroid plexus epithelium and is thought to be involved in both the secretion and reabsorption of CSF [66]. Due to its role in CSF reabsorption, AQP1 may be an important factor in edema resolution under ischemic conditions. However, a role for AQP1 in ischemia has yet to be determined. In addition, at least six AQPs have been found in the rodent brain at various locations or expressed in brain-derived cells [50]. The role of these AQPs in postischemic edema formation and resolution is also unclear, but likely to have a significant impact due to their water fluxing properties.

25.4
Cellular Regulation of Cerebrovascular Permeability

25.4.1
Role of Actin

Morphologically, brain ECs are more similar to epithelial cells than to ECs in peripheral blood vessels in that there is limited molecular transport due to a low rate of fluid-phase endocytosis (that limits transcellular flux) and coupling by high electrical resistance tight junctions (that limits paracellular flux) [3–5]. Significant steps have been made in understanding the molecular structure of tight junctions in brain ECs that affords these features. It is becoming increasingly apparent that tight junctions are not passive structures, but can be rapidly modulated by signaling pathways that affect the structure of the tight junction and the organization of the EC actin cytoskeleton [4, 5, 14, 67, 68]. At least ten proteins have been localized to the tight junction, some of which are involved in signal transduction processes that regulate permeability and are directly attached to the actin cytoskeleton [67, 68]. For example, within the tight junction, actin filaments bind to the zona occludens, forming a direct connection from the paracellular space through the membrane to the cytoplasm [4, 5, 69], providing a mechanism of regulation by factors that influence the actin cytoskeleton (see Chapters 1 and 4).

The contractile activity of EC actin stress fibers within the cytoplasm is of central importance to regulating tight junction permeability. Agonists that promote relaxation of stress fibers (e.g., cyclic-AMP) decrease permeability by cell spreading that strengthens cell-cell contact and reduces paracellular transport [15]. Alternatively, agonists that promote stress fiber contraction (e.g., PKC, VEGF) promote increased permeability by causing cell rounding that decreases cell-cell contact [70]. This hypothesis is supported by several studies that have demonstrated that inhibition of myosin light-chain (MLC) phosphorylation, which inhibits actin stress fiber contraction, decreases agonist-induced permeability [71–73]. Stress fiber activity as an underlying mechanism of paracellular transport is an important consideration for the numerous mediators of cerebral edema formation that are produced by the injured brain, e.g., histamine, bradykinin, arachidonic acid, etc. (see Section 25.6).

Further evidence for the involvement of EC actin in maintaining barrier function is demonstrated by numerous studies that have shown that depolymerization of actin filaments with cytochalasin increased permeability in both epithelial and ECs and that stabilization of the cytoskeleton with either phalloidin or jasplakinolide (compounds that specifically bind to and stabilize F-actin) prevented the increased permeability [14, 16, 74–77]. In addition to affecting paracellular flux, the polymerization state of actin in ECs has also been shown to affect pinocytotic vesicle formation and transcellular flux. Rats infused with cytochalasin B, a cell-permeable compound that disrupts actin filaments, had increased pinocytotic vesicle formation in cerebral ECs and enhanced transcellular transport of electron-dense horseradish peroxidase (HRP) into the brain cortex [16]. However, these conditions likely mimic endothelial cell damage and possibly necrosis that occurs after longer periods of cerebral ischemia or due to severe reperfusion injury.

25.4.2
Ischemia and Hypoxia Effects on EC Actin and Permeability

Hypoxia has been shown to affect the state of actin in ECs and influence permeability in vitro. For example, in cultured brain endothelial cells exposed to hypoxia, a dramatic increase in permeability to sucrose was noted that correlated with increased filamentous (F-)actin and stress fiber formation [13, 78]. Other studies have found that hypoxia with subsequent reoxygenation increased F-actin by 41% in cultured ECs and changed the morphology of ECs to be more spindle-shaped [74]. Alternative to these findings, several studies have suggested that EC damage, including during hypoxia/reoxygenation, disrupts F-actin and leads to increased permeability [13, 79], suggesting that the duration of ischemia or hypoxia impacts the cellular processes that influence permeability. In cultured brain ECs, long periods of hypoxia were associated with a marked increase in permeability that was attributed to ATP depletion that caused depolymerization of EC actin filaments [13]. It therefore appears that there is a gradient of damage to the EC actin cytoskeleton, such that short periods of hypoxia increase stress fiber formation and enhance paracellular permeability, whereas longer periods cause depolymerization and overall EC permeability that may also include transcellular transport. The underlying cellular mechanisms by which I/R affects the EC actin cytoskeleton are not clear, but may relate to ATP depletion during ischemia [13] and/or the production of autocoids (NO, O_2^-) during reperfusion (see Section 25.5) [13, 80].

In a cell culture system, the role of EC actin in reperfusion injury was studied by exposure of cerebral microvascular endothelial cells to posthypoxic reoxygenation. Mark and Davis found that hypoxia induced a 2.6-fold increase in paracellular permeability to sucrose that was reduced by 58% during reoxygenation [78], suggesting that reperfusion may actually decrease permeability. While these in vitro studies importantly demonstrate that hypoxia has cellular effects on the cerebral endothelium that promotes changes in permeability, the true impact in the brain where there are complex interactions between hemodynamic factors and chemical mediators are less clear.

25.5
Reperfusion Injury

The mechanisms by which ischemia promotes brain tissue damage and cerebral edema are likely different from those during reperfusion. During ischemia, there is reduced oxygen delivery that diminishes ATP and causes metabolic disturbances, whereas during reperfusion there is repletion of ATP and release of damaging O_2^- [3]. Increased permeability of cerebral vessels after brain ischemia is thought to involve the opening of EC tight junctions and increased vesicular transport, together with F-actin and tight junction protein rearrangements that promote paracellular opening. During reperfusion, the release of various substances, including O_2^-, can affect EC structure and function to increase permeability, as well as promote hyperemia that has been shown to cause vasogenic edema [81–83] (see Section 25.9). Therefore, there is clearly a differential effect of ischemia versus reperfusion on vascular permeability and BBB properties.

25.5.1
Nitric Oxide and Other Reactive Oxygen Species as Mediators of EC Permeability During Ischemia and Reperfusion

Nitric oxide (NO) is a ubiquitous intercellular signaling molecule synthesized in several cell types, including neurons, endothelium, and macrophages, from the amino acid L-arginine by the enzyme nitric oxide synthetase (NOS) [84]. Large amounts of NO are produced within minutes after MCA occlusion within the cortex and vascular endothelium [85, 86]. The initial increase in NO during ischemia is thought to be neuroprotective by improving cerebral blood flow (CBF). However, the sustained increase in NOS activity is thought to contribute to ischemic injury [87, 88]. The detrimental effects of NO production during ischemia are demonstrated in numerous studies showing reduced infarct size in animals genetically deficient in the neuronal isoform of NOS, and by inhibiting NOS activity during reperfusion [87–90].

One major mechanism of injury associated with NO production in vivo is due to its reaction with superoxide anion in an irreversible, diffusion-limited reaction to form peroxynitrite ($ONOO^-$) [84]. Reactive oxygen species (ROS), such as O_2^- and hydrogen peroxide (H_2O_2) are produced in significant amounts during the reperfusion phase following ischemia and have been implicated in reperfusion injury by their ability to incur oxidative damage to lipids, proteins and nucleic acids, as well as their ability to interact with other compounds such as NO to form more reactive adducts [84, 91–95]. The production of $ONOO^-$ during reperfusion seems likely since large amounts of NO produced during ischemia are available to combine with O_2^- generated during reperfusion. Several studies have demonstrated $ONOO^-$ production during ischemia and reperfusion which has led to the hypothesis that it is a mediator of reperfusion injury [95–97]. As an anion, peroxynitrite is remarkably stable and is particularly effective at oxidizing electron-rich groups such as thiols, iron/sulfur centers and zinc fingers [84].

Increased vascular permeability in response to increased NO, O_2^-, and $ONOO^-$ production has been demonstrated in numerous studies using cultured cells [79, 98–101], many of which noted altered EC F-actin and damage as well [100]. For example, significant amounts of O_2^- and H_2O_2 were measured in cultured cells within minutes of reoxygenation after being exposed to hypoxia that was associated with EC damage [99]. This effect was blocked by allopurinol, a compound that inactivates O_2^-. Similarly, Lum et al. found that hypoxia/reoxygenation increased permeability of ECs to albumin two- to five-fold [79]. The increased permeability was associated with significant F-actin reorganization in ECs. Treatment of the cells with superoxide dismutase (SOD) prior to hypoxia prevented the increase in O_2^- and 50% of the increase in permeability. Other in vitro studies have shown that damage to cerebral ECs due to posthypoxic reoxygenation was attenuated by the presence on both SOD and L-NAME, suggesting that $ONOO^-$ is a primary mediator of EC injury during hypoxia/reoxygenation [101]. In vivo studies have also demonstrated a role for $ONOO^-$ in mediating reperfusion-induced cerebral edema. Inhibition of NO with L-NAME after 3 h of ischemia reduced BBB disruption by 65% and improved stroke outcome in a mouse model of transient focal ischemia [102]. A similar study in rats found that NO inhibition with L-NAME reduced ischemic brain damage when given just before reperfusion after 2 h of ischemia that was associated with diminished immunoreactivity for nitrosotyrosine, a marker for $ONOO^-$, suggesting that $ONOO^-$ formation contributes to reperfusion-induced cerebral edema [95].

25.5.2
Thresholds of Injury

It is well known that the depth and duration of ischemia affect the magnitude of stroke outcome, giving rise to the threshold concept of cerebral ischemia [103–105]. Flow and time thresholds for neuronal injury and edema formation have been established in humans and many experimental animal models. For example, one study in baboons found that when CBF fell below 35–40% of normal, the time threshold for edema formation was 30 min [105]. The cellular mechanisms that underlie vascular permeability are also likely to be affected by the duration of ischemia and reperfusion. For example, in cultured brain ECs, hypoxia was shown to affect F-actin in a threshold-dependent manner [13]. Short periods of hypoxia were shown to cause increased F-actin, but longer periods caused actin depolymerization that correlated with apoptosis and ATP depletion. This suggests there may be a threshold duration of ischemia for actin cytoskeletal changes that underlies vascular permeability. Understanding these thresholds is important when considering the window of opportunity for therapeutic intervention.

25.6
Transcellular Transport as a Mechanism of BBB Disruption During Ischemia

The passage of blood-borne substances into the brain parenchyma via vesicular or transcellular transport is considerably less studied during ischemia and reperfusion, but is by no means a less important route of passage than the paracellular route. The paucity of information regarding transcellular transport as a mechanism of enhanced cerebrovascular permeability, compared to that of the paracellular (tight junction) opening, is likely due to the fact that cell culture studies do not easily lend themselves to measuring transcellular transport. A great deal of information regarding how tight junctions are structured and affected by conditions such as hypoxia and reoxygenation have been gained from cell culture studies that measure transendothelial electrical resistance (TEER) as an indicator of barrier disruption. Because TEER is a measure of ionic flux, this information is limited to the paracellular route. A few studies have used cultured endothelium to investigate transcellular uptake of protein tracers during hypoxia, which is distinguished from the transcellular route by inhibition at low temperature [106].

Evidence for a role of pinocytosis in vascular permeability is largely circumstantial, but has suggested that this mode of transport is important for enhancing the permeability of the BBB under pathologic conditions. Under normal conditions, the majority of pinocytosis is from apical to basolateral [9, 107], which argues against bidirectional trafficking through the endothelium and supports the concept of a BBB. Electron microscopy studies have shown that the number of apical vesicles that are apparent in endothelium correlate with how permeable the vessel is. Vesicles are found in large numbers in vessels known to have high permeability (e.g., skeletal muscle arterioles) and low numbers in vessels with low permeability (e.g., cerebral) [108]. In cerebral endothelium that has BBB properties, vesicle formation is low under normal conditions, but increases substantially under experimental and pathologic conditions in which barrier permeability is enhanced [9, 109, 110], suggesting that this mode of transport contributes to enhanced permeability and BBB disruption.

There is evidence that transcellular transport from blood to brain is enhanced during ischemia and reperfusion [9, 106, 111–114]. Using a coculture model of brain capillary endothelial cells and astrocytes, Plateel demonstrated hypoxia-induced permeability and uptake of albumin was temperature-sensitive, suggesting that enhanced vesicular transport is a means of albumin accumulation during hypoxia [106]. However, in a model of focal ischemia without reperfusion, transfer constants for sucrose and inulin were measured and showed that convective, fluid-phase endocytosis transport was not increased [115]. The difference in these studies may be due to measuring vesicular transport (pinocytosis) versus fluid-phase endocytosis (membrane turnover).

The polarized nature of the cerebral endothelium not only contributes to barrier properties, but may also provide a mechanism of protection. Enzymatic substances on the abluminal or basolateral membrane inactivate neurotransmitters and remove toxins from the brain into the bloodstream [116, 117]. Studies from our laboratory have shown that ischemia and reperfusion increase basolateral pinocytosis in cerebral endothelium, suggesting an efflux mechanism is invoked to rid the brain of toxic substances that may accumulate during ischemia, or to reduce intracranial pressure that has been elevated during edema formation, since vesicular transport has been shown to be pressure-induced [9].

There are several mechanisms by which transcellular transport can increase apical to basolateral and promote edema formation. First, mediators of cerebral edema such as bradykinin and histamine that are released during ischemia have been shown to enhance permeability by vesicular (receptor-mediated) transport [118] (see Section 25.7). Second, prolonged ischemia causes actin depolymerization that could also increase pinocytosis. Nag et al. showed a significant increase in pinocytosis in the cerebral cortex of anesthetized rats infused with cytochalasin B, a cell-permeable compound that causes depolymerization of actin [16]. A role for actin in mediating transcellular transport is likely given that the endocytotic machinery is directly linked to the actin cytoskeleton [119]. Last, changes in cerebral hemodynamics due to postischemic reperfusion likely contribute to an increase in apical pinocytosis, since pressure has been shown to increase pinocytotic vesicle formation and the passage of HRP through the cerebral endothelium [9]. It is important to note that many studies that show enhanced transcellular transport or pinocytosis during ischemia/reperfusion or hypoxia/reoxygenation have not always shown an increase in transport into the brain parenchyma. While this mode of transport likely contributes to enhanced vascular permeability, more evidence is needed to ascertain its specific role in cerebral edema formation.

25.7
Mediators of EC Permeability During Ischemia

The increase in BBB permeability and edema formation during stroke is also partly due to the production of autocoids by the injured brain tissue and endothelium. In general, a mediator has the properties that it can: (a) increase BBB permeability, (b) have the capacity to induce brain edema, (c) cause vasodilation that augments permeability due to increased surface area, and (d) increase its concentration in the brain during ischemia and/or reperfusion [118, 120]. These criteria provide a framework for which therapeutic interaction may influence cerebral edema formation. The following is a list of some of the known mediators of cerebral edema during ischemia and reperfusion. Some of these are inflammatory mediators and it is important to note that there is an inflammatory response associated with stroke that is thought to cause secondary injury to brain tissue following ischemia. Because inflammatory mediators of BBB permeability are nicely presented in Chapter 24, they will only be mentioned here in the context of how they are involved in BBB disruption and cerebral edema during ischemia.

1. Bradykinin (BK) is an endogenous nonapeptide produced by activation of the kallikrein-kinin system that has been shown to increase in brain tissue after ischemia and cause brain edema [121, 122]. It has been shown to increase BBB permeability by both vesicular transport and paracellular mechanisms (i.e., decreased TEER) [123–125]. It is a potent vasorelaxant by endothelium-dependent vasodilation due to release of NO, free radicals, leukotrienes, and prostanoids [126–128]. Inhibition of the BK B_2 receptor significantly reduces postischemic brain injury in both rat and mouse models [129, 130].

2. Histamine (HA) is an inflammatory mediator that is increased in brain tissue after ischemia, likely due to release from mast cells [131, 132]. It causes significant brain edema formation and BBB permeability by both vesicular transport and paracellular tight junction opening (measured by TEER) [46, 125, 133, 134]. It dilates pial vessels by activation of H2 receptors on smooth muscle and H1 receptors on endothelium [118, 135, 136]. Histamine antagonists have been shown to reduce accumulation of water and sodium in brain tissue after bilateral common carotid artery occlusion in rats [137].

3. Arachidonic acid (AA) is increased after focal ischemia and causes nonspecific opening of the BBB [138, 139]. Its vasomotor activity has been assessed in pial arterioles and found to both cause constriction and dilation [141, 142]. Application of glucocorticoids to decrease phospholipase A2 successfully decreased brain edema after ischemia [143, 144].

4. Endothelin-1 (ET-1) increases in brain tissue and plasma after cerebral ischemia [145] and has been shown to increase BBB permeability [146, 147]. While ET-1 may not fit the category of mediator because it causes potent vasoconstriction that is thought to further promote ischemia instead of dilation [148, 149], its inhibition by ET-A receptor antagonists has been shown to be effective at decreasing postischemic brain edema and improving stroke outcome [150, 151].

5. Vascular endothelial growth factor (VEGF) is a secreted mitogen whose expression is increased by hypoxia and ischemia [152–154]. It is a potent permeability factor that increases BBB permeability by both transcellular and paracellular mechanisms [155, 156]. It also produces potent vasorelaxation [157] and can therefore be considered a mediator of cerebral edema. Its role in stroke outcome is complex. Because of its mitogen activity, it can increase angiogenesis of cerebral microvessels after stroke and improve outcome, but its permeability effects preclude it from being an effective or beneficial treatment. However, due to the fact that ischemia-induced permeability increases in the acute stage (within minutes), while angiogenesis develops much later in the ischemic brain, Zhang et al. demonstrated that late (48 h) administration of recombinant VEGF to rats after ischemia significantly improved stroke outcome due to increased microvessel density [158].

25.8
Hyperglycemic Stroke

Serum glucose concentration is a known determinant of stroke outcome [159–163]. Diabetic patients have a significantly increased magnitude of stroke that has been attributed to hyperglycemia and not to other complicating factors associated with diabetes, such as atherosclerosis or microangiopathy [161, 164, 165]. In animal models, preexisting hyperglycemia has been shown to triple infarct size as a result of permanent MCA occlusion and to increase the incidence of fatal edema seven-fold during transient ischemia [166, 167]. Therefore, the pathologic processes that increase vascular permeability and promote brain edema during ischemia and reperfusion are severely aggravated by the presence of high glucose (see Chapter 26). Hyperglycemia has been shown to increase lactate production [168] and impair ionic hemostasis during ischemia, which is thought to contribute to brain tissue damage [169]. While these factors may have a role in enhancing edema formation during ischemia, the mechanisms of enhanced edema formation during reperfusion, when ATP levels are restored and production of autocoids and second messengers is pronounced, is likely very different.

25.8.1
Role of PKC Activity in Mediating Enhanced BBB Permeability During Hyperglycemic Stroke

There is extensive literature implicating a role for protein kinase C in regulating tight junction permeability [3, 4, 170–174]. Agonists that are known to increase vascular permeability both in vivo and in vitro, including bradykinin and thrombin, activate PKC through the phospholipase C (PLC)-diacylglycerol (DAG) pathway [172, 173]. This thrombin- and bradykinin-induced increase in permeability did not occur in ECs depleted of PKC [172–174], suggesting a central role for PKC activity in modulating permeability. In addition, specific inhibitors of PKC, including calphostin C and H-7, prevented agonist-induced permeability [172, 173, 175] in ECs; and several studies have shown that PKC activation by phorbol esters increases paracellular permeability [176].

The involvement of PKC in increasing permeability during hyperglycemia may relate to PKC-induced affects on EC actin. PKC activation phosphorylates cytoskeletal proteins, leading to F-actin rearrangements [177]. PKC has been shown to colocalize with ZO-1 at the tight junction and is thought to affect permeability through its ability to phosphorylate tight junction proteins such as ZO-1 [178], directly altering the structure of the tight junction and attachment to the cytoskeleton. In addition, PKC activation, through its ability to promote calcium/calmodulin-dependent actin-myosin interaction, is thought to increase permeability by causing EC contraction [174, 175, 179, 180] and increasing intercellular gap formation. This effect was demonstrated by one study which showed that PKC-induced permeability was prevented by the inhibition of MLC phosphorylation by ML-7 (a specific inhibitor of MLC kinase), suggesting PKC increases permeability through affecting actin-myosin interaction [176].

The importance of PKC activation in increasing EC permeability is that elevated glucose is known to augment PKC activation in a number of cell types, including ECs [181, 182]. Cells exposed to high glucose have an increased flux of glucose through the glycolytic pathway, causing de novo synthesis of DAG [181, 183]. DAG is the primary activator of PKC and its enhanced activation in response to elevated glucose is thought to be a contributor to many of the vascular complications associated with diabetes, including edema formation [181, 182, 184]. It is likely that enhanced PKC activity predisposes the cerebral endothelium to increased permeability, providing a mechanism by which hyperglycemia worsens postischemic edema formation. Hyperglycemia-induced PKC activation has been recognized for some time and has provided a therapeutic target for treatment of diabetic peripheral vascular disease. A similar therapeutic target likely exists for the cerebral circulation as well.

25.9
Hemodynamic Changes During Ischemia and Reperfusion and Its Role in Cerebral Edema

Ischemia is associated with significant alterations in hemodynamic forces on the cerebral endothelium that have been shown to influence vascular permeability. Flow cessation causes a decrease in both pressure and circumferential wall tension in the cerebral circulation, as well as alterations in shear stress, all of which have been shown to influence barrier properties. Reduced flow in anesthetized female rats causes both decreased pressure and transport of insulin and potassium across the BBB, suggesting that ionic homeostasis may be disturbed under conditions of reduced flow [185]. Shear stress alterations due to flow cessation have been studied in a dynamic model of BBB in which cerebral microvascular endothelial cells were cocultured with astrocytes in a tube in which flow and shear stress were controlled [10]. One hour of normoxic, normoglycemic flow cessation caused a drop in nitric oxide production that was restored upon reperfusion. These changes were not accompanied by measurable changes in TEER, suggesting that paracellular permeability was not changed during alterations in shear stress. Transcellular permeability was not measured in this study.

The reperfusion period following the opening of an occluded artery is associated with autoregulatory failure and diminished myogenic tone in the cerebral arteries and arterioles, all factors contributing to reperfusion-induced hyperemia [186–188]. While hyperemia increases CBF and delivers substrates to an ischemic region, the loss of tone during this postischemic period has been shown to cause BBB disruption and edema [118, 186, 189]. The effect of postischemic hyperemia on BBB disruption and edema formation was investigated by Kuroiwa et al., who demonstrated that ischemic edema was severely exacerbated by hyperemia during reperfusion and that suppression of postischemic hyperemia significantly reduced brain edema and BBB opening [189]. These hemodynamic alterations during postischemic reperfusion may predispose cerebral ECs to damage, producing an additive effect that significantly increases permeability.

One effect that ischemia-induced loss of myogenic tone may have is to increase pinocytotic vesicle formation and transcellular transport. Our own studies in pressurized and perfused cerebral (pial) arterioles demonstrated a significant increase (78%) in apical pinocytosis after a step increase in pressure to cause forced dilatation of myogenic tone, as determined by electron microscopy [9]. Since this study was done in vitro, without the influence of pharmacologic agents to induce acute hypertension, it was the first study to demonstrate that pressure is a primary stimulus for pinocytosis. These results are similar to a study that showed that hypertension after postischemic reperfusion accelerated BBB disruption by increasing pinocytosis in cerebral endothelium [190].

References

1 Williams GR, Jiang J, Matchar DB, Samsa GP **1999**, Incidence and occurrence of total (first-ever and recurrent) stroke, *Stroke* 30, 2523–2529.
2 Hacke W, Steiner T, Schwab S **1997**, Critical management of acute stroke: medical and surgical therapy, in *Cerebrovascular Disease*, ed. HH Batjer, Lippincott-Raven, Philadelphia.
3 Dirnagl U, Lindauer U **1998**, Microcirculatory disturbances in cerebral ischemia, in *Cerebrovascular Disease: Pathophysiology, Diagnosis and Management*, vol. 1, eds. MD Ginsberg, J Bogousslavsky, Blackwell Science, Malden.
4 Kniesel U, Wolburg H **2000**, Tight junctions of the blood-brain barrier, *Cell Mol Neurobiol* 20, 57–76.
5 Rubin LL, Standdon JM **1999**, The cell biology of the blood-brain barrier, *Annu Rev Neurosci* 22, 11–28.
6 Betz AL, Dietrich WD **1998**, Blood-brain barrier dysfunction in cerebral ischemia, in Cerebrovascular Disease: Pathophysiology, Diagnosis and Management, vol. 1, eds. MD Ginsberg, J Bogousslavsky, Blackwell Science, Malden.
7 Wahl M, Unterberg A, Baethmann A, Schilling L **1988**, Mediators of blood-brain barrier dysfunction and formation of vasogenic brain edema, *J Cereb Blood Flow Metab* 8, 621–634.
8 Johansson B **1974**, Blood-brain barrier dysfunction in acute arterial hypertension after papaverine-induced vasodilatation, *Acta Neurol Scand* 50, 573–580.
9 Cipolla MJ, Crete R, Vitullo L, Rix RD **2004**, Transcellular transport as a mechanism of blood-brain barrier disruption during stroke, *Front Biosci* 9, 777–785.
10 Krizanac-Bengez L, Kapural M, Parkinson F, Cucullo L, Hossain M, Mayberg M, Janigro D **2003**, Effects of transient loss of shear stress on blood-brain barrier endothelium: role of nitric oxide and IL-6, *Brain Res* 977, 239–246.

11 Mark KS, Burroughs AR, Brown RC, Huber JD, Davis TP **2003**, Nitric oxide mediates hypoxia-induced changes in paracellular permeability of cerebral microvasculature, *Am J Physiol* 286, H174–H180.

12 Ramirez MM, Kim DD, Duran WN **1996**, Protein kinase C modulates microvascular permeability through nitric oxide synthase, *Am J Physiol* 271, H1702–H1705.

13 Plateel M, Dehouck P, Torpier G, Cecchelli R, Teissier E **1995**, Hypoxia increases the susceptibility to oxidant stress and the permeability of the blood-brain barrier endothelial cell monolayer, *J Neurochem* 65, 2138–2145.

14 Blum MS, Toninelli E, Anderson JM, Balda MS, Zhou J, O'Donnel L, Pardi R, Bender JR **1997**, Cytoskeletal rearrangement mediates human microvascular endothelial tight junction modulation by cytokines, *Am J Physiol* 273, H286–H294.

15 Goeckeler ZM, Wysolmerski RB **1995**, Myosin light chain kinase-regulated endothelial cell contraction: relationship between isometric tension, actin polymerization and myosin phosphorylation, *J Cell Biol* 130, 613–627.

16 Nag S **1995**, Role of endothelial cytoskeleton in blood-brain barrier permeability to protein, *Acta Neuropathol* 90, 454–460.

17 Ropper AH, Shafran B **1984**, Brain edema after stroke: clinical syndrome and intracranial pressure, *Arch Neurol* 41, 26–29.

18 Oxbury JM, Greenhall RC, Grainger KM **1975**, Predicting outcome of stroke: acute stage after cerebral infarction, *Br Med J* 3, 125–127.

19 Bounds JV, Wiebers DO, Whisnant JP, Okazaki H **1981**, Mechanisms and timing of deaths from cerebral infarction, *Stroke* 12, 474–477.

20 Iannotti F, Hoff JT, Schielke GP **1985**, Brain tissue pressure in focal cerebral ischemia, *J Neurosurg* 62, 83–89.

21 Koudstaal PJ, Stibbe J, Vermeulen M **1988**, Fatal ischaemic brain oedema after early thrombolysis with tissue plasminogen activator, *Br Med J* 297, 1571–1574.

22 Bell BA, Symon TD, Branston NM **1985**, CBF and time thresholds for the formation of ischemic cerebral edema and effect of perfusion in the baboon, *J Neurosurg* 62, 31–41.

23 Brott TG, Hacke W **1998**, Thrombolytic and defibrinogenating agents for ischemic and hemorrhagic stroke, in *Stroke: Pathophysiology, Diagnosis and Management*, ed. Barnett E, Churchill Livingstone, Philadelphia, pp. 1164–1165.

24 Klatzo I **1967**, Neuropathologic aspects of brain edema, *J Neuropathol Exp Neurol* 26, 1–14.

25 Yang GY, Chen SF, Kinouchi H, et al. **1992**, Edema, cation content, and ATPase activity after middle cerebral artery occlusion in rats, *Stroke* 23, 1331–1336.

26 Betz AL, Keep RF, Beer ME, Ren X-D **1994**, Blood-brain barrier permeability and brain content of sodium, potassium and chloride during focal ischemia, *J Cereb Blood Flow Metab* 14, 29–37.

27 Young W, Rappaport ZH, Chalif DJ, Flamm ES **1987**, Regional brain sodium, potassium, and water changes in the rat middle cerebral artery occlusion model of ischemia, *Stroke* 18, 751–759.

28 Betz AL, Ennis SR, Schielke GP **1989**, Blood-brain barrier sodium transport limits development of brain edema during partial ischemia in gerbils, *Stroke* 20, 1253–1259.

29 Gotoh O, Asano T, Koide T, Takakura K **1985**, Ischemic brain edema following occlusion of the middle cerebral artery in the rat, I: the time courses of the brain water, sodium and potassium contents and blood-brain barrier permeability to ^{125}I-albumin, *Stroke* 16, 101–109.

30 Dietrich WD, Prado R, Watson BD, Nakayama H **1988**, Middle cerebral artery thrombosis: acute blood-brain barrier alterations, *J Neuropath Exp Neurol* 47, 443–451.

31 Hassel B, Iversen EG, Fonnum F **1986**, Neurotoxicity of albumin in vivo, *Neurol Res* 8, 146–151.

32 Fenstermacher JD **1984**, Volume regulation of the central nervous system, in *Edema*, eds. NC Staub, AE Taylor, Raven Press, New York, pp. 383–404.

33 Dodson RF, Chu LW, Welch KM, Achar VS **1977**, Acute tissue response to cerebral ischemia in the gerbil: an ultrastructural study, *J Neurol Sci* 33, 161–170.

34 Chen Y, Swanson RA **2003**, Astrocytes and brain injury, *J Cereb Blood Flow Metab* 23, 137–149.

35 Anderson CM, Swanson RA **2000**, Astrocyte glutamate transport: review of properties, regulation and physiological functions, *Glia* 32, 1–14.

36 Walz W, Hertz L **1983**, Functional interactions between neurons and astrocytes, part II: potassium homeostasis at the cellular level, *Prog Neurobiol* 20, 133–183.

37 Tanaka J, Toku K, Zhang B, Ishihara K, Sakanaka M, Maeda N **1999**, Astrocytes prevent neuronal death induced by reactive oxygen and nitrogen species, *Glia* 28, 85–96.

38 Reese TS, Karnovsky MJ **1976**, Fine structural localization of a blood-brain barrier to exogenous peroxidase, *J Cell Biol* 34, 207–217.

39 Brightman MW, Reese TS **1969**, Junctions between intimately apposed cell membranes in the vertebrate brain, *J Cell Biol* 40, 648–677.

40 Davson H, Oldendorf WH **1967**, Transport in the central nervous system, *Proc R Soc Med* 60, 326–328.

41 Reinhart CA, Gloor SM **1997**, Co-culture blood-brain barrier models and their use for pharmatoxicologic screening, *Toxicol Vitro* 11, 513–518.

42 Bauer HC, Bauer H **2000**, Neural induction of the blood-brain barrier: still an enigma, *Cell Mol Neurobiol* 20, 13–28.

43 Quick AM, Cipolla MJ **2005**, Pregnancy-induced upregulation of aquaporin 4 in brain and its role in eclampsia, *FASEB J* 19, 170–175.

44 Rash JE, Yasumura T, Hudson CS, Agre P, Neilsen S **1998**, Direct immunogold labeling of aquaporin-4 in square arrays of astrocyte and ependymocyte plasma membrane in rat brain and spinal cord, *Proc Natl Acad Sci USA* 95, 11981–11986.

45 Butt AM, Jones HC **1992**, Effect of histamine and antagonists on electrical resistance across the blood-brain barrier in rat brain-surface microvessels, *Brain Res* 569, 100–105.

46 Abbott NJ **2002**, Astrocyte-endothelial interactions and blood-brain barrier permeability, *J Anat* 200, 629–638.

47 Fischer S, Wobben M, Kleinstuck J, Renz D, Schaper W **2000**, Effect of astroglial cells on hypoxia-induced permeability in PBMEC cells, *Am J Physiol* 279, C935–C944.

48 Kondo T, Kinouchi H, Kawase M, Yoshimoto T **1996**, Astroglial cells inhibit the increasing permeability of brain endothelial cell monolayer following hypoxia/reoxygenation, *Neurosci Lett* 208, 101–104.

49 Fischer S, Wobben M, Marti HH, Renz D, Schaper W **2002**, Hypoxia-induced hyperpermeability in brain microvessel endothelial cells involves VEGF-mediated changes in the expression of zona occludens-1, *Microvasc Res* 63, 70–80.

50 Badaut J, Lasbennes F, Magistretti PJ, Regli L **2002**, Aquaporins in the brain: distribution, physiology, and pathophysiology, *J Cereb Blood Flow Metab* 22, 367–378.

51 Ke C, Poon WS, NG HK, Pang JC, Chan Y **2001**, Heterogeneous responses of aquaporin-4 in oedema formation in replicated severe traumatic brain injury model in rats, *Neurosci Lett* 301, 21–24.

52 Ishibashi K, Kuwahara M, Sasaki S **2000**, Molecular biology of aquaporins, *Rev Physiol Biochem Pharmacol* 141, 1–32.

53 Agre P, Bonhivers M, Borgnia MJ **1998**, The aquaporins, blueprints for cellular plumbing systems, *J Biol Chem* 273, 14659–14662.

54 Verkman AS **2002**, Aquaporin water channels and endothelial cell function, *J Anat* 200, 617–627.

55 Amiry-Moghaddam M, Ottersen OP **2003**, The molecular basis of water transport in the brain, *Nat Rev Neurosci* 4, 991–1001.

56 Hasegawa H, Ma T, Skach W, Matthay MA, Verkman AS **1994**, Molecular cloning of a mercurial-insensitive water channel expressed in selected water-transporting tissues, *J Biol Chem* 269, 5497–5500.

57 Jung JS, Bhat RV, Preston GM, Guggino WB, Baraban JM, Agre P **1994**, Molecular characterization of an aquaporin cDNA from brain: candidate osmoreceptor and regulator of water balance, *Proc Natl Acad Sci USA* 91, 13052–13056.

58 Nielsen S, Nagelhus EA, Amiry-Moghaddam M, Bourque C, Agre P, Ottersen OP **1997**, Specialized membrane domains for water transport in glial cells: high-resolution immunogold cytochemistry of aquaporin-4 in rat brain, *J Neurosci* 17, 171–180.

59 Amiry-Moghaddam M, Xue R, Haug F-M, Neely JD, Bhardwaj A, Agre P, Adams ME, Froehner SC, Mori S, Ottersen OP **2004**, Alpha syntrophin deletion removes the perivascular but not the endothelial pool of aquaporin-4 at the blood-brain barrier and delays the development of brain edema in an experimental model of acute hyponatremia, *FASEB J* 18, 542–544.

60 Amiry-Moghaddam M, Otuska T, Hurn PD, Traystman RJ, Haug F-M, Froehner SC, Adams ME, Neely JD, Agre P, Ottersen OP, Bhardwaj A **2003**, An alpha-syntrophin-dependent pool of AQP4 in astroglial end-feet confers bidirectional water flow between blood and brain, *Proc Natl Acad Sci USA* 100, 2106–2111.

61 Saadoun S, Papadopoulos MC, Davies DC, Krishna S, Bell BA **2002**, Aquaporin-4 expression is increased in oedematous human brain tumors, *J Neurol Neurosurg Psychiat* 72, 262–265.

62 Taniguchi M, Yamashita T, Kumura E, Tamatani M, Kobayashi A, Yokawa T, Maruno M, Kato A, Ohnishi T, Kohmura E, Tohyama M, Yoshimine T **2000**, Induction of aquaporin-4 water channel mRNA after focal cerebral ischemia in rat, *Mol Brain Res* 78, 131–137.

63 Manley GT, Fujimura M, Ma T, Noshita N, Filiz F, Bollen AW, Chan P, Verkman AS **2000**, Aquaporin-4 deletion in mice reduces brain edema after acute water intoxication and ischemic stroke, *Nat Med* 6, 159–163.

64 Vizuete ML, Venero JL, Vargas C, Ilundain AA, Echevarria M, Machado A, Cano J **1999**, Differential upregulation of aquaporin–4 mRNA expression in reactive astrocytes after brain injury: potential role in brain edema, *Neurobiol Dis* 6, 245–258.

65 Pardridge WM **1997**, Blood-brain barrier transport mechanisms, in *Primer on Cerebrovascular Disease*, eds. KMA Welch, LR Caplan, RJ Reis, BK Siesjo, B Weir, Academic Press, New York, pp. 17–20.

66 Nielsen S, Smith BL, Christensen EI, Agre P **1993**, Distribution of the aquaporin CHIP in secretory and resorptive epithelia and capillary endothelia, *Proc Natl Acad Sci USA* 90, 7275–7279.

67 Stevenson BR, Keon BH **1998**, The tight junction: morphology and molecules, *Annu Rev Cell Biol* 14, 89–109.

68 Staddon J, Ratcliffe M, Morgan L, Hirase T, Smales C, Rubin L **1997**, Protein phosphorylation and the regulation of cell-cell junctions in brain endothelial cells, *Heart Vessels Suppl* 12, 106–109.

69 Ito M, et al. **1999**, Direct binding of three tight junction-associated MAGUKs ZO-1, ZO-2, and ZO-3, with the COOH termini of claudins, *J Cell Biol* 147, 1351–1363.

70 Lum H, Malik AB **1994**, Regulation of vascular endothelial barrier function, *Am J Physiol* 267, L223–L241.

71 Yuan Y, Huang Q, Wu HM **1997**, Myosin light chain phosphorylation: modulation of basal and agonist-stimulated venular permeability, *Am J Physiol* 272, H1437–H1443.

72 Yuan SY **2000**, Signal transduction pathways in enhanced microvascular permeability, *Microcirculation* 7, 395–405.

73 Garcia JG, Davis HW, Patterson CE **1995**, Regulation of endothelial cell gap formation and barrier dysfunction: role of myosin light chain phosphorylation, *J Cell Physiol* 163, 510–522.

74 Crawford LE, Milliken EE, Iran K, Zweier JL, Becker LC, Johnson TM, Eissa NT, Crystal RG, Finkel T, Goldschmidt-Clermont PJ **1996**, Superoxide-

mediated actin response in post-hypoxic endothelial cells, *J Biol Chem* 271, 26863–26867.

75 Hippenstiel S, Tannert-Otto S, Vollrath N, Krull M, Just I, Aktories K, von Eichel-Streiber C, Suttorp N **1997**, Glucosylation of small GTP-binding Rho proteins disrupts endothelial barrier function, *Am J Physiol* 272, L38–L43.

76 Stevenson BR, Begg DA **1994**, Concentration-dependent effects of cytochalasin D on tight junctions and actin filaments in MDCK epithelial cells, *J Cell Sci* 107, 367–375.

77 Alexander JS, Hechtan HB, Shepro D **1988**, Phalloidin enhances endothelial barrier function and reduces inflammatory permeability in vitro, *Microvasc Res* 35, 308–315.

78 Mark KS, Davis TP **2002**, Cerebral microvascular changes in permeability and tight junctions induced by hypoxia-reoxygenation, *Am J Physiol* 282, H1485–H1492.

79 Lum H, Barr DA, Shaffer JR, Gordon RJ, Ezrin AM, Malik AB **1992**, Reoxygenation of endothelial cells increases permeability by oxidant-dependent mechanisms, *Circ Res* 70, 991–998.

80 Lagrange P, Romero I, Minn A, Revest PA **1999**, Transendothelial permeability changes induced by free radicals in an in vitro model of the blood-brain barrier, *Free Rad Med Biol* 27, 667–672.

81 Olsson Y, Crowell RM, Klatzo I **1971**, The blood-brain barrier to protein tracers in focal ischemia and infarction caused by occlusion of the middle cerebral artery, *Acta Neuropathol* 18, 89–102.

82 Kuroiwa T, Ting P, Martinez H, Klatzo I **1985**, The biphasic opening of the blood-brain barrier to proteins following temporary middle cerebral artery occlusion, *Acta Neuropathol* 68, 122–129.

83 Shigeno T, Teasdale GM, McCulloch J, Graham D **1985**, Recirculation model following MCA occlusion in rats: Cerebral blood flow, cerebrovascular permeability and brain edema, *J Neurosurg* 63, 272–277.

84 Beckman JS, Koppenol WH **1996**, Nitric oxide, superoxide and peroxynitrite: the good, the bad and the ugly, *Am J Physiol* 271, C1424–C1437.

85 Malinski T, Bailey F, Zhang ZG, Chopp M **1993**, Nitric oxide measured by a porphyrinic microsensor in rat brain after transient middle cerebral artery occlusion, *J Cereb Blood Flow Metab* 13, 355–358.

86 Nagafuji T, Sugiyama M, Matsui T, Muto A, Naito S **1995**, Nitric oxide synthase in cerebral ischemia: possible contribution of nitric oxide synthase activation in brain microvessels to cerebral ischemic injury, *Mol Chem Neuropathol* 26, 107–157.

87 Dalkara T, Morikawa H, Moskowitz MA, Panahian N **1994**, Blood flow-dependent functional recovery in a rat model of cerebral ischemia, *Am J Physiol* 267, H678–H683.

88 Huang Z, Huang PL, Panahian N, Dalkara T, Fishman MC, Moskowitz MA **1994**, Effects of cerebral ischemia in mice deficient in neuronal nitric oxide synthase, *Science* 265, 1883–1885.

89 Huang Z, Huang PL, Ma J, Meng W, Ayata C, Fishman MC, Moskowitz MA **1996**, Enlarged infarcts in endothelial nitric oxide synthase knockout mice are attenuated by nitro-L-arginine, *J Cereb Blood Flow Metab* 16, 981–987.

90 Panahian N, Yoshida T, Huang PL, Hedley-Whyte ET, Dalkara T, Fishman M, Moskowitz MA **1996**, Attenuated hippocampal damage after global cerebral ischemia in knock-out mice deficient in neuronal nitric oxide synthase, *Neuroscience* 72, 343–354.

91 Clemens JA, Panetta JA **1994**, Neuroprotection by antioxidants in models of global and focal ischemia, in *The Neurobiology of NO and OH*, eds. CC Chiueh, DL Gilbert, CA Colton, New York Academy of Sciences, New York.

92 Sakomoto A, Ohnishi ST, Ohnishi T, Ogawa R **1991**, Relationship between free-radical production and lipid peroxidation during ischemia-reperfusion injury in the rat brain, *Brain Res* 554, 186–192.

93 Oliver CN, Starke-Reed PE, Stadtman ER, Lui GJ, Carney JM, Floyd RA **1990**, Oxidative damage to brain proteins, loss of glutamate synthetase activity, and production of free radicals during ischemia/reperfusion-induced injury to gerbil brain, *Proc Natl Acad Sci USA* 87, 5144–5147.

94 Kumura E, Yoshimine T, Iwatsuki KI, Yamanaka K, Tanaka S, Hayakawa T, Shiga T, Kosaka H **1996**, Generation of nitric oxide and superoxide during reperfusion after focal cerebral ischemia in rats, *Am J Physiol* 270, C748–C752.

95 Gursoy-Ozdemir Y, Bolay H, Saribas O, Dalkara T **2000**, Role of endothelial nitric oxide generation and peroxynitrite formation in reperfusion injury after focal cerebral ischemia, *Stroke* 31, 1974–1981.

96 Greenacre S, Ridger V, Wilsoncroft P, Brain SD **1997**, Peroxynitrite: a mediator of increased microvascular permeability, *Clin Exp Pharmacol Physiol* 24, 880–882.

97 Gobbel GT, Chan TYY, Chan PH **1997**, Nitric oxide and superoxide-mediated toxicity in cerebral endothelial cells, *J Pharmacol Exp Ther* 282, 1600–1607.

98 Gorodeski GI **2000**, NO increases permeability of cultured human cervical epithelia by cGMP-mediated increase in G-actin, *Am J Physiol* 278, C942–C952.

99 Beetsch JW, Park TS, Dugan LL, Shah AR, Gidday JM **1998**, Xanthine oxidase-derived superoxide causes reoxygenation injury of ischemic cerebral endothelial cells, *Brain Res* 786, 89–95.

100 Kuhne W, Besselmann N, Noll T, Muhs A, Watanabe H, Piper HM **1993**, Disintegration of cytoskeletal structure of actin filaments in energy-depleted endothelial cells, *Am J Physiol* 264, H1599–H1608.

101 Nagashima T, Wu S, Ikeda K, Tamaki N **2000**, The role of nitric oxide in reoxygenation injury of brain microvascular endothelial cells, *Acta Neurochir Suppl* 76, 471–473.

102 Ding-Zhou L, Marchand-Verrecchia C, Croci N, Plotkine M, Margaill I **2002**, L-NAME reduces infarction, neurologic deficit and blood-brain barrier disruption following cerebral ischemia in mice, *Eur J Pharmacol* 457, 137–146.

103 Jones TH, Morawetz RB, Crowell RM, Marcoux FW, Fitzgibbon SJ, DeGirolami U, Ojemann RG **1981**, Thresholds of focal cerebral ischemia in awake monkeys, *J Neurosurg* 54, 773–782.

104 Hossman K-A **1998**, Thresholds of ischemic injury, in *Cerebrovascular Disease: Pathophysiology, Diagnosis and Management*, eds. MD Ginsberg, J Bogousslavsky, Blackwell Science, Malden, Mass., pp. 193–204.

105 Bell BA, Symon TD, Branston NM **1985**, CBF and time thresholds for the formation of ischemic cerebral edema and effect of perfusion in the baboon, *J Neurosurg* 62, 31–41.

106 Plateel M, Teissier E, Cecchelli R **1997**, Hypoxia dramatically increases the nonspecific transport of blood-borne proteins to the brain, *J Neurochem* 68, 874–877.

107 Broadwell RD **1989**, Transcytosis of macromolecules through the blood-brain barrier: a cell biological perspective and critical appraisal, *Acta Neuropathol* 79, 117–128.

108 Coomber BL, Stewart PA **1986**, Three-dimensional reconstruction of vesicles in endothelium of blood-brain barrier versus highly permeable microvessels, *Anat Rec* 215, 256–261.

109 Nag S, Robertson DM, Dinsdale HB **1979**, Quantitative estimate of pinocytosis in experimental acute hypertension, *Acta Neuropathol (Berl)* 46, 107–116.

110 Kapadia SE **1984**, Ultrastructural alterations in blood vessels of the white matter after experimental spinal cord trauma, *J Neurosurg* 61, 539–544.

111 Westergaard E, Go G, Klatzo I **1976**, Increased permeability of cerebral vessels to horseradish peroxidase induced by ischemia in Mongolian gerbils, *Acta Neuropathol* 35, 307–325.

112 Cole DJ, Matsumura JS, Drummond JC, Schultz RL, Wong MH **1991**, Time- and pressure-dependent changes in blood-brain barrier permeability after temporary middle cerebral artery occlusion in rats, *Acta Neuropathol* 82, 266–273.

113 Ayata C, Ropper AH **2002**, Ischaemic brain edema, *J Clin Neurosci* 9, 113–124.

114 Spatz M, Klatzo I **1976**, Ischaemic brain edema, *Adv Exp Med Biol* 69, 479–495.

115 Preston E, Webster J **2002**, Differential passage of [^{14}C]sucrose and [3H]inulin across the rat blood-brain barrier after cerebral ischemia, *Acta Neuropath (Berl)* 103, 237–242.

116 Drewes L **2001**, Molecular architecture of the brain microvasculature: perspective on blood-brain barrier transport, *J Mol Neurosci* 16, 93–98.

117 Hardebo JE, Owman C **1979**, Barrier mechanisms for neurotransmitter monoamines and their precursors at the blood-brain barrier, *Ann Neurol* 8, 1–11.

118 Schilling L, Wahl M **1997**, Brain edema: pathogenesis and therapy, *Kidney Int* 51, S69–S75.

119 Orth JD, McNiven MA **2003**, Dynamin at the actin-membrane interface, *Curr Opin Cell Biol* 15, 31–39.

120 Abbott NJ **2000**, Inflammatory mediators and modulation of blood-brain barrier permeability, *Cell Mol Neurobiol* 2, 131–147.

121 Kamiya T, Katayama Y, Kashiwagi F, Terashi A **1993**, The role of bradykinin in mediating ischemic brain edema in rats, *Stroke* 24, 571–576.

122 Zausinger S **2003**, Bradykinin receptor antagonists in cerebral ischemia and trauma, *IDrugs* 6, 970–975.

123 Raymond JJ, Robertson DM, Dinsdale HB **1986**, Pharmacological modification of bradykinin induced breakdown of the blood-brain barrier, *Can J Neurol Sci* 13, 214–220.

124 Olesen S-P, Crone C **1986**, Substances that rapidly augment ionic conductance of endothelium in cerebral venules, *Acta Physiol Scand* 127, 233–241.

125 Butt AM **1995**, Effect of inflammatory agents on electrical resistance across the blood-brain barrier in pial microvessels of anaesthetised rats, *Brain Res* 696, 145–150.

126 Wahl M, Young AR, Edvinsson, Wagner F **1983**, Effects of bradykinin on pial arteries and arterioles in vitro and in situ, *J Cereb Blood Flow Metab* 3, 231–237.

127 Gorlach C, Wahl M **1996**, Bradykinin dilates rat middle cerebral artery and its large branches via endothelial B2 receptors and release of nitric oxide, *Peptides* 17, 1373–1378.

128 Onoue H, Kaito N, Tomii M, Tokudome S, Nakajima M, Abe T **1994**, Human basilar and middle cerebral arteries exhibit endothelium-dependent responses to peptides, *Am J Physiol* 267, H880–H886.

129 Ding-Zhou L, Margaill I, Palmier B, Pruneau D, Plotkine M, Marchand-Verrecchia C **2003**, LF 16-0687 Ms, a bradykinin B2 receptor antagonist, reduces ischemic brain injury in a murine model of transient cerebral ischemia, *Br J Pharm* 139, 1539–1547.

130 Zausinger S, Lumenta DB, Pruneau D, Schmid-Elsaesser R, Plesnila N, Baethmann A **2003**, Effects of LF 16-0687 MS, a bradykinin B_2 receptor antagonist, on brain edema formation and tissue damage in a rat model of temporary focal cerebral ischemia, *Brain Res* 950, 268–278.

131 Subramanian N, Theodore D, Abraham J **1981**, Experimental cerebral infarction in primates: regional changes in brain histamine concentration, *J Neural Trans* 50, 225–232.

132 Adachi N, Oishi R, Saeki K **1991**, Changes in the metabolism of histamine and monoamines after occlusion of the middle cerebral artery in rats, *J Neurochem* 57, 61–66.

133 Dux E, Joo F **1982**, Effects of histamine on brain capillaries: fine structural and immunohistochemical studies after intracarotid infusion, *Exp Brain Res* 47, 252–258.

134 Schilling L, Wahl M **1994**, Opening of the blood-brain barrier during cortical superfusion with histamine, *Brain Res* 653, 289–296.

135 Toda N **1990**, Mechanism underlying responses to histamine of isolated monkey and human cerebral arteries, *Am J Physiol* 258, H311–H317.

136 Wahl M, Kuschinsky W **1979**, The dilating effects of histamine on pial arteries of cats and its mediation by H_2 receptors, *Circ Res* 44, 161–165.

137 Joo F, Kovacs J, Szerdahelyi P, Temesvari P, Tosaki A **1994**, The role of histamine in brain oedema formation, *Acta Neurochir* 60, 76–78.

138 Bhakoo KK, Crockard HA, Lascelles PT **1984**, Regional studies of changes in brain fatty acids following experimental ischemia and reperfusion in the gerbil, *J Neurochem* 43, 1025–1031.

139 Tomimoto H, Shibata M, Ihara M, Akiguchi I, Ohtani R, Budka H **2002**, A comparative study on the expression of cyclooxygenase and 5-lipoxygenase during cerebral ischemia in humans, *Acta Neuropathol (Berl)* 104, 601–607.

140 Rao AM, Hatcher JF, Kindy MS, Dempsey RJ **1999**, Arachidonic acid and leukotriene C4: role in transient cerebral ischemia in gerbils, *Neurochem Res* 24, 1225–1232.

141 Unterberg A, Wahl M, Hammersen F, Baethmann A **1987**, Permeability and vasomotor responses of cerebral vessels during exposure to arachidonic acid, *Acta Neuropathol* 73, 209–219.

142 Rosenblum WI, Nelson GH **1988**, Endothelium-dependent constriction demonstrated in vivo in mouse cerebral arterioles, *Circ Res* 63, 837–843.

143 Barbosa-Coutinho LM, Hartmann A, Hossmann K-A, Rommel T **1985**, Effect of dexamethasone on serum protein extravasation in experimental brain infarcts of monkeys: An immunohistochemical study, *Acta Neuropathol* 65, 255–260.

144 Dux E, Ismail M, Szerdahelyi P, Joo F, Dux L, Koltai M, Draskoczy M **1990**, Dexamethasone treatment attenuates the development of ischaemic brain oedema in gerbils, *Neuroscience* 34, 203–207.

145 Bian LG, Zhang TX, Zhao WG, Shen JK, Yang GY **1994**, Increased endothelin-1 in the rabbit model of middle cerebral artery occlusion, *Neurosci Lett* 174, 47–50.

146 Matsuo Y, Mihara S, Ninomiya M, Fujimoto M **2001**, Protective effect of endothelin type A receptor antagonist on brain edema and injury after transient middle cerebral artery occlusion in rats, *Stroke* 32:2143–2148.

147 Miller RD, Monsul NT, Vender JR, Lehman JC **1996**, NMDS- and endothelin–1-induced increases in blood-brain barrier permeability quantitated with Lucifer Yellow, *J Neurol Sci* 136, 37–40.

148 Stanimirovic DB, Bertrand N, McCarron R, Uematsu S, Spatz M **1994**, Arachidonic acid release and permeability changes induced by endothelins in human cerebromicrovascular endothelium, *Acta Neurochir Suppl (Wien)* 60, 71–75.

149 Yanagisawa M, Kurihara H, Kimura S, Tomobe Y, Kobayashi M, Mitsui Y, Yazaki Y, Goto K, Masaki T **1988**, A novel potent vasoconstrictor peptide produced by vascular endothelial cells, *Nature* 332, 411–415.

150 Inoue A, Yanagisawa M, Kimura S, Kasuya Y, Miyauchi T, Goto K, Masaki T **1989**, The human endothelin family: three structurally and pharmacologically distinct isopeptides predicted by three separate genes, *Proc Natl Acad Sci USA* 86, 2863–2867.

151 Lehmberg J, Putz C, Furst M, Beck J, Baethmann A, Uhl E **2003**, Impact of the endothelin-A receptor antagonist BQ 610 on microcirculation in global cerebral ischemia and reperfusion, *Brain Res* 961, 277–286.

152 Kovacs Z, Ikezaki K, Samoto K, Inamura T, Fukui M **1996**, VEGF and flt. expression time kinetics in rat brain infarct, *Stroke* 27, 1865–1872.

153 Hayashi T, Abe K, Suzuki H, Itoyama Y **1997**, Rapid induction of vascular endothelial growth factor gene expression after transient middle cerebral artery occlusion in rats, *Stroke* 28, 2039–2044.

154 Lennmyr F, Ata KA, Funa K, Olsson Y, Terent A **1998**, Expression of vascular endothelial growth factor (VEGF) and its receptors (Flt-1 and Flk-1) following permanent and transient occlusion of the middle cerebral artery in the rat, *J Neuropathol Exp Neurol* 57, 874–882.

155 Fischer S, Wiesnet M, Marti HH, Renz D, Schaper W **2004**, Simultaneous activation of several second messengers in hypoxia-induced hyperpermeability of brain derived endothelial cells, *J Cell Physiol* 198, 359–369.

156 Hofman P, Blaauwgeers HG, Tolentino MJ, Adamis AP, Nunes Cardozo BJ, Vrensen GF, Schlingemann RO **2000**, VEGF-A induced hyperpermeability of blood-retinal barrier endothelium in vivo is predominantly associated with pinocytotic vesicular transport and not with formation of fenestrations, *Curr Eye Res* 21, 637–645.

157 Kusters B, de Waal RM, Wesseling P, Verrijp K, Maass C, Heerschap A, Barentsz JO, Sweep F, Ruiter DJ, Leenders WP **2003**, Differential effects of vascular endothelial growth factor A isoforms in a mouse brain metastasis model of human melanoma, *Cancer Res* 63, 5408–5413.

158 Zhang ZG, Zhang L, Jiang Q, Zhabg R, Davies K, Powers C, van Bruggen N, Chopp M **2000**, VEGF enhances angiogenesis and promotes blood-brain barrier leakage in the ischemic brain, *J Clin Invest* 106, 829–838.

159 Sacco R, Zamanillo MC, Kargman D **1994**, Predictors of mortality and recurrence after hospitalized cerebral infarction in an urban community: the northern Manhattan stroke study, *Neurology* 44, 626–634.

160 Oppenheimer S, Halfbraid BI, Oswald GA, Yudkin JS **1985**, Diabetes mellitus and early mortality from stroke, *Br Med J* 291, 1014–1015.

161 Pulsinelli W, Levy DE, Sigsbee B **1983**, Increased damage after ischemic stroke in patients with hyperglycemia with or without established diabetes mellitus, *Am J Med* 74, 540–544.

162 Candelise L, Landi G, Orazio EN, Boccardi E **1985**, Prognostic significance of hyperglycemia in acute stroke, *Arch Neurol* 42, 661–663.

163 Matchar DB, Divine GW, Heyman A, Feussner JR **1992**, The influence of hyperglycemia on outcome of cerebral infarction, *Ann Intern Med* 117, 449–456.

164 Plum F **1983**, What causes infarction in ischemic brain? *Neurology* 33, 222–233.

165 Kiers L, Davis SM, Larkins R **1992**, Stroke topography and outcome in relation to hyperglycemia and diabetes, *J Neurol Neurosurg Psychiatry* 55, 263–270.

166 de Courten-Myers G, Myers RE, Schoolfield L **1988**, Hyperglycemia enlarges infarct size in cerebrovascular occlusion in cats, *Stroke* 19, 623–630.

167 de Courten-Myers G, Kleinholz M, Wagner KR, Myers RE **1989**, Fatal strokes in hyperglycemic cats, *Stroke* 20, 1707–1715.

168 Frizzell TR, Batjer HH **1997**, Cerebral ischemia: insights from animal models, in *Cerebrovascular Disease*, ed. HH Batjer, Lippincott-Raven, Philadelphia.

169 Venables GS, Miller SA, Gibson G, Hardy JA, Strong AJ **1985**, The effect of hyperglycemia on changes during reperfusion following focal cerebral ischemia in the cat, *J Neurol Neurosurg Psychiatry* 48, 663–669.

170 Clarke H, Marano CW, Soler AP, Mullin JM **2000**, Modification of tight junction function by protein kinase C isoforms, *Adv Drug Deliv Rev* 41, 283–301.

171 Wolburg H, Lippoldt A **2002**, Tight junctions of the blood-brain barrier: Development, composition and regulation, *Vasc Pharm* 38, 323–337.

172 Murray MA, Heistad DD, Mayhan WG **1991**, Role of protein kinase C in bradykinin-induced increases in microvascular permeability, *Circ Res* 68, 1340–1348.

173 Aschner JL, Lum H, Fletcher PW, Malik AB **1993**, The differential effects of protein kinase C on bradykinin- and thrombin-mediated phospholipase C activation, FASEB J 7, A719.

174 Lynch J, Ferro T, Blumenstock AM, Brockenauer AM, Malik AB **1990**, Increased endothelial albumin permeability mediated by protein kinase C, *J Clin Invest* 85, 1991–1998.

175 Silinger-Birnboim AM, Goligorsky MS, Del Vecchio PJ, Malik AB **1992**, Activation of protein kinase C pathway contributes to hydrogen peroxide-induced increase in endothelial permeability, *Lab Invest* 67, 24–30.

176 Yuan SY **2000**, Signal transduction pathways in enhanced microvascular permeability, *Microcirculation* 7, 395–405.

177 Stuart RO, Nigam SK **1995**, Regulated assembly of tight junctions by protein kinase C, *Proc Natl Acad Sci USA* 92, 6072–6076.

178 Lapierre LA **2000**, The molecular structure of the tight junction, *Adv Drug Deliv Rev* 41, 255–264.

179 Yuan Y, Huang Q, Wu HM **1997**, Myosin light chain phosphorylation: modulation of basal and agonist-stimulated venular permeability, *Am J Physiol* 272, H1437–H1443.

180 Lapierre LA **2000**, The molecular structure of the tight junction, *Adv Drug Deliv Rev* 41, 255–264.

181 Lee T-S, Saltsman K, Ohashi H, King G **1989**, Activation of protein kinase C by elevation of glucose concentration: proposal for a mechanism in the development of diabetic vascular complications, *Proc Natl Acad Sci USA* 86, 5141–5144.

182 Williams B, Schrier R **1992**, Characterization of glucose-induced in situ protein kinase C activity in cultured vascular smooth muscle, *Diabetes* 41, 1464–1472.

183 Wolf B, Williamson J, Easom R, Chang K, Sherman W, Turk J **1991**, Diacylglycerol accumulation and microvascular abnormalities induced by elevated glucose levels, *J Clin Invest* 87, 31–38.

184 Craven P, Davidson C, DeRubertis F **1990**, Increase in diacylglycerol mass in isolated glomeruli by glucose from de novo synthesis of glycerolipids, *Diabetes* 39, 667–674.

185 Hom S, Egleton RD, Huber JD, Davis TP **2001**, Effect of reduced flow on blood-brain barrier transport systems, *Brain Res* 890, 38–48.

186 Iadecola C **1998**, Cerebral circulatory dysregulation in ischemia, in *Cerebrovascular Disease: Pathophysiology, Diagnosis and Management*, Vol. 1, eds. MD Ginsberg, J Bogousslavsky, Blackwell Science, Malden, Mass., pp. 319–357.

187 Cipolla M, Lessov N, Hammer ES, Curry AB **2001**, The threshold duration of ischemia for myogenic tone in middle cerebral arteries: effect on vascular smooth muscle actin, *Stroke* 32, 1658–1664.

188 Cipolla MJ, McCall AL, Lessov N, Porter JP **1997**, Reperfusion decreases myogenic reactivity and alters middle cerebral artery function after focal cerebral ischemia in rats, *Stroke* 28, 176–180.

189 Kuroiwa T, Shibutani M, Okeda R **1989**, Nonhyperemic blood flow restoration and brain edema in experimental focal cerebral ischemia, *J Neurosurg* 70, 73–80.

190 Ito U, Ohno K, Yamaguchi T, Takei, H, Tomita H, Inaba Y **1980**, Effect of hypertension on blood- brain barrier; change after restoration of blood flow in post-ischemic gerbil brains; an electronmicroscopic study, *Stroke* 11, 606–611.

26
Diabetes and the Consequences for the Blood-Brain Barrier

Arshag D. Mooradian

26.1
Introduction

The prevalence of diabetes is rapidly increasing worldwide. This change in the epidemiology of diabetes is fuelling the surge in studies related to the pathogenesis and complications of this disease.

Diabetes is a systemic metabolic disorder that affects various tissues of the body, albeit to a variable degree. At the present time, there are several potential biochemical changes that contribute to the emergence of diabetes complications. These include increased oxidative load, increased glycation of important proteins, activation of protein kinase C pathway and possibly increased accumulation of intracellular sorbitol and depletion of myoinositol [1, 2]. These biochemical changes cause, at least partly, the various long-term complications of diabetes, notably retinopathy, nephropathy, neuropathy and macrovascular disease. The central nervous system (CNS) complications of diabetes have not been widely appreciated, primarily because the most overt complication, namely stroke, is viewed as a vascular complication while the other more subtle changes in the CNS are usually not clinically relevant. Nevertheless, in the past decade there has been increasing evidence to indicate that diabetes may have significant effects on the CNS that extend beyond clinically appreciable cerebrovascular accidents [3–6]. There are several potential mechanisms that contribute to the diabetes-related changes in the CNS [5]. These include alterations in cerebral microvessels with poor auto-regulation, blood distribution, and altered blood-brain barrier (BBB) function, neurochemical changes, alterations in neurotransmitter receptor activity and other contributing factors, such as repeated hypoglycemic reactions, peripheral neuropathy and nephropathy with uremia. In this chapter, the changes in cerebral microvasculature will be reviewed and the potential clinical manifestations of diabetes-related changes in the CNS will be discussed from the perspective of the BBB function.

26.2
Histological Changes in the Cerebral Microvessels

In addition to the well characterized ischemic changes that occur following a cerebrovascular accident (see Chapter 25), the microvasculature within the CNS, like the microvasculature in a variety of tissues in a diabetic animal, undergoes significant histological changes [7–10]. Although the overall number of cerebral capillaries may not be altered [7], the density of capillaries within the cortical areas may be reduced and the capillary basement membrane thickening is increased in streptozotocin (STZ)-induced diabetic animals [8–10]. In addition, calcium depositions have been documented in microvessel walls of these animals [9].

Studies using intravital microscopy have also observed the formation of arterio-venous shunting in cerebral cortical tissue after 5 months of uncontrolled experimental diabetes [9]. These changes, although significant, are not of the same magnitude as those observed in retinal tissue where the microvasculature is also endowed with tight junctions between endothelial cells. Thus, local tissue factors appear to be an important determinant of the alterations in the microvessels that occur in diabetes. It is noteworthy that some of the histological changes in cerebral microvessels in diabetes, namely the paucity of cortical capillaries and establishment of arterio-venous shunts are similar to the changes described in aging rat models [11]. These similarities in histological changes in cerebral microvasculature support the notion that diabetes may often simulate premature aging [12].

The structural changes of cerebral microvessels predispose the animal to cerebral hypoxic injury during periods of reduced blood flow or poor ventilation and may contribute to changes in the BBB function.

26.3
Functional Changes in the Blood-Brain Barrier

The principal known functions of the BBB are to assure the adequate delivery of nutrients and substrates to the CNS, eliminating metabolic end-products through efflux and protecting the CNS from environmental toxins circulating in the blood. The latter is accomplished by the dynamic barrier that is formed by the tight junctions between endothelial cells. The molecular anatomy of these tight junctions may change in diabetes. In a recent study, Western and Northern blot analyses were carried out to measure the steady-state level of occludin and zonula occludens-1 (ZO-1) proteins and mRNA levels in cerebral tissue of STZ-induced diabetic rats and the results were compared to insulin-treated diabetic rats and vehicle-injected control rats [13]. The cerebral occludin content in diabetic rats was significantly reduced compared to insulin-treated diabetic rats or nondiabetic control rats. The ZO-1 content of cerebral tissue from diabetic rats was not significantly altered compared to controls [13]. The cerebral occludin or ZO-1 mRNA content relative to G3PDH mRNA was not altered in diabetes. Thus, select structural proteins in the tight junctions may be altered in diabetes. It is noteworthy that occludin content of rat retina is reduced in experimental diabetes [14]. In addition, high glucose or low insulin concentrations reduce the ZO-1 protein content of bovine retinal capillary endothelial cells in culture [15]. Considering the similarities in the tight junctions of the BBB and blood-retinal barrier, it is not surprising that the changes in occludin content of the two vascular barriers in diabetic rats may be similar. However, the physiological consequences of these changes are not readily demonstrable.

Studies on the barrier function have not always been consistent. Using immuno-histochemical techniques, Stauber et al. [16] found that albumin, but not immunoglobulin G or complement C, selectively enters the cerebral cortex after 2 weeks of experimental diabetes. In another study, extravasations of Evan's blue albumin into the cerebral tissue of diabetic rats after adrenalin induced acute hypertension was significantly increased compared to controls [17]. Furthermore, studies using intravital microscopy and fluorescein-labeled albumin found increased BBB permeability in diabetes, although they did not find any increase in the permeability of the BBB with acute hypertension [18]. These results should be interpreted with caution since extravasation of ligands through the BBB is often hard to quantitate with intravital microscopy.

The increased permeability of the cerebral microvessels to albumin is suspected to be the result of enhanced microvascular uptake of glycosylated albumin [19] and is considered to be independent of alteration in the BBB. In this regard, it is noteworthy that the BBB permeability to albumin in long-term diabetic rats could be normalized following insulin treatment [20].

The BBB permeability to a variety of markers has been studied. The volume of distribution of inulin in select regions of the cerebrum, such as mediobasal hypothalamus, mediodorsal hypothalamus and periaqueductal gray [21], is increased in rats after 4 weeks of uncontrolled hyperglycemia. However, the permeability of the cerebral microvessels to other markers such as horseradish peroxidase, sucrose, or cytochrome c is not altered in diabetic rats [21, 22].

Overall, it appears that the barrier function of the BBB in diabetic rats is well maintained, but it may be impaired, especially in certain regions of the cerebrum. Future studies should employ a host of markers with a wide spectrum of size and polarity to more comprehensively study the effect of diabetes on the BBB permeability profile.

26.3.1
Diabetes-Related Changes in the BBB Transport Function

In addition to being a protective barrier, the cerebral microcirculation plays a critical role in regulating the availability of nutrients and substrates for the production of neurotransmitters. To maintain an optimal physiologic milieu, the cerebral microcirculation regulates the efflux of byproducts of the cerebral metabolism. These transport functions are highly adaptable in response to alterations in blood glucose levels. Thus, sustained hypoglycemia or hyperglycemia is associated with significant changes in various transport parameters. One of the key alterations is the up-regulation or down-regulation of glucose transport as a consequence of chronic hypoglycemia and hyperglycemia, respectively. Some studies in human subjects have found that sustained hypoglycemia for several days induces an increase in BBB transport of glucose [23–25]. However, studies by Segel et al. [26] could not confirm these findings. In this study, there was no increase in blood-to-brain glucose transport after 24 h of experimental sustained hypoglycemia in eight healthy subjects. These investigators did not exclude the possibility of a change in regional blood-to-brain glucose transport and could not address the possibility that alterations in transport might occur beyond the BBB [26].

The effect of hypoglycemia on BBB transport function and cerebral metabolism is more comprehensively studied in animal models. In chronically hypoglycemic rats bearing insulinomas, glucose transport is up-regulated while the transport of large neutral amino acids and choline is unaffected and the transport of lactate and pyruvate is diminished [27]. Furthermore, in this rat model of hypoglycemia, cerebral tissue content of glucose-6-phosphate dehydrogenase (G6PD), creatine phosphate and ATP are not significantly decreased. This suggests that the enhanced glucose extraction at the BBB during hypoglycemia preserves brain energy consumption [27]. However, despite such adaptations to hypoglycemia [27–30], the mitochondrial respiration within the cerebral tissue remains abnormal [31].

The effect of sustained hyperglycemia on the BBB transport of nutrients has been studied in various animal models of diabetes and in human subjects with diabetes. Most, but not all studies have found that chronic hyperglycemia is associated with down-regulation of the BBB transport of glucose [32–37]. Insulin treatment with amelioration of hyperglycemia normalizes BBB glucose transport [33–35]. However, some studies have questioned this finding [36, 37]. Differences in the experimental models used and the degree or duration of hyperglycemia as well as differences in methodology employed, may explain the discrepancy in the reported results.

In addition, cerebral microvessels isolated from diabetic animals have altered glucose metabolism [38]. The latter observation could have been because of reduced cellular uptake of glucose, presumably as a result of reduced glucose transporter function or could be secondary to alterations in metabolic pathways of glucose metabolism beyond the glucose transporter.

There are few clinical studies on BBB glucose transport during hyperglycemia. Fanelli and his colleagues [39] found that neither blood-to-brain glucose transport nor cerebral glucose metabolism is measurably reduced in people with poorly controlled type 1 diabetes. Brooks et al. [40] observed 18% reduction in blood-to-brain glucose transport in poorly controlled diabetic patients, but this difference was not statistically significant. In a small study of six human subjects, Hasselbalch et al. [41] found that 2 h of acute hyperglycemia did not cause significant changes in the maximal transport velocity or affinity in the glucose transporter of the BBB. In addition, a short duration of hyperglycemia did not change the global cerebral blood flow (CBF) or cerebral glucose metabolism rate (CMR_{glc}), except in the white matter where a significant increase in CMR_{glc} could be demonstrated [41].

Inconsistencies in the published literature on diabetes-related changes in BBB glucose transport are further compounded by the inconsistencies in studies evaluating the molecular basis of changes in BBB glucose transport. Choi et al. [42] found increased GLUT-1 mRNA levels in the brain capillaries of rats with STZ-induced diabetes but decreased transport and GLUT-1 protein (the major BBB glucose transporter). To determine the molecular mechanisms of diabetes-related changes in the expression of GLUT-1 in cerebral tissue, we studied STZ-induced diabetic rats and vehicle-injected controls after 4 weeks of diabetes [43]. The GLUT-1 mass in cerebral microvessels was reduced in diabetic rats by approximately 38%. The GLUT-1 concentration in the insulin-treated diabetic group was not significantly different from controls. The GLUT-1 mRNA content of cerebral tissue in diabetic rats was significantly reduced compared to control rats or insulin-treated diabetic rats. The in vitro translation of GLUT-1 mRNA of diabetic rats was also significantly lower than that in nondiabetic control rats or insulin-treated diabetic rats [43]. These changes occurred in association with a reduction in poly(A) tail length of GLUT-1 mRNA, which decreased from a control value of 200–350 nt to only 50–100 nt in diabetic rats [43]. Shortening of the poly(A) tail of mRNAs is a novel mechanism of diabetes-related changes in the expression of specific genes that are regulated at a translational level.

In addition to the changes in BBB glucose transport, diabetes is associated with a host of changes in the transport of other nutrients and metabolites. Since the glucose transporter mediates the BBB transport of various hexoses, it is not surprising that diabetes is associated with a reduction in brain uptake of hexoses other than glucose [33]. In addition, the hexose transporter is also the key cellular transporter of dehydroascorbate [44]. In previously published studies, we demonstrated that dehydroascorbate, but not ascorbate, competes with glucose for entry into the brain [45]. Whether sustained hyperglycemia in diabetes deprives the brain of vitamin C and whether vitamin C deficiency in brain tissues leads to cerebral dysfunction is not known [45]. Nevertheless, several studies in animal models of diabetes or in human subjects have suggested that tissue deficiency in ascorbic acid may occur [46, 47], and may partially account for increased oxidative load in diabetes [48, 49].

Another example of adaptive changes in BBB related to diabetes is the increased BBB transport of ketones, such as hydroxybutyrate. This increase in ketone body transport is normalized after insulin treatment [33]. A similar up-regulation of BBB transport of ketones is reported in other models of ketosis. The exact mechanism of this change is not known. It is tempting to suggest that it may be related to alterations in the expression of ketone body transporter within the BBB. Indeed, ketosis and independently acidosis has been shown to alter the expression of several genes through a pH response element in the regulatory regions of genes [50].

Unlike the changes in BBB glucose and ketone transport, diabetes does not appear to alter the transport of amino acids across the BBB [33]. However, the permeability of phenylalanine was decreased by an average of 40% throughout the entire brain after 4 weeks of STZ diabetes in rats [51] and influx was depressed by 35%. The permeability of lysine was increased by an average of 44%, but the influx was decreased by 27%. Several plasma neutral amino acids (branched chain) were increased, whereas all basic amino acids were decreased. Brain tryptophan, phenylalanine, tyrosine, methionine and lysine contents were markedly decreased. The transport changes were attributed to alterations in the concentrations of the plasma amino acids that compete for the neutral and basic amino acid carriers at the BBB [51].

The BBB transport of choline, the substrate for neurotransmitter acetylcholine, is also reduced in STZ-induced diabetic rats [52]. Insulin treatment for 5 days did not rectify the abnormal choline transport. This change may contribute to the cognitive and memory changes related to diabetes [5, 6]. It is of note that the BBB choline transport is also reduced in aging animals and humans where memory deficits are common [53].

The BBB permeability to minerals such as sodium and potassium, but not chloride or calcium, is also decreased in diabetic rats [54, 55]. These changes may be related to alterations in the activity of key ATPase-dependent pumps. However, in our studies of cerebral microvascular ATPases, we found that the sodium/potassium or magnesium ATPase activity of the endothelial cells of the BBB was not altered in diabetic rats [56]. This finding was unexpected since glucotoxicity, via activation of the polyol pathway, is often associated with reduced activity of Na^+/K^+-ATPase [57]. However, the role of the polyol pathway in diabetes-related changes in Na^+/K^+-ATPase activity is not clear, since treatment of diabetic rats with an aldose reductase inhibitor does not normalize BBB permeability to sodium, although dietary supplementation with myoinositol does [55].

The effect of diabetes on the BBB transport of various lipophilic hormones or drugs is not well studied. The cerebral transport of thyroid hormone may be reduced in diabetes [58]. However, the brain uptake of benzodiazepines with different lipophilic and protein-binding characteristics is not altered in diabetes [59]. Because of conceptual importance and potential clinical implications, more studies are needed to determine whether diabetes alters the brain uptake of various centrally acting drugs.

Overall, it appears that the *BBB transport of certain substrates is altered by diabetes*. Most studies are done in animal models of diabetes with 2–4 weeks of uncontrolled diabetes, giving results like these:

1. *Barrier function*
    - increased permeability to albumin, inulin;
    - unaltered permeability to immunoglobulin G, complement 3, horseradish peroxidase, ethylenediaminetetraacetic acid (EDTA), sucrose.

2. *Transport function*
- increased transport of monocarboxylic acids, e.g. beta hydroxybutyrate;
- decreased transport of hexoses (e.g. glucose), dehydroascorbate, choline, sodium, potassium, protons (H^+);
- unaltered transport of neutral amino acids (e.g. tryptophan, tyrosine), basic amino acids (e.g. lysine), lactate, calcium, chloride.

In addition, repeated hypoglycemic episodes and hyperinsulinemia in obese type 2 diabetic individuals, or insulin deficiency, may contribute to the diabetes-related changes in BBB transport.

26.4
Potential Mechanisms of Changes in the BBB

The barrier function of the cerebral microvasculature is determined by both structural and functional characteristics, such as tight junctions between endothelial cells, the ionic charges on the surface of endothelial cells and a host of enzymes and transporters that limit the access of circulating peptides or potentially toxic substrates into the CNS. Thus, diabetes-related changes in the BBB can be the result of alterations in the hemodynamics of cerebral circulation, changes in the physico-chemical properties of key constituents of the cerebral microvessels or changes in neuronal control of the BBB. These changes are usually the result of chronic uncontrolled hyperglycemia, although co-morbidities such as hypertension [60, 61], transient cerebral ischemic events [62, 63] and hyperosmolality [64] would further aggravate the changes in the BBB. These potential mechanisms will be discussed briefly.

26.4.1
Hemodynamic Changes

Studies using intravital microscopy have found significant changes in the cerebral microcirculation of diabetic animals [65, 66]. As early as 1 month, the pial vessels of diabetic rats were dilated and tortuous and the vasoconstrictive response of arterioles and venules to local increases of P_{O_2} in the artificial cerebrospinal fluid (CSF) bathing these vessels was reduced [9]. In addition, the linear velocity of blood flow was reduced in diabetic rats and increased arterio-venous shunting was also observed. Electron microscopic studies found that, after 5 months of diabetes, focal changes in the thickness and density of the vascular basement membrane were noted [9]. In addition, astrocytic endfeet were swollen and contained mitochondria having a longitudinal rearrangement of their cristae. Basement membranes contained nodules of electron-lucent material that impinged on degenerating smooth muscle cells, pericytes and astrocytes. The tight interendothelial junctions appeared to be intact even though dramatic changes had taken place perivascularly. Although the vasoconstrictive response to in-through different signal transduction mechanisms [90]. Endothelins stimulate sodium uptake into rat brain capillary endothelial cells through endothelin A-like receptors [91]. Some of these effects of endothelin are mediated through phosphorylation of key proteins in endothelial cells [92]. Multiple signaling pathways are involved in the regulation of tight junction permeability [93, 94]. These include the cyclic nucleotides that decrease paracellular permeability, protein kinase C, G proteins, MAP kinases and other protein kinases. The effect of diabetes on these signaling pathways within the cerebral microcirculation is not well characterized.

26.5
Potential Clinical Consequences of Changes in the BBB

A host of diabetes-related changes in the CNS has been recognized. The pathogenesis of CNS diseases is often related to cerebrovascular accidents, repeated hypoglycemic attacks, ketosis, hyperglycemia and hyperosmolarity. In addition, peripheral or autonomic neuropathy, impaired vision, renal failure and medications further aggravate CNS-related diseases in diabetes [3–5]. Although there are multiple potential causes underlying these changes, the BBB function is likely to be an important contributor to the changes in the CNS. Clinical evidence for loss of BBB integrity in diabetic subjects is scarce. This is the result of the limitations in the current technology available to study the BBB function in human subjects. In this context, the availability of reliable serum biomarkers of the barrier function would be helpful. One such marker of BBB openings is the serum level of S100B protein and/or the presence of anti-S100B auto-antibodies [95–97]. The S100B is an astroglial protein that belongs to the calcium receptor superfamily. Serum levels of S100B protein are increased when BBB is disrupted with mannitol [96]. When the BBB disruption is severe, neuronal injury is reflected in the increased levels of serum neuron-specific enolase (NSE) [96]. The measurement of serum concentrations of NSE and S100B may have a prognostic role in evaluating clinical outcome following severe hypoglycemia [98]. In one case of diabetic ketoacidosis, serum S100B concentration rose coinciding with the onset of cerebral edema [99].

To determine whether clinical diabetes is associated with disruption of the BBB and/or brain injury, serum levels of S100B and NSE as well as auto-antibodies against them were measured in subjects with type 1 and type 2 diabetes [100]. Serum S100B concentrations in type 2 diabetic subjects, but not in type 1 diabetic subjects, were significantly lower than those found in healthy controls. There were no significant differences in serum NSE levels of either type 1 or type 2 diabetics compared to healthy controls. However, there was a significant increase in antibodies to NSE in both type 1 and type 2 diabetic subjects compared to controls, whereas diabetics and controls had equally very low levels of anti S100B auto-antibodies. These studies suggest that diabetes in humans may be associated with alterations in the BBB integrity that allow the emergence of antibodies against neuronal antigens [100].

Cerebrovascular accidents are widely recognized as the most serious complication of diabetes. However, diabetes is also associated with less dramatic disease of the CNS, including the slow deterioration of cognitive function, increased incidence of depression and reduced pain tolerance.

Several clinical studies have found that type 1 diabetic patients have modest impairment in learning, memory and problem solving, as well as impairments in mental and motor speed [101, 102]. The cognitive changes of type 2 diabetic patients are more consistently demonstrable. Moderate cognitive impairment is reported particularly in tasks involving verbal memory and complex information processing, while basic attention process, motor reaction time and short-term memory are relatively unaffected [101, 102]. Several studies have reported that diabetic patients are significantly inferior to the nondiabetic control group in the serial learning task and Benton's visual retention test [103–105]. The digit span test usually shows no significant difference between the groups. Electroencephalogram (EEG) frequency band analysis has shown a slowing over the central cortex and reduction of alpha activity over the parietal area in diabetic patients [105]. The checkerboard-elicited P100 wave did not reveal a significant increase in latency nor were the P300 wave latencies significantly different in diabetic patients. However, a trend toward longer latencies in diabetics was evident at Fz and Cz recording sites [105]. These observations suggest that diabetic patients have an impairment in the retrieval of recently learned material with preservation of auditory attention and immediate recall.

It is also possible that the diabetes-related changes in the BBB contribute to the increased incidence of Alzheimer's disease in diabetes [106]. The changes observed in diabetes are often similar to those found in aging [107]; and disruptions in the BBB have been suspected to be a contributing cause of Alzheimer's disease [108]. Indeed, BBB changes in the GLUT-1 content of cerebral tissue from patients with Alzheimer's disease are similar to those found in diabetes [109].

Similar changes in cognition are found in animal models of diabetes [110, 111]. In one study, mice with STZ-induced diabetes mellitus had normal acquisition for relatively simple tasks but showed problems in learning more complex tasks such as shuttle box avoidance. However, other studies have found increased learning in diabetes. In the latter studies, enhanced learning shown in simple passive avoidance tasks appears to be due to increased foot-shock sensitivity [112]. Diabetic mice also show a marked memory retention deficit after learning an active avoidance T-maze task. This retention deficit was reversed by a single injection of insulin, suggesting that it may be related to hyperglycemia per se. Furthermore, diabetic mice were found to have a shift to the left in the inverted U-shaped dose-response curve for memory retention produced by an acetylcholine agonist [113]. Responses to several other pharmacological memory enhancers are also altered in diabetic mice. These studies suggest that this mouse model of diabetes mellitus demonstrates a deficit in memory retention and retrieval similar to that seen in humans with diabetes mellitus.

In addition to cognitive changes in diabetes, affective disorders and notably the incidence of depression are increased in diabetes [114]. Major depression has a higher recurrence rate and depressive episodes may last longer in individuals with type 1 or type 2 diabetes [115–119]. It is not entirely clear whether this is the result of neurobiochemical changes associated with diabetes, or the result of psychological factors related to chronic disease state or its treatment. It is noteworthy that depressive disorders may increase the risk for diabetes, possibly because of increased insulin resistance presumably secondary to increased plasma corticosteroid levels [120].

The onset of major depression typically precedes the diagnosis of type 2 diabetes by many years, while in type 1 diabetes it seems to follow the diagnosis [115–119]. A significant relationship between poor glycemic control and major depression has been reported, although this relationship was not supported by other studies [120–123].

Another clinical manifestation of diabetes-related changes in the CNS is alteration in pain perception. Pain intolerance is a common feature of diabetes, independent of peripheral neuropathy [124, 125]. Animal studies have suggested an altered response to opiate agonists and antagonists as well as an altered pain threshold in diabetic animals [126]. STZ-induced diabetic rats have a higher spontaneous pain threshold but reduced sensitivity to morphine analgesia [126]. The exaggerated painful response to stimuli, as well as spontaneous pain in diabetic rats was attributed to hyper-responsiveness and increased conduction veloci-ty of a subset of C-nerve fibers [126]. In clinical studies, infusion of 50 g glucose in normal subjects resulted in a significant decrease in both the threshold level of pain and the maximal level of pain tolerated, as measured by responses to electrical pain induced by a Grass stimulator [127]. In addition, patients with diabetes mellitus were hyperalgesic when compared with normal subjects. These findings have potential clinical implications in the pathophysiology and management of painful diabetic neuropathy and the use of narcotic agents in diabetes mellitus.

The sensory threshold of pain is elevated in diabetic subjects. Major risk factors for the elevation of sensory threshold for pain were age, duration of diabetes and tall stature [127]. Finally, diabetic subjects have a prolongation of anginal perceptual threshold (the time from the onset of 0.1 mV ST segment depression to the onset of chest pain during treadmill exercise). This prolongation correlates positively with the somatic pain threshold [125]. The changes in anginal threshold may delay the diagnosis of myocardial infarction in these patients.

Another potential consequence of the changes in BBB is decreased responsiveness to certain hormones, notably leptin [128]. It appears that obesity-related resistance to leptin is partially the result of reduced transport of leptin across the BBB in obese individuals. Finally, it is possible that the aggravation of hypoglycemia unawareness in individuals with long-standing diabetes as a result of very strict blood glucose control [129] may partly be related to up-regulation of the glucose transporters in response to repeated hypoglycemic attacks [27]. The up-regulation of glucose transporters in these patients may theoretically retard the CNS response to hypoglycemia. Alternatively, the down-regulation of glucose transporters in chronic hyperglycemia may account for the emergence of hypoglycemic symptoms when blood glucose levels are rapidly reduced from elevated levels to near normal range. In addition, sustained hyperglycemia may compete with cerebral ascorbate uptake through competitive inhibition of transport and may result in cerebral tissue deficiency in vitamin C content [45]. These speculations are tantalizing but have not been proven with experimental data.

26.6
Conclusions

These observations indicate that uncontrolled hyperglycemia significantly alters the biochemical composition and physiology of cerebral microvessels. These alterations may have subtle clinical implications. It is possible that the down-regulation of glucose transport at the BBB may protect the CNS against the ravages of hyperglycemia.

There are multiple mechanisms involved in the changes observed at the BBB. The relative importance of these mechanisms to the overall alterations in BBB function is not known. Future studies should address the unique responses of the BBB to hyperglycemia and altered metabolic milieu commonly seen in diabetes. In addition, pharmacologic interventions that can protect the BBB function independent of the degree of hyperglycemic control would help ameliorate some of the complications of diabetes.

References

1 A. D. Mooradian, J. Thurman **1999**, Glucotoxicity: potential mechanisms, *Clin. Geriatr. Med.* 15, 255–263.
2 M. Brownlee **2001**, Biochemistry and molecular cell biology of diabetic complications, *Nature* 414, 813–820.
3 A. D. Mooradian **1988**, Diabetic complications of the central nervous system, *Endocrin. Rev.* 9, 346–356.
4 A. D. Mooradian **1997**, Central nervous system complications of diabetes mellitus – a perspective from the blood-brain barrier, *Brain Res. Brain Res. Rev.* 23, 210–218.
5 A. D. Mooradian **1997**, Pathophysiology of central nervous system complications in diabetes mellitus, *Clin. Neurosci.* 4, 322–326.
6 M. H. Horani, A. D. Mooradian **2003**, Effect of diabetes on the blood brain barrier, *Curr. Pharm. Des.* 9, 833–840.
7 J. Jacobsen, P. Sidenius, H. J. G. Gunderson, R. Osterby **1987**, Quantitative changes of cerebral neocortical structure in insulin treated long term streptozocin diabetic rats, *Diabetes* 36, 597–601.
8 W. G. Mayhan **1993**, Cerebral circulation during diabetes mellitus, *Pharm. Ther.* 57, 377–391.

9 P. A. McCuskey, R. S. McCuskey **1984**, In vivo and electron microscopic study of the development of cerebral diabetic microangiopathy, *Microcirc. Endo. Lymph.* 1, 221–244.

10 G. M. Knudsen, U. Gobel, O. B. Paulson, W. Kuschinsky **1991**, Regional density of perfused capillaries and cerebral blood flow in untreated short-term and long-term streptozocin diabetes, *J. Cereb. Blood. Flow Metabol.* 11, 361–365.

11 A. D. Mooradian, R. S. McCuskey **1992**, In vivo microscopic studies of age-related changes in the structure and the reactivity of cerebral microvessels, *Mech. Ageing Dev.* 64, 247–254.

12 A. D. Mooradian **1988**, Tissue specificity of premature aging in diabetes mellitus, the role of cellular replicative capacity, *J. Am. Geriatr. Soc.* 36, 831–839.

13 J. M. Chehade, M. J. Haas, A. D. Mooradian **2002**, Diabetes-related changes in rat cerebral occludin and zonula occludens-1 (ZO-1) expression, *Neurochem. Res.* 27, 249–252.

14 D. A. Antonetti, A. J. Barker, S. Khin **1998**, Vascular permeability in experimental diabetes is associated with reduced endothelial occludin content, vascular endothelial growth factor decreases occludin in retinal endothelial cells, *Diabetes* 47, 1953–1959.

15 T. W. Gardner **1995**, Histamine, ZO-1 and increased blood-retinal barrier permeability in diabetic retinopathy, *Trans. Am. Ophthalmol. Soc.* 93, 583–621.

16 W. T. Stauber, S. H. Ong, R. S. McCuskey **1981**, Selective extravascular escape of albumin into the cerebral cortex of the diabetic rat, *Diabetes* 30, 500–503.

17 B. Oztaz, M. Kucuk **1995**, Influence of acute arterial hypertension on blood-brain barrier permeability in streptozocin-induced diabetic rats, *Neurosci. Lett.* 188, 53–56.

18 W. G. Mayhan **1990**, Effect of diabetes mellitus on the disruption of the blood brain barrier during acute hypertension, *Brain Res.* 534, 106–110.

19 S. K. Williams, J. J. deVenny, M. W. Bilensky **1981**, Micropinocytic ingestion of glycosylated albumin by isolated microvessels: possible role in pathogenesis in diabetic microangiopathy, *Proc. Natl. Acad. Sci. USA* 78, 2393–2397.

20 J. Jajobsen, L. Malmgren, Y. Olsson **1978**, Permeability of the blood-nerve barrier in streptozocin-induced diabetic rats, *Exp. Neurol.* 60, 277–285.

21 B. Oztaz, M. Kucuk **1982**, Blood-brain barrier permeability in streptozocin-induced diabetic rats, *Med. Sci. Res.* 15, 645–646.

22 M. Lorenzi, D. P. Healy, R. Hawkins, J. M. Printz, M. P. Printz **1986**, Studies on the permeability of the blood-brain barrier in experimental diabetes, *Diabetologia* 29, 58–62.

23 T. G. Christensen, N. H. Diemer, H. Laursen, A. Gjedde **1981**, Starvation accelerates blood-brain glucose transfer, *Acta. Physiol. Scand.* 112, 221–223.

24 D. A. Pelligrino, L. J. Segil, R. F. Albrecht **1990**, Brain glucose utilization and transport and cortical function in chronic vs acute hypoglycemia, *Am. J. Physiol.* 259, E729–E735.

25 S. G. Hasselbalch, G. M. Knudsen, J. Jakobsen, L. P. Hageman, S. Holm, O. B. Paulson **1995**, Blood brain barrier permeability of glucose and ketone bodies during short-term starvation in humans, *Am. J. Physiol.* 268, E1161–E1166.
26 S. A. Segel, C. G. Fanelli, C. S. Dence, J. V. Markham, D. S. Paramore, W. J. Powers, P. E. Cryer **2001**, Blood-to-brain glucose transport, cerebral glucose metabolism, and cerebral blood flow are not increased after hypoglycemia, *Diabetes* 50, 1911–1917.
27 A. L. McCall, L. B. Fixman, N. Fleming, K. Tornheim, W. Chick, N. B. Ruderman **1986**, Chronic hypoglycemia increases brain glucose transport, *Am. J. Physiol.* 251, E442–E447.
28 D. Pelligrino, B. K. Siesjo **1981**, Regulation of extra- and intracellular PH in the brain in severe hypoglycemia, *J. Cereb. Blood Flow Metab.* 1, 85–96.
29 D. Pelligrino, L. O. Almquist, B. K. Siesjo **1981**, Effect of insulin-induced hypoglycemia on intracellular PH and impedance in the cerebral cortex of the rat, *Brain Res.* 221, 129–147.
30 D. A. Pelligrino, L. J. Segil, R. F. Albrecht **1990**, Brain glucose utilization and transport and cortical function in chronic versus acute hypoglycemia, *Am. J. Physiol* 259, E729–E735.
31 D. A. Pelligrino, G. L. Becker, D. J. Mimetic, R. F. Albrecht **1989**, Cerebral mitochondria respiration in diabetic and chronically hypoglycemic rats, *Brain Res* 479, 241–246.
32 A. Gjedde, C. Crone **1981**, Blood-brain glucose transfer: repression in chronic hyperglycemia, *Science* 214, 456–457.
33 A. L. McCall, W. R. Millington, R. J. Wurtman **1982**, Metabolic fuel and amino acid transport into the brain in experimental diabetes mellitus, *Proc. Natl. Acad. Sci. USA* 79, 5406–5410.
34 W. M. Pardridge, D. Triguero, C. R. Farrell **1990**, Down-regulation of blood-brain barrier glucose transporter in experimental diabetes, *Diabetes* 39, 1040–1044.
35 A. D. Mooradian, A. M. Morin **1991**, Brain uptake of glucose in diabetes mellitus: the role of glucose transporters, *Am. J. Med. Sci.* 301, 173–177.
36 R. B. Duckrow, R. M. J. Bryan **1987**, Regional cerebral glucose utilization during hyperglycemia, *J. Neurochem.* 48, 989–993.
37 S. I. Harik, J. C. LaManna **1988**, Vascular perfusion and blood-brain glucose transport in acute and chronic hyperglycemia, *J. Neurochem.* 51, 1926–1929.
38 A. L. McCall, J. B. Gould, N. B. Ruderman **1984**, Diabetes induced alterations of glucose metabolism in rat cerebral microvessels, *Am. J. Physiol.* 247, E462–E467.
39 C. G. Fanelli, C. S. Dence, J. Markham, T. O. Videen, D. S. Paramore, P. E. Cryer, W. J. Powers **1998**, Blood to brain glucose transport and cerebral glucose metabolism are not reduced in poorly controlled type 1 diabetes, *Diabetes* 47, 1444–1450.
40 D. J. Brooks, J. S. Gibbs, P. Sharp, S. Herold, D. R. Turton, S. K. Luthra, E. M. Kohner, S. R. Bloom, T. Jones **1986**, Regional cerebral glucose transport in

74 A. D. Mooradian, J. L. Pinnas, C. C. Lung, M. D. Yahya, K. Meredith **1994**, Diabetes related changes in the protein composition of rat cerebral microvessels, *Neurochem. Res.* 19, 123–128.

75 A. D. Mooradian, G. Grabau, A. Uko-eninn **1993**, In-situ quantitative estimates of the age-related and diabetes-related changes in cerebral endothelial barrier antigen, *Brain Res.* 309, 41–44.

76 A. D. Mooradian, P. J. Scarpace **1992**, Beta adrenergic receptor activity of cerebral microvessels in experimental diabetes mellitus, *Brain Res.* 583, 155–160.

77 M. S. Magnoni, H. Kobayashi, E. Trezzi, A. Catapano, P. F. Spano, M. Trabucchi **1984**, Beta adrenergic receptors in brain microvessels of diabetic rats, *Life Sci.* 34, 1095–1100.

78 D. Bates, R. M. Weinshiltboum, R. J. Campbell, M.T. Sundt **1977**, The effect of lesions in the locus coeruleus on the physiological responses of the cerebral blood vessels in rats, *Brain Res.* 136, 431–443.

79 R. L. Grubb, M. E. Richle, J. O. Eichling **1978**, Peripheral sympathetic regulation of brain water permeability, *Brain Res.* 144, 204–207.

80 E. T. Mackenzie, J. McCulloch, M. O'Kean, J. D. Pickard, A. M. Harper **1976**, Cerebral circulation and norepinephrine: relevance of blood-brain barrier, *Am. J. Physiol.* 231, 483–488.

81 M. Raichle, B. K. Hartman, J. O. Eichling, L. G. Sharp **1975**, Central noradrenergic regulation of cerebral blood flow and vascular permeability, *Proc. Natl Acad. Sci. USA* 72, 3726–3730.

82 P. Lass, G. M. Knudsen, E. V. Pederson, D. I. Barry **1989**, Impaired beta-adrenergic mediated cerebral blood flow response in streptozocin diabetic rats, *Pharmacol. Toxicol.* 65, 318–350.

83 A. D. Mooradian, B. Bastani **1993**, Identification of proton-translocating adenosine triphosphatases in rat cerebral microvessels, *Brain Res.* 629, 128–132.

84 A. D. Mooradian, B. Bastani **1997**, The effect of metabolic acidosis and alkalosis on the H^+ ATPase of rat cerebral microvessels, *Life Sci.* 61, 2247–2253.

85 W. G. Mayhan, S. P. Didion **1996**, Glutamate-induced disruption of the blood-brain barrier in rats: role of nitric oxide, *Stroke* 27, 965–969.

86 J. Zheng, H. Benveniste, B. Klitzman, C. A. Piantadosi **1995**, Nitric oxide synthase inhibition and extracellular glutamate concentration after cerebral ischemia/reperfusion, *Stroke* 26, 298–304.

87 O. Durieu-Trautmann, C. Federeci, C. Cerminon, N. Foignanat-Chaverot, F. Roux, M. Claire, A. D. Strosberg, P. O. Couraud **1993**, Nitric oxide and endothelin secretion by brain microvessel endothelial cells: regulation by cyclic nucleotides, *J. Cell Physiol.* 155, 104–111.

88 N. Shyamaladevi, A. R. Jayakumar, R. Sujatha, V. Paul, E. H. Subramanian **2002**, Evidence that nitric oxide production increases gamma-amino butyric acid permeability of blood-brain barrier, *Brain Res. Bull.* 57, 231–236.

89 D. B. Stanimirovic, N. Bertrand, R. McCarron, S. Uematsu, M. Spatz **1994**, Arachidonic acid release and permeability changes induced by endothelins in human cerebromicrovascular endothelium, *Acta Neurochir. Suppl.* 60, 71–75.

90 N. Kawai, R. M. McCarron, M. Spatz **1997**, The effect of endothelins on ion transport systems in cultured rat brain capillary endothelial cells, *Acta Neurochir. Suppl.* 70, 138–140.

91 N. Kawai, R. M. McCarron, M. Spatz **1995**, Endothelins stimulate sodium uptake into rat brain capillary endothelial cells through endothelin A-like receptors, *Neurosci. Lett.* 190, 85–88.

92 R. E. Catalan, A. M. Martinez, M. D. Aragones, F. Hernandez, G. Diaz **1996**, Endothelin stimulates protein phosphorylation in blood-brain barrier, *Biochem. Biophys. Res. Commun.* 219, 366–369.

93 I. A. Krizbai, M. A. Deli **2003**, Signalling pathways regulating the tight junction permeability in the blood-brain barrier, *Cell. Mol. Biol.* 49, 23–31.

94 W. G. Mayhan **2001**, Regulation of blood-brain barrier permeability. *Microcirculation* 8, 89–104.

95 M. Kapural, L. J. Krizanac-Bengez, G. Barnett, J. Perl, T. Masaryk, D. Apollo, P. Rasmussen, M. R. Mayberg, D. Janigro **2002**, Serum S-100 beta as a possible marker of blood-brain barrier disruption, *Brain Res.* 940, 102–104.

96 N. Marchi, P. Rasmussen, M. Kapural, V. Fazio, K. Kight, M. R. Mayberg, A. Kanner, B. Ayumar, B. Albensi, M. Cavaglia, D. Janigro **2003**, Peripheral markers of brain damage and blood-brain barrier dysfunction, *Restor. Neurol. Neurosci.* 21, 109–121.

97 P. Mecocci, L. Parnetti, G. Romano, A. Scarelli, F. Chionne, R. Cecchetti, M. C. Polidori, B. Palumbo, A. Cherubini, U. Senin **1995**, Serum anti-GFAP and anti-S100 autoantibodies in brain aging, Alzheimer's disease and vascular dementia, *J. Neuroimmunol.* 57, 65–170.

98 M. W. Strachan, H. D. Abraha, R. A. Sherwood, G. A. Lammie, I. J. Deary, F. M. Ewing, P. Perros, B. M. Frier **1999**, Evaluation of serum markers of neuronal damage following severe hypoglycaemia in adults with insulin-treated diabetes mellitus, *Diabetes Metab. Res. Rev.* 15, 5–12.

99 E. A. McIntyre, H. D. Abraha, P. Perros, R. A. Sherwood **2000**, Serum S-100beta protein is a potential biochemical marker for cerebral edema complicating severe diabetic ketoacidosis, *Diabetic Med.* 17, 807–809.

100 M. R. Hovsepyan, M. J. Haas, A. S. Boyajyan, A. A. Guevorkyan, A. A. Mamikonyan, S. E. Myers, A. D. Mooradian **2004**, Astrocytic and neuronal biochemical markers in the sera of subjects with diabetes mellitus, *Neurosci. Lett.* 369, 224–227.

101 G. J. Biessels, A. C. Kappelle, B. Bravenboer, D. W. Erkelens, W. H. Gispen **1994**, Cerebral function in diabetes mellitus, *Diabetologia* 37, 643–650.

102 C. M. Ryan **1988**, Neurobehavioral complication of type 1 diabetes. Examination of possible risk factors, *Diabetes Care* 11, 86–93.

103 R. Stewart, D. Liolitsa **1999**, Type 2 diabetes mellitus, cognitive impairment and dementia, *Diabetic Med.* 16, 93–112.

104 M. W. Starchan, I. J. Deary, F. M. Ewing, B. M. Frier **1997**, Is type 2 diabetes associated with an increased risk of cognitive dysfunction? A critical review of published studies, *Diabetes Care* 20, 438–445.

105 A. D. Mooradian, K. Perryman, J. Fitten, G. Kavonian, J. E. Morley **1988**, Cortical function in elderly non-insulin dependent diabetic patients: behavioral and electrophysiological studies, *Arch. Int. Med.* 148, 2369–2372.

106 D. Galasko **2003**, Insulin and Alzheimer's disease: an amyloid connection, *Neurology* 60, 1886–1887.

107 G. N. Shah, A. D. Mooradian **1997**, Age-related changes in the blood-brain barrier, *Exp. Gerontol.* 32, 501–519.

108 M. O. Chaney, J. Baudry, C. Esh, J. Childress, D. C. Luehrs, T. A. Kokjohn, A. E. Roher **2003**, Diabetes, aging, and Alzheimer's disease: a tale, models, and hypotheses, *Neurol. Res.* 25, 581–589.

109 A. D. Mooradian, H. C. Chung, G. N. Shah **1997**, GLUT-1 expression in the cerebra of patients with Alzheimer's disease, *Neurobiol. Aging* 18, 469–474.

110 J. F. Flood, A. D. Mooradian, J. E. Morley **1990**, Characteristics of learning and memory in streptozotocin-induced diabetic mice, *Diabetes* 39, 1391–1398.

111 G. J. Biessels, A. Kamal, I. J. Urban, B. M. Spruijt, D. W. Erkelens, W. H. Gispen **1998**, Water maze learning and hippocampal synaptic plasticity in streptozocin-diabetic rats: effects of insulin treatment, *Brain Res.* 800, 125–135.

112 J. A. Gavard, P. J. Lustman, R. E. Clouse **1993**, Prevalence of depression in adults with diabetes, *Diabetes Care* 16, 1167–1178.

113 P. J. Lustman, L. S. Griffith, R. E. Clouse **1988**, Depression in adults with diabetes: results of a 5 year follow-up study, *Diabetes Care* 11, 605–612.

114 P. J. Lustman, L. S. Griffith, R. E. Clouse, K. E. Freedland, S. A. Eisen, E. H. Rubin, R. M. Carney, J. B. McGill **1997**, Effects of nortriptyline on depression and glycemic control in diabetes: results of a double-blind, placebo-controlled trial, *Psychosom. Med.* 59, 241–250.

115 P. J. Lustman, L. S. Griffith, K. E. Freedland, R. E. Clouse **1997**, The course of major depression in diabetes, *Gen. Hosp. Psychiatry* 19, 138–143.

116 M. Kovacs, D. S. Obrosky, D. Goldston, A. Drash **1997**, Major depressive disorder in youths with IDDM: a controlled prospective study of course and outcome, *Diabetes Care* 20, 45–51.

117 S. Kalmijn, E. J. Feskens, L. J. Launer, T. Stijnen, D. Kromhout **1995**, Glucose tolerance, hyperinsulinemia and cognitive function in a general population of elderly men, *Diabetologia* 38, 1096–1102.

118 P. J. Lustman, L. S. Griffith, R. E. Clouse, P. E. Cryer **1986**, Psychiatric illness in diabetes mellitus: relation to symptoms and glucose control, *J. Nerv. Ment. Dis.* 174, 736–746.

119 M. Peyrot, R. R. Rubin **1999**, Persistence of depressive symptoms in diabetic adults, *Diabetes Care* 22, 448–452.

120 S. J. Niemczyk, M. A. Speers, L. B. Travis, H. E. Gary **1990**, Psychosocial correlates of hemoglobin A 1c in young adults with type I diabetes, *J. Psychosom. Res.* 34, 617–627.

121 A. M. Jacobson, A. G. Adler, J. I. Wolfsdorf, B. Anderson, L. Derby **1990**, Psychological characteristics of adults with IDDM: comparison of patients in poor and good glycemic control, *Diabetes Care* 13, 375–381.

122 U. Rajala, S. Keinänen-Kiukaanniemi, S.-L. Kivelä **1997**, Non-insulin dependent diabetes mellitus and depression in a middle-aged Finnish population, *Soc. Psychiatry Psychiatr. Epidemiol.* 32, 363–367.

123 M. Peyrot, R. R. Rubin **1997**, Levels and risks of depression and anxiety symptomatology among diabetic adults, *Diabetes Care* 20, 585–589.

124 T. Bergenheim, B. Borssen, F. Lithner **1992**, Sensory thresholds for vibration, perception and pain in diabetic patients aged 15–50 years, *Diabetes Res. Clin. Pract.* 16, 47–52.

125 V. Umachandran, K. Ranjadayalan, G. Ambepityia, B. Marchant, P. G. Kopelman, A. D. Timmis **1991**, The perception of angina in diabetes: relation to somatic pain threshold and autonomic function, *Am. Heart J.* 121, 1649–1654.

126 X. Chen, J. D. Levine **2001**, Hyper-responsivity in a subset of C-fiber nociceptors in a model of painful diabetic neuropathy in the rats, *Neuroscience* 102, 185–192.

127 G. K. Morley, A. D. Mooradian, A. S. Levine, J. E. Morley **1984**, Why is diabetic peripheral neuropathy painful? The effect of glucose on pain perception in humans, *Am. J. Med.* 77, 79–82.

128 W. A. Banks **2003**, Is obesity a disease of the blood-brain barrier? Physiological, pathological, and evolutionary considerations, *Curr. Pharm. Des.* 9, 801–809.

129 P. E. Cryer, S. N. Davis, H. Shamoon **2003**, Hypoglycemia in diabetes, *Diabetes Care* 26, 1902–1912.

27
Human Parasitic Disease in the Context of the Blood-Brain Barrier – Effects, Interactions, and Transgressions

Mahalia S. Desruisseaux, Louis M. Weiss, Herbert B. Tanowitz, Adam Mott, and Danny A. Milner

27.1
Introduction *(by D. Milner)*

Human parasitic diseases currently comprise the largest group of infections in humans worldwide including the nematodes ("roundworms") at $\sim 3.7 \times 10^{12}$, trematodes ("flatworms" or "flukes") at $>200 \times 10^6$, cestodes ("tapeworms") at $>50 \times 10^6$, the protozoa at $>500 \times 10^6$, and amoeba at $>50 \times 10^6$. Although the population at risk for certain diseases such as malaria has decreased over the past 40 years, the number of individuals infected has increased, as has the mortality rate. With the exception of dracunculiasis (i.e. guinea worm), the incidence and prevalence of the vast majority parasitic diseases has steadily grown with the population, although accurate estimates of these diseases are difficult at best. The growing problem of human parasitic diseases is largely one of socioeconomic inadequacies, compounded by specific environmental and political situations. The evolution of drug resistance complicates the treatment of malaria (i.e. *Plasmodium falciparum* malaria, rapidly evolving resistance to a regiment of drugs), but the bulk of parasites (e.g. the intestinal nematodes, the filaria worms, the trematodes) remain susceptible to existing treatments, but require recurrent treatment and aggressive management plans if eradication is desired. When programs of prevention and elimination have been successful, the incidence and prevalence of some extremely common diseases have dropped drastically (e.g. dracunculiasis). Conversely, parasitic diseases have emerged as a result of the rising population of immunocompromised patients; and these carry higher mortality due to the lack of knowledge and drug treatments available (e.g. disseminated amebiasis).

Parasitic diseases have, almost by definition, evolved to survive in harmony with their hosts. However, some degree of host morbidity is often a consequence of this coexistence and is a leading cause of massive economic loss worldwide (e.g. chronic anemia in China from hookworms, blindness in Africa from river blindness). Host mortality is a negative result for the parasite and, thus, stems from inappropriate or abnormal perturbations of the parasite's life cycle. In some cases, humans are not the intended (or usual biological) hosts of infecting parasites, but rather an inappropriate or inadequate host, and this can lead to severe consequences (e.g. African trypanosomiasis, cysticercosis). In most cases, however, the human-parasite relationship appears to be perturbed (e.g. immunocompromise, human and parasite genetic influences, other infections), resulting in fatal outcomes (e.g. toxoplasmosis, severe malaria). Central nervous system involvement by parasitic diseases, especially infections which affect, disrupt, and/or destroy the blood brain barrier are more commonly fatal than parasitic infections which do not involve the central nervous system. Overall, death of the host in parasitic diseases is a relatively rare event when compared to the pool of infected individuals. Tables 27.1 and 27.2 summarize the majority of parasites interacting with the blood-brain barrier either as part of their normal life cycles or as part of human disease.

The life cycles of human parasitic diseases deserve ample attention in order to clarify why disease of the central nervous system (CNS) and breakdown of the blood-brain barrier (BBB) occur, given the prevailing concept of ultimate parasite survival. As the integrity of the CNS and the BBB are important for host survival, it seems unlikely that a disease process impairing either of these would serve an evolutionary advantage for the parasite. Figure 27.1 is a diagrammatic representation of life cycles for the majority of human parasites. Many parasitic organisms are transmitted to humans by the bites or fecal exposure near a bite from mosquitos, biting flies, and other arthropods such as *Ixodes*

Table 27.1 Organisms affecting the blood-brain barrier.

Organism type	Genus and species
Amoeba	*Acanthamoeba* spp
	Balamuthia mandrillaris
	Entamoeba histolytica
	Naegleria fowleri
Cestode	*Taenia solium*
Nematode	*Angiostrongylus cantonensis*
	Onchocerca volvulus
	Strongyloides stercoralis
	Toxocara canis
	Trichinella spp
Protozoa	*Leishmania* spp
	Plasmodium falciparum
	Toxoplasma gondii
	Trypanosoma brucei gambiense, T.b. rhodesiense
	Trypanosoma cruzi
Trematode	*Schistosoma mansoni, S. haematobium, S. japonicum*

Table 27.2 Parasitic diseases known to affect the blood-brain barrier (all data 2005, www.cdc.gov).

Organism	Disease	Prevalence/incidence/mortality	Clinical outcome
Acanthamoeba spp (*A. culbertsoni, A. polyphaga, A. castellanii, A. healyi, A. astronyxis, A. hatchetti, A. rhysodes*)	Granulomatous amoebic encephalitis (and/or disseminated disease); "contact" conjunctivitis	Uncommon but worldwide	"Contact" conjunctivitis (from contaminated contact lenses) usually treatable; disseminated disease (especially in immunocompromised) almost universally fatal
Angiostrongylus cantonensis	Eosinophilic meningitis	Rare (majority in Southeast Asia and Pacific Basin)	Mostly a product of the host response to the parasite but larvae may be found in the brain; rarely fatal and usually does not require treatment unless CNS symptoms are severe
Balamuthia mandrillaris	Granulomatous amoebic encephalitis	100 cases reported worldwide	Spectrum includes granulomatous amoebic encephalitis; nasopharyngeal, cutaneous, and disseminated infections
Entamoeba histolytica	Amoebiasis, disseminated amoebiasis	Luminal (asymptomatic/symptomatic) is common worldwide; extraluminal (especially CNS involvement) is rare	Spectrum includes asymptomatic carriage, severe intraluminal intestinal disease, severe extraluminal disease with abscess of the liver and CNS

Table 27.2 (continued)

Organism	Disease	Prevalence/incidence/mortality	Clinical outcome
Leishmania spp: *L. donovani* complex (*L. donovani, L. infantum, L. chagasi*); *L. mexicana* complex (*L. mexicana, L. amazonensis, L. venezuelensis*); *L. tropica*; *L. major*; *L. aethiopica*; and subgenus *Viannia* [*L. (V.) braziliensis, L. (V.) guyanensis, L. (V.) panamensis*, and *L. (V.) peruviana*]	Leishmaniasis (Kala-azar)	350 million at risk worldwide with 1.5 million cases per year of cutaneous leishmaniasis and 500 000 cases per year of visceral leishmaniasis	Manifestations of leishmaniasis are dependent on the species. Cutaneous leishmaniasis causes sores on the skin where sandflies have fed with localized lymphadenopathy and often spontaneous resolution. Visceral leishmaniasis results in fever, weight loss, and hepatosplenomegaly. Rarely, leishmania can cross the blood brain barrier
Naegleria fowleri	Primary amoebic meningo-encephalitis	<30 cases in United States, but worldwide	Due to brain destruction, rapid progression with degenerating clinical signs; almost universally fatal
Onchocerca volvulus	Onchocerciasis (river blindness)	17.7 million infections (99% in Africa), 270,000 blind, 500,000 with visual impairment	Patients may first develop skin rash and subcutaneous nodules (onchocercomas) which contain adult female worms; eye lesions develop when microfilaria from gravid female begin circulating in the bloodstream and demonstrate a tropism for ocular tissues; disease is almost never fatal
Plasmodium falciparum	Cerebral malaria and severe malarial anemia (severe malaria)	2.5 billion at risk, annually 300–500 million infections, 1.5–2.0 million deaths	Mortality is 20–50% for severe malaria (approximately 1.5–2.5 million deaths yearly)

Table 27.2 (continued)

Organism	Disease	Prevalence/incidence/mortality	Clinical outcome
Schistosoma mansoni, S. haematobium, and *S. japonicum*	Schistosomiasis (Bilharzia), neuroschistosomiasis	200 million infected worldwide	Typical infections involves migration of adult worms to species specific human venous regions with production of eggs which migrate through tissues locally; rarely, eggs migrate to the brain and spinal cord resulting in neuroschistosomiasis
Strongyloides stercoralis	Strongyloidiasis, hyperinfection	70 million worldwide; severe disease (hyperinfection), although very rare, has mortality up to 80%	Acute/chronic strongyloidiasis manifests with cutaneous, pulmonary or gastrointestinal symptoms; severe disease (immunocompromised) leads to dissemination including the CNS
Taenia solium	Taeniasis, cysticercosis, neurocysticercosis	Estimated to 50 million worldwide; prevalence higher in regions consuming pork; approximately 1–2% of cases develop neurocysticercosis	Cysticercosis: dissemination of larval cysts throughout the body, usually asymptomatic (80%) unless trauma, involution, or mass effect brings them to clinical attention. Neurocysticercosis: cyst involving the central nervous system which are usually asymptomatic unless trauma, involution, or progression to mass effect brings them to clinical attention usually with seizures, hydrocephalus, and disability

Table 27.2 (continued)

Organism	Disease	Prevalence/incidence/mortality	Clinical outcome
Toxocara canis	Ocular larva migrans and visceral larva migrans	Worldwide; in United States, 10,000 cases occur yearly with 700 individuals developing eye damage	OLM occurs when a microscopic worm enters the eye; it may cause inflammation and formation of a scar on the retina. Symptoms of VLM, which are caused by the movement of the worms through the body, include fever, coughing, asthma, or pneumonia.
Toxoplasma gondii	Acute localized lymphadenopathy, congenital toxoplasmosis, cerebral toxoplasmosis	60 million (US data) but worldwide distribution	Immunocompetent: rarely "flu-like" symptoms with swollen lymph nodes. Immunocompromised: acute or reactivation causes serious (often fatal) disease of the CNS. Congenital: most asymptomatic at birth with CNS problems developing later, small per cent with serious damage at birth
Trichinella spp (*T. spiralis, T. pseudospiralis, T. nativa, T. nelsoni, T. britovi*)	Trichinellosis, trichonosis	Worldwide; prevalence decreasing due to meat regulations; less than 15 cases yearly in United States	Infections may be asymptomatic but intestinal invasion may have diarrhea, abdominal pain, and vomiting. Muscular larval migration causes periorbital and facial edema, conjunctivitis, fever, myalgias, rashes, and eosinophilia. Fatal manifestations may include myocarditis, central nervous system involvement, and pneumonitis

Table 27.2 (continued)

Organism	Disease	Prevalence/incidence/mortality	Clinical outcome
Trypanosoma brucei gambiense and Trypanosoma brucei rhodesiense	West African [*T. (b) gambiense*] and East African trypanosomiasis [*T. (b) rhodesiense*], "African sleeping sickness"	40 000–100 000 new cases yearly in Africa	Universally fatal without medical treatment
Trypanosoma cruzi	American trypanosomiasis (Chaga's disease)	16–18 million infected worldwide (majority South America) with 50 000 deaths yearly	Acute (1%) can result in death with CNS involvement; chronic (20–30%) results in systemic muscular damage (e.g. cardiac, esophagus, bowel)

ticks (Fig. 27.1 A). Examples include the human malarias (*Plasmodium falciparum, P. vivax, P. ovale, P. malariae*) and the nematodes responsible for the human filarial infections. The overwhelming majority of these infections do not result in a fatal outcome (e.g. less than 1% for malaria), though disabling morbidity is common. A recurring theme of these infections is that no known animal reservoir exists – humans are the only host. Clinically significant disease, therefore, manifests either rarely (e.g. in *P. falciparum* malaria probably due to a combination of factors including human genetics, parasite burden, immune status, comorbidities) or with only great morbidity but limited mortality (e.g. lymphatic filariasis when the worm burden and inflammatory component reach a threshold). This is in stark contrast to diseases carrying large animal reservoirs (Fig. 27.1 B) such as African trypanosomiasis, in which case infections in humans are fatal if untreated, while cattle may remain asymptomatic. Other parasites with a very similar life cycle to *Trypanosoma* and *Plasmodia*, namely American trypanosomiasis and *Babesia* spp, have rare or little clinical consequence for the CNS. *Toxoplasma gondii* and *Echinococcus* spp are excellent examples of a life cycle where humans are incidental "end hosts" which cannot transmit the infection (Fig. 27.1 C). In toxoplasmosis, disease occurs either acutely in the infection (congenital toxoplasmosis or focal lymphadenitis) or as a result of reactivation during a period of immunosuppression (cardiac or neural toxoplasmosis). Echinococcosis results from human incidental exposure to canine or other carnivorous feces, with development of intrahost infecting cysts, which can rarely invade the central nervous system. Other less problematic diseases of a similar pattern include North American *Brugia, Anisakis* spp ingestion, sparganosis, *Toxocara canis* infection, and *Dirofilaria immitis* infection. The trematodes such as *Schistosoma* spp demonstrate yet another life cycle pattern (Fig. 27.1 D) in which humans become infected with a free-swimming stage (i.e. cercariae), harbor the worms to adulthood, and release eggs into the environment. At that point, secondary hosts (e.g. snails) are infected by the miracidia from eggs and ultimately release the sporocysts, which become the cercariae. A vulnerable point in all of these life cycles is the water source; and in each case, water purification (as was used for the nematode *Dracunculus medinensis* globally) and/or snail extermination (as was implemented in China under Mao-Tse Tong for the *Schistosoma* spp) could eradicate transmission.

The amoebae and geohelminths (i.e. ascariasis, hookworms, whipworms, *Strongyloides stercoralis*) are examples of a continuous life cycle (Fig. 27.1 E), in which the parasitic organisms are either somewhat or completely free-living and therefore can continually reinfect humans. The vast majority of these organisms, with the exception of *Entamoeba histolytica*, do so with "innocent" colonization of the human host (i.e. remaining within the gastrointestinal and/or respiratory tract, without wandering). The other clinically significant amebic (discussed below) and geohelminth infections (e.g. the hyperinfection of *S. stercoralis*) are almost always the result of immunosuppression.

Lastly, there are those parasites (e.g. cysticercosis with *Taenia solium*) which are usually in cyclic harmony with the host and reservoir but can enter the

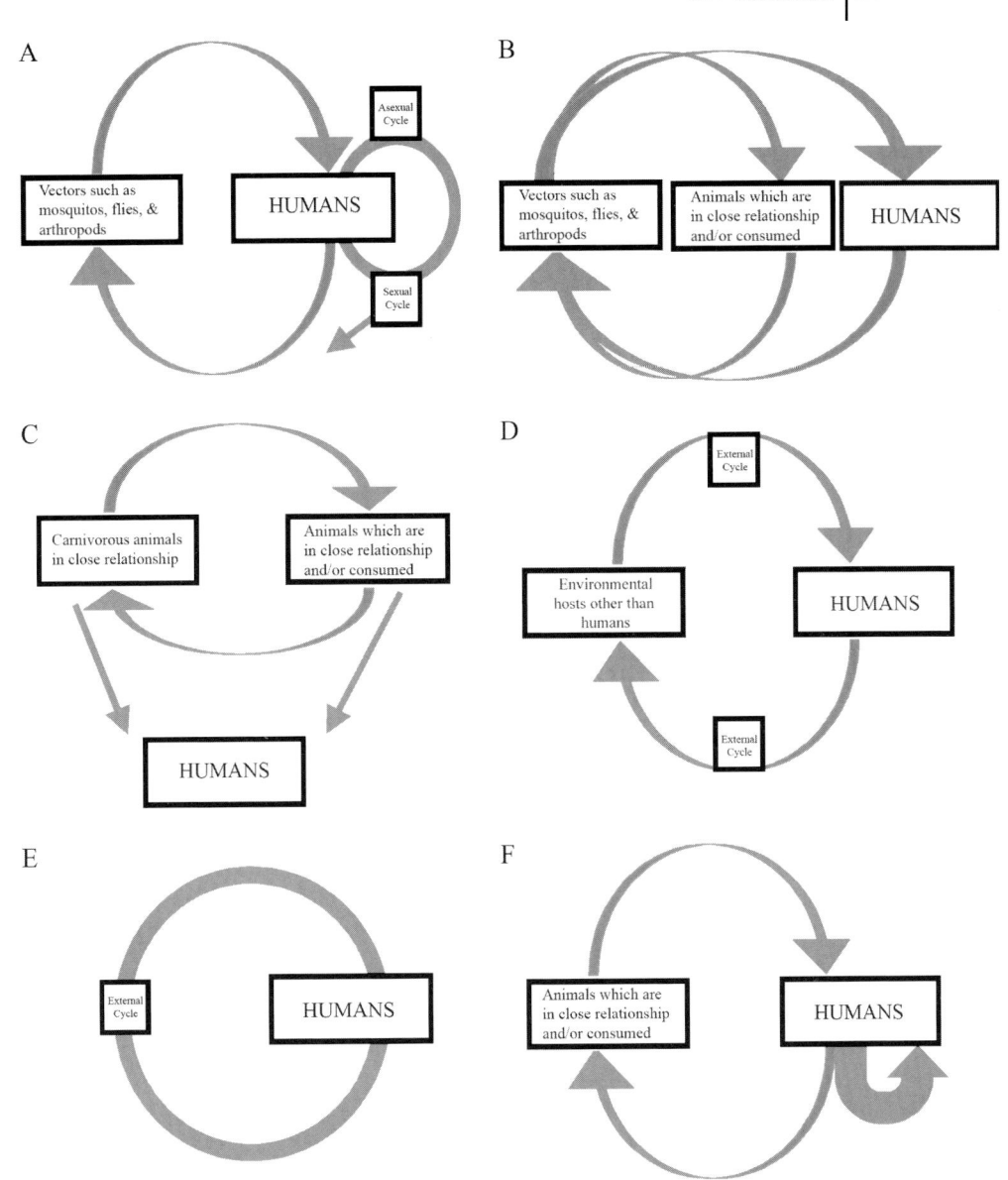

Fig. 27.1 Sets of diagrams are depicted, representing the general patterns of life cycles seen in parasites, which can interact with the blood-brain barrier. See text for details.

cycle abnormally (e.g. consumption of fecal material of pigs or humans rather than consumption of infected uncooked pig meat), resulting in dissemination (Fig. 27.1 F).

The spectrum of parasite interactions with the BBB is well characterized for several organisms that will be explored in more detail below. Little is known about the majority of the remaining parasite-human interactions, but a review of the anatomic pathology findings as well as the indirect evidence of involvement are helpful in understanding the resulting human diseases that are seen clinically.

The anatomic pathology of the CNS as observed under light microscopy is also a useful tool for understanding the degree to which the BBB is involved by these parasitic infections and why treatment at such stages is often futile. Figure 27.2 depicts several examples of different types of pathology encountered. *Plasmodium falciparum* with its unique ability to sequester in human tissue can be found to "stuff" and distend cerebral vessels in patients dying of cerebral malaria. The majority of the life cycle from early trophozoites to late schizonts (Fig. 27.2 A) occurs in close proximity to human endothelium as part of every human infection (see Section 27.2 "Malaria", below). The rate of clinically evident cerebral involvement, however, remains very rare, provoking one of the most frequently argued/long-standing controversies in the malaria community.

Both African and American trypanosomiasis can been visualized by blood smear for clinical diagnosis of the infections (Fig. 27.2 B). The central nervous system pathology of the lesions are quite distinct, with African trypanosomiasis causing meningeal and perivascular mononuclear infiltration, reactive gliosis, and the classic morula cells, while American trypanosomiasis demonstrates microglial nodules and occasionally amastigotes within glial cells.

Toxoplasma gondii remains quiescent in human tissues after primary infection, which can include the acute syndromes of either localized lymphadenitis in nonpregnant humans or the catastrophic infection of gestating infants seen in mothers exposed to the organism for the first time in pregnancy. When patients become immunosuppressed, particularly with falling T-cell counts, the bradyzoites (*Gr.* slow animals) resting within cysts reactivate to tachyzoites (*Gr.* fast animals), resulting in minimal inflammation, tissue necrosis, and neurologic sequelae (Fig. 27.2 C). Treatment, therefore, is directed at prevention by monitoring the serology/exposures of expectant mothers and prophylactic drug use in the immunosuppressed.

The free-living amoebae generally cause massive tissue necrosis, resulting in large abscess cavities, as with *Entamoeba histolytica* (seen in Fig. 27.2 D with an ingested red blood cell), or diffuse granulomatous inflammation of the cerebral cortex, as with *Acanthamoeba* spp (seen in Fig. 27.2 E at the border of the granular and molecular layers of the cerebellum). Presentation with most amoebic infections is late, hence the high clinical failure rate in the immunosuppressed.

Clinical disease and the resulting pathology of trematode infections depend on the venous bed to which the parasite couple localizes. The trematodes do not usually directly invade the CNS, but occasionally the eggs metastasize to the spinal cord and rarely to the cerebral cortex. *Schistosoma haematobium* male and female adult worms can be seen coupling within a bladder vessel, while their eggs (higher magnification inset) move towards mucosal surfaces, to return to the environment inducing a surrounding inflammation (Fig. 27.2 F). Traditionally, drugs targeting these parasites only kill the mature adult worms and thus must be taken within a precise window after known exposure to present egg-related disease. Recent data from several clinical trials demonstrates effective preventative treatment shortly after exposure for all three species with subsequent reduction in disease burden and clinical symptoms [127–129].

Cysticercosis results when humans ingest the proglottids and/or eggs of *Taenia solium* rather than the infected meat of a pig. In this situation, the human becomes akin to the pig, and human muscle can be encysted with metacestode larvae (Fig. 27.2 G). These viable larval cysts are almost always clinically silent unless trauma or mass effect reveals them. Conversely, when the larva dies, the immune system responds dramatically and the resultant inflammation (Fig. 27.2 H) can be a focal point of neurologic symptoms, most commonly seizures. As these signs are of dead worms, the role for parasite treatment is minimal while symptomatic treatment is paramount.

The majority of filarial diseases manifest due to the adult worm burden in the host. River blindness, however, results from relatively few adult worms of *Onchocerca volvulus*, often embedded within a subcutaneous nodule and filled with microfilaria (Fig. 27.2 I) releasing massive amounts of progeny into the bloodstream and lymphatics. These microfilariae migrate into the eye tissues through the BBB and die. Treatment with ivermectin only suppresses the microfilarial (or larval) stages to prevent blindness while subcutaneous nodulectomy offers a definitive but disfiguring cure.

Indirect evidence of CNS invasion by demonstration of the production of IgG within the cerebrospinal fluid has been seen in cysticercosis, schistosomiasis, onchocerciasis, and granulomatous amoebic encephalitis [1–9]. These findings suggest parasite antigen exposure to the responding human immune system within the CNS (and, thus, organisms having crossed the BBB). Perhaps one finds this a logical conclusion, as the pathology of these diseases (shown above) can involve large and often destructive central nervous system lesions. However, cerebrospinal fluid markers of disease prove very helpful in making definitive diagnoses, especially in locations where imaging is not available and coinfection is expected. Additional clinical markers, which are directly related to violation of the BBB by the invading parasite, include tissue-specific cytokines and neuron-specific enolase, as demonstrated by *Taenia solium* infection [10–13]. Although these are not specific, within the correct clinical and radiographic context, they prove extremely useful for diagnostic purposes. Other parasitic organisms may venture to the central nervous system without harm to the host. For example, in experimental mouse models, *Leishmania* spp were demonstrated to cross the BBB. Transgression by the same organism has been definitely reported in human patients [14, 15]. These findings are of interest because the human clinical disease of leishmaniasis does not include clinical manifestations of CNS involvement. Other disease conditions, such as the eosinophilic meningitis of *An-*

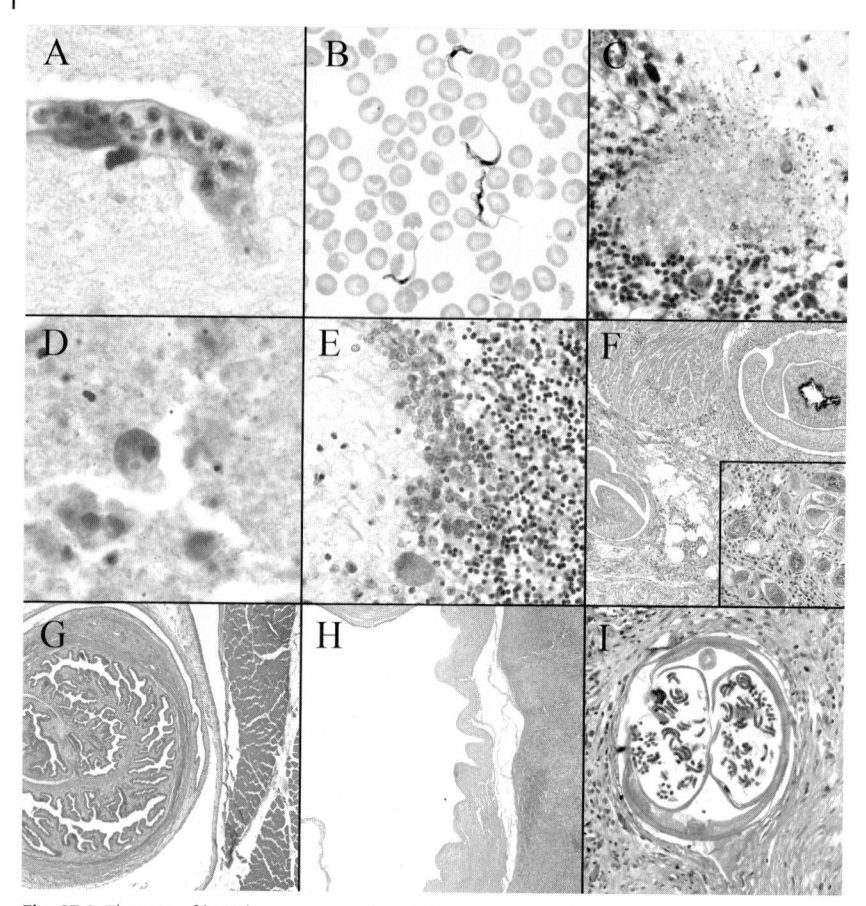

Fig. 27.2 This set of histologic images depicts the patterns of injury seen in the central nervous system and elsewhere resulting from parasitic infections in humans. These images are hematoxylin and eosin stained sections taken under the following relative magnifications: (A) 1000× in frontal lobe, (C) 1000× in cerebellum, (D) 1000× in cerebral cortex, (E) 400× in cerebellum, (F) 20× in bladder wall (inset, 400× in bladder submucosa), (G) 20× in skeletal muscle, (H) 20× in cerebral cortex, (I) 200× in subcutaneous nodule. Section (B) is 1000× on a Giemsa stained blood film. All images supplied by D. Milner, Boston.

giostrongylus cantonensis infection or the response of heat shock proteins in *Trichinella spiralis* infections are immune and/or systemic responses to the organism without its entrance into the CNS [16, 17]. Though these responses may ultimately play no role in the clinical outcome, the neurobiology of this interaction remains fascinating within the context of the BBB.

Free-living amoebae represent an interesting subgroup of human parasites because of their tropism for the human CNS. In the case of *Entamoeba histolytica*, one of the only pathologic *Entamoeba* spp found in the gastrointestinal tract of humans, dissemination is rare to various organs of the body but leads to large necrotic abscess formation, especially in the liver and lungs. Occasionally, *E. histolytica* will form a cerebral abscess presenting with headache and nonspecific features of encephalitis, meningoencephalitis, or mass lesions. The mechanism by which *E. histolytica* traverses the BBB is unknown, but the resulting infection is usually fatal. *Naegleria fowleri* and *Acanthamoeba castellanii* cause similar diseases resulting in granulomatous encephalitis. *A. castellanii* usually disseminates to skin and other organs concomitantly with CNS infection. Cytokine priming experiments have revealed that release of endogenous tumor necrosis factor (TNF-α), interleukin-6 (IL-6), and IL-1β are important in defense against *A. castellanii* infections in mice [18]. The T1 genotype of *A. castellanii* in an experimental model induces programmed cell death (i.e. apoptosis) in human endothelial cells by specifically activating phosphatidylinositol 3-kinase [19]. The T3, T4, and T11 genotypes bind with higher affinity to human endothelial cells in culture and demonstrate increased cytotoxicity, while the effects of T2 and T7 genotypes on the same human endothelial cell system are significantly less [20]. As *A. castellanii* represents but one species of the *Acanthamoeba* spp capable of dissemination within humans, further studies are needed to determine the significance and importance of genotypes and their tropisms. *Balamuthia mandrillaris*, a causative agent of granulomatous amoebic meningoencephalitis, has been successfully cultured on human endothelial cells from the cerebrospinal fluid of an infected patient and demonstrated cytopathic effects, suggesting important interactions between the amoeba and the endothelial barrier during infection [21]. Phosphatidylinositol 3-kinase is also an important mediator of IL-6 release from human endothelial cells exposed to *B. mandrillaris*, suggesting again that the human response to free-living amoebae is normally preventative of infection [22]. For this reason, immunocompromised patients almost exclusively suffer from disseminated disease. However, the mystery of the CNS tropism of free-living amoebae remains.

27.2
Malaria: The *Plasmodium berghei* Mouse Model and the Severe Falciparum Malaria in Man *(by M.S. Desruisseaux and D. Milner)*

Plasmodium spp are parasitic organisms across the animal kingdom, with recent evolutionary studies suggesting that human parasites arose from lateral transfer events from animals or birds to humans. The four human malarias, therefore, are less related to each other than they are to other animal and bird malarias. It is, therefore, not so surprising that *P. falciparum* infections in humans are very different from the other three species.

The life cycle (see Fig. 27.1 A) of these four species involves a mosquito vector transferring infective sporozoites to humans during a blood meal. This is followed by an obligatory liver stage to form the red blood cell-infecting merozoites. Upon release of merozoites from the liver, these parasites begin an asexual reproduction stage with continual production of new merozoites which reinfect red blood cells.

investigative efforts, the exact mechanisms of these processes have never been clearly elucidated. Several authors have tried to clarify the progression of neuropathology that occurs in African trypanosomiasis.

27.3.1
CNS Pathology in African Trypanosomiasis

Early parasitic infiltration of the brain in animal models occurs in areas where there is paucity in the BBB development. Early damage is thought to occur in the choroid plexus and the periventricular regions, with parasites penetrating these tissues via some unclear mechanism and spilling into the cerebral spinal fluid [39, 40, 46, 47]. This results in chronic meningitis whereby inflammatory cells and Mott cells, plasma cells containing immunoglobulin, are able to permeate throughout the meninges [38, 39, 45, 46]. Access to the perivascular space is then afforded to the parasite and the inflammatory cells by direct extension, leading to the formation of perivascular cuffing [39, 45, 51].

One of the hallmarks of sleeping sickness is autoantibody production. Antibodies against erythrocytes, muscle cells, and single-stranded DNA, as well as against neurons, gangliosides, and galactocerebrosides have been observed [39, 46]. There are several implications to this, including demyelination observed in late stages of the disease, tissue damage and hemolytic anemia among others. The autoantibody production might be important in the autonomic dysfunction as well as the myocarditis observed in the disease [45].

27.3.2
Cytokines and Endothelial Cell Activation

In the presence of inflammatory stimuli activated endothelial cells express cellular adhesion molecules which allow sequestration and transmigration of leukocytes across the endothelium. *In vitro* studies have demonstrated that trypanosome extracts, as well as the trypanosomal variable surface glycoproteins (VSGs) are able to significantly induce a transient increase in the expression of ICAM-1, VCAM-1, and E-selectin on the surface of human bone marrow endothelial cells, comparable to levels obtained upon stimulation with TNF-α [52]. These molecules are able to mediate changes in the extracellular matrix and the cytoskeleton of the endothelium, which may lead to increased permeability [52]. *T. brucei* mediates the activation of endothelial cells by inducing an increase in cytokines such as TNF-α and IFN-γ.

Several studies demonstrate that cytokines are markedly elevated in infected subjects at all stages of infection, and that their levels continue to increase as the disease progresses [53]. Plasma levels of IFN-γ increase as early as day 7 post-infection. The levels of IFN-γ and TNF-α in the brain increase exponentially with the severity of the neuropathology [53]. A similar relationship exists with

IL-1β and brain pathology, though the correlation is not as strong as with TNF-α and IFN-γ [53, 54]. These cytokines may act synergistically in altering the immune response and in modulating the expression of cell adhesion molecules [55]. In addition, when exposed to endothelial cells, they may be implicated in promoting increased permeability [40]. However, a negative relationship exists between the anti-inflammatory cytokines, such as IL-10 and IL-6, and brain pathology. The levels of these factors are elevated at stages when brain pathology is mild in infected animals [53]. It should be noted that, while the CNS levels of IFN-γ increase with disease progression, the plasma levels of the cytokine decrease. The same pattern is observed with systemic TNF-α, where concentrations decrease in late disease, suggesting that the CNS concentration of these cytokines originates from an endogenous source [53]. It is postulated that an early increase of systemic IFN-γ and other cytokines is responsible for astrocyte and microglial activation in the brain. These cells in themselves are able to produce inflammatory cytokines; and this may provide an explanation for the endogenous production of cytokines in the brain [53].

27.3.3
Astrocytosis and Microglial Cell Activation

Astrocyte activation is early evidence of CNS histopathology in sleeping sickness, with morphological changes evident even prior to any major brain lesions [39, 56]. Early in the disease process, activation occurs in a diffuse manner. However, as the infection progresses, the distribution of activated clusters of microglia is more prevalent in regions like the hypothalamus, the cerebral cortex, and the septum [56, 57]. There is a marked hypertrophy in these cells at later stages of disease; and these changes appear to correlate with the changes in sleep cycle characteristic of sleeping sickness [56, 57].

Microglia and astrocytes express certain chemokines involved in recruitment of systemic T lymphocytes into the CNS. Notably, the parasites are able to stimulate CD8[+] T lymphocytes to secrete IFN-γ, a cytokine known to promote parasite growth and proliferation [56]. Trypanosomal factors are also able to mediate inducible nitric oxide synthase production within astrocytes and microglia in the brain [39, 56].

27.3.3.1 Nitric Oxide
Nitric oxide (NO) under physiological conditions causes smooth muscle cell relaxation and inhibits platelet aggregation, thus ensuring adequate blood supply to various organs. Excessive NO production may be important in vascular permeability and may contribute to BBB breakdown and the transmigration of inflammatory cells [46, 51, 58]. Under physiologic situations, NO is constitutively synthesized by nitric oxide synthase in endothelial cells (eNOS or NOS3) and neuronal cells (BNOS or NOS1) and is dependent upon calcium-calmodulin interactions

[51]. In inflammatory settings, in the presence of certain cytokines and microorganisms, the production of NO is increased via activation of inducible nitric oxide synthase (iNOS or NOS2) in macrophages and brain microglia in a process that is independent of calcium [51]. It is believed that there is a dual action exerted by NO. The constitutively synthesized NO, when found in excess, has a cytotoxic role and is speculated to have some effect in the permeabilization of the BBB with trypanosomal infection. The concentration of constitutive NOS is found to be directly correlated with the evolution of disease [51]. However, when surplus NO is synthesized in the macrophages, there appears to be an attenuation of T-helper cell response and immunosuppression which aids in the survival of the parasite [58].

Though the specific role of NO and its toxic effects on CNS events in trypanosomal infection remains to be elucidated, *in vitro* studies using purified mouse brain astrocytes and microglia demonstrate an increase of iNOS when these cells are cocultured with trypanosomes, leading to increased NO release. This implies that there is direct contact of these cells with trypanosomal components in the brain of infected subjects [59]. Indeed, in mice infected with *T.b. brucei*, increased iNOS activity can be observed in mononuclear cells in the meninges, the choroid plexus, and at times in the parenchyma [60]. The presence of iNOS in these regions appears to precede neurological signs [60]. IFN-γ is also able to induce iNOS production in those cells and acts synergistically with trypanosomes to induce these changes [59].

27.3.4
American Trypanosomiasis (Chagas' Disease)

Chagas' disease is most often associated with acute myocarditis and chronic dilated cardiomyopathy, as well as the mega syndromes. Although the brain is involved in *T. cruzi* infection, it has not been extensively investigated. The CNS is involved in congenital Chagas' disease which is most prevalent in highly endemic areas. Neonates may suffer from seizures, jaundice, and be clinically indistinguishable from the manifestations of congenital toxoplasmosis or other etiologies in the TORCH syndromes.

The acute infection phase occurs in approximately 1% of cases and occurs mostly in young children, less than 2 years old [29, 43, 52]. It is characterized by parasitic penetration into the CNS, with resultant inflammatory foci of glial nodules. Trypomastigotes at that stage can sometimes be present in the cerebral spinal fluid; and amastigotes can be found in glial cells and macrophages [39, 43]. The parasites are postulated to enter the brain parenchyma through the pial vessels; and invasion of the CNS may occur because of immaturity of the BBB in that age group [43]. It is important to note that acute *T. cruzi* infection is associated with increased levels of TNF-α which can modulate alterations in the BBB [42, 43]. There is then an intermediate phase in the majority of patients where there may be waning of the inflammation [39]. This is followed by the chronic stage which may occur years after infection. This includes a wide array of CNS abnormalities, including damage to the peripheral autonomic and skeletal muscle innervations via some autoimmune mechanism [39].

More recently, Chagas' meningoencephalitis has been observed in the setting of HIV/AIDS, where the clinical and radiographic manifestations are indistinguishable from toxoplasmosis in the same setting [61]. The involvement of the CNS is rare, though the presence of the parasite in the CNS and especially the CSF during the acute phase of the disease has been demonstrated [111–114]. The first report of reactivation in an AIDS patient was in 1990 [115] and there have been several reports since that time [116–119]. In patients with advanced AIDS, cerebral infection is the prevalent form of reactivation disease. The clinical features of these infections are similar to those of CNS toxoplasmosis. The CNS disease normally presents as multifocal, necrotizing encephalitis [113, 115, 118]. In some cases, the trypomastigote form of the parasite can be visualized in CSF, but generally the diagnosis is confirmed by intracranial biopsy. The amastigote form of the parasite can be observed in many cell types of the brain tissue by light or electron microscopy [111]. At this time, no efficacious treatment protocol has been defined and the number of case fatalities is high, with a short duration of survival. An acute CNS reactivation of chronic Chagasic disease may also occur in immunosuppressed patients, such as individuals with leukemia and post-transplantation [39, 44]. In these patients, there is clear evidence of encephalitis and inflammatory foci with multiple amastigotes [39, 44]. Imaging of the brain is significant for the presence of ring-enhancing lesions in the white matter, subcortical regions, and the cerebellum, with very little edema [44]. Necrosis and hemorrhage is apparent both histologically and by MRI [39, 44].

Trypomastigotes of *T. cruzi* invade endothelial cells, resulting in an increased synthesis of cytokines such as TNF-α and IL-1β, and in addition, an increased synthesis of the vasoactive peptide, endothelin-1 [62, 63]. These findings could be responsible in part for vascular spasm and loss of the integrity of the BBB. Moreover, *T. cruzi* infection of endothelial cells have been reported to alter gap junctions between endothelial cells which also lead to alterations in the BBB [64]. Finally, recent evidence indicates that, like African trypanosomiasis, *T. cruzi* infection causes an increase in inflammatory mediators in the brain [65].

Trypanosomal disease of the CNS involves a complex process in which the details are not altogether clear. There is evidence of direct trypanosomal penetration in the CNS in both African and American disease, mostly in areas of decreased BBB stability. However, as disease progresses, the distribution of parasites in the brain parenchyma may not follow a specific pattern. There is a surge of proinflammatory cytokines with infection, which can help to stimulate certain adhesion molecules on the surface of endothelial cells, thereby bringing about compromise in the BBB permeability. There is also an increase in the synthesis of NO, which can bring about cell damage and can lead to increased permeability of the endothelium. There is demonstration of glial cell activation, and in the case of *T. cruzi*, there are trypanosomal forms within these glial cells. There is also an autoimmune phenomenon in which anti-host cell antibodies are synthesized, causing a lot of destruction to many structures inside and out of the CNS. Moreover, at the later stages of disease, there is evidence of hemorrhage (*T. cruzi*) and edema (*T. brucei*), which further strengthens the hypothesis that there is widespread compromise of the BBB in these disease entities.

27.4
Toxoplasmosis: Transgression, Quiescence, and Destructive Infections
(by L. Weiss)

Toxoplasma gondii is an obligate intracellular protozoan parasite found worldwide that can infect any mammal. Infection results from ingestion of either the tissue cyst stage or oocysts. Initial infection in the gastrointestinal tract is followed by dissemination of tachyzoites and acute disease, which is often asymptomatic and self-limiting. Studies in mice and rats after oral infection with oocysts suggest that *T. gondii* first infects mesenteric lymph nodes, then the spleen, and subsequently the brain, heart, and other organs via the bloodstream [66]. Recent observations in mice suggest that *T. gondii* may transverse the BBB inside CD11c and CD11b expressing leukocytes, prior to replication in the brain [67]. Other studies have demonstrated that *T. gondii* microneme protein 2 (MIC2) can bind to ICAM-1 and that, following this, the organism actively penetrates endothelial cell monolayers [68]. Interestingly, the integrity of the endothelial cell barrier was not altered during parasite transmigration. ICAM-1 is upregulated during the course of *T. gondii* infection [68]. Following this acute dissemination, a latent phase develops with tissue cysts containing bradyzoites being present in brain, muscle, and other tissues, which lasts for the lifetime of the host. Tissue cysts persist within astrocytes and neurons of the brain and elicit little to no inflammatory reaction [69]. Tissue cyst rupture and bradyzoite differentiation to the tachyzoite stage are thought to occur intermittently, but disease is limited due to the immune response of the host [70, 71].

T. gondii is an important opportunistic infection of the CNS in AIDS patients as well as other immunocompromised patients, such as transplant recipients and those receiving chemotherapy. In the setting of immune suppression, encephalitis often results from a reactivation of the tissue cyst stage in the brain [72]. In these immunocompromised hosts when tissue cysts rupture, tachyzoites can replicate freely in the brain, resulting in lysis of their host cells and in encephalitis with necrotic foci in the brain.

IFN-γ is the primary cytokine controlling both the acute and chronic stages of infection [73–75]. Elevated levels of IFN-γ and the cytokines IL-1β and TNF-α are present in the brains of *T. gondii* infected mice [76–78], and the administration of anti-IFN-γ results in an increase in the amount of encephalitis in infected mice [79, 80]. Protection against both the acute and chronic stages of *T. gondii* requires cells from both the hematopoietic and nonhematopoietic lineages [81]. The effector cell populations in the brain include CD4$^+$ and CD8$^+$ T cells, microglia, astrocytes, dendritic cells, B cells and an IFN-γ producing a non-T cell population [82, 83]. Neurons, astrocytes, and microglia have all been shown to produce IFN-γ in *vitro* and these cells may be important sources of IFN-γ [84, 85]. IFN-γ stimulates the microglia to produce TNF-α which upregulates ICAM-1 [86].

Neutrophils are the first cells to be recruited to sites of infection and they play a critical role in retarding intracellular parasite growth and lysing extracellular parasites [87]. In mouse models of toxoplasmosis, depletion of neutrophils before infection leads to an increase in parasite load and an impaired ability of the host to produce IFN-γ, TNF-α and IL-12 [88]. Likewise, mice in which neutrophils had impaired migration had increased parasite loads and decreased levels of IFN-γ [89, 90]. Dendritic cells serve as antigen-presenting cells (APC) early in infection and produce IL-12, priming the differentiation of $CD4^+$ T cells to produce IFN-γ and thus ensuring an adaptive immune response [91]. Infected astrocytes produce factors that enable dendritic cells to differentiate into APC and thus help initiate the intracerebral immune response. Brain dendritic cells represent a novel type of APC, because both astrocytes and microglia (the other known APCs) require IFN-γ, whereas brain dendritic cells do not. This feature may explain how intracerebral T cell activation occurs. IL-12 in conjunction with TNF-α, IL-1β and IL-15 also activates natural killer (NK) cells to produce IFN-γ [92–94], priming macrophages to produce IL-12, thereby amplifying local IFN-γ production [95] with IFN-γ and IL-12, driving a T cell Th1 response [92]. Studies indicate the major sources of IFN-γ are $CD4^+$ and $CD8^+$ T cells that are recruited into the brain during infection [96, 97]. $CD8^+$ T cells are thought to be the primary effector cell against *T. gondii* in the brain, with $CD4^+$ T cells playing a synergistic role [72, 92, 98, 99]. $CD8^+$ T cells exhibit cytotoxic lymphocyte (CTL) activity in a MHC-restricted manner [100–103]. In the chronic phase of infection, prevention of TE depends upon $CD8^+$ T cells and IFN-γ [96]. During the chronic stage of the infection, the Th1 cytokines IL1, IL6, TNF-α, and IFN-γ decrease, while IL-10 increases in the brain [96].

Microglia, the resident macrophage population in the brain, are probably a major effector cell in the prevention of *T. gondii* tachyzoite proliferation in the brain [86, 104, 105]. Astrocytes can also inhibit growth of *T. gondii* in the brain [106] and produce proinflammatory cytokines [107]. These, together with the cytokines produced by activated microglia, probably play an important role in inducing the infiltration of immune cells into the brain. The proinflammatory by-products of activation such as NO, while inhibiting parasite replication in the brain, are themselves detrimental to neurons and may also contribute to defects in the BBB. *T. gondii* can infect microvascular endothelial cells and IFN-γ has been demonstrated to inhibit *T. gondii* replication in these cells [108]. Since one of the first steps in the development of cerebral toxoplasmosis is penetration of the BBB, IFN-γ-induced inhibition in these cells may be important for limiting the replication of *T. gondii* in the brain.

27.5
Conclusion

Human parasitic infections remain crucially important within the scope of human disease. The very complex interactions of the BBB with a subset of these parasites seem to lead to the greatest morbidity and highest mortality in each disease group. Further understanding of these relatively rare events within the context of a global understanding of the diseases as a whole is a vital aspect of reducing disease mortality in human populations – especially for those who suffer most. Their suffering is not due to ignorance, but is a reflection of the larger socioeconomic or political circumstance.

References

1 Estanol Vidal B, Corona Vazquez T **1989**, *Rev. Invest. Clin.* 41, 327–330.
2 Jay A **1987** *Arq. Neuropsiquiatr.* 45, 261–275.
3 Miller BL, Tourtellotte WW, Shapshak P, Goldberg M, Heiner D, Weil M **1985**, *Arch. Neurol.* 42, 782-784.
4 Ferrari TC, Xavier MA, Gazzinelli G, Cunha AS **1999**, *Trans. R. Soc. Trop. Med. Hyg.* 93, 558–559.
5 Ferrari TC, Oliveira RC, Ferrari ML, Gazzinelli G, Cunha AS **1995**, *Trans. R. Soc. Trop. Med. Hyg.* 89, 469–500.
6 Duke BO, Moore PJ **1976**, *Tropenmed. Parasitol.* 27, 123–132.
7 Dorta-Contreras AJ, Escobar-Perez X, Duenas-Flores A, Mena-Lopez R **2003**, *Rev. Neurol.* 36, 506–509.
8 Dorta Contreras AJ, Plana Bouly R, Diaz Martinez AG **1989**, *Rev. Cubana Med. Trop.* 41, 242–249.
9 Dorta-Contreras AJ, **1998**, *Clin. Diagn. Lab. Immunol.* 5, 452–455.
10 Alvarez JI, Castano CA, Trujillo J, Teale JM, Restrepo BI **2002**, *J. Neuroimmunol.* 127, 139–144.
11 Lima JE, Garcia LV, Leite JP **2004**, *J. Neurol. Sci.* 217, 31–35.
12 Lima JE, Garcia LV, Leite JP **2004**, *Braz. J. Med. Biol. Res.* 37, 19–26.
13 Fernandez-Bouzas A, Fernandez T, Ricardo-Garcell J, Casian G, Sanchez-Conde R **2001**, *Clin. Neurophysiol.* 112, 2281–2287.
14 Abreu-Silva AL, Tedesco RC, Mortara RA, da Costa SC **2003**, *Am. J. Trop. Med. Hyg.* 68, 661–665.
15 Prasad LS, **1996**, *Am. J. Trop. Med. Hyg.* 55, 652–654.
16 Hou RF, Lee HH, Chen KM, Chou HL, Lai SC **2004**, *Int. J. Parasitol.* 34, 1355–1364.
17 Martinez J, Bernadina WE, Rodriguez-Cabeiro F **1999**, *Parasitology* 118, 605–613.
18 Benedetto N, Auriault C **2002**, *Eur. Cytokine Netw.* 13, 447–455.
19 Sissons J, Stins M, Jayasekera S, Alsam S, Khan NA **2005**, *Infect. Immun.* 73, 2704–2708.

20 Alsam S, Stins M, Rivas AO, Sissons J, Khan NA **2003**, *Microb. Pathogen.* 35, 235–241.
21 Jayasekera S, Tucker J, Rogers C, Nolder D, Warhurst D, Alsam S, White JM, Khan NA **2004**, *J. Med. Microbiol.* 53, 1007–1012.
22 Jayasekera S, Sissons J, Maghsood AH, Khan NA **2005**, *Microbes Infect.* in press.
23 Hunt NH, Grau GE **2003**, *Trends Immunol.* 24, 491–499.
24 Thumwood CM, Hunt NH, Clark IA, Cowden WB **1988**, *Parasitology* 96, 579–589.
25 Lou J, Lucas R, Grau GE **2001**, *Clin. Microbiol.* 14, 810–820.
26 Ma N, Madigan MC, Chan-Ling T, Hunt NH **1997**, *Glia* 19, 135–151.
27 Lucas R, Lou J, Juillard P, Moore M, Bluethmann H, Grau GE **1997**, *J. Neuroimmunol.* 72, 143–148.
28 Dietrich J **2002**, *J. Neuroimmunol.* 128, 58–68.
29 Piguet PF, Kan CD, Vesin C **2002**, *Lab. Invest.* 82, 1155–1166.
30 Adams S, Brown H, Turner G **2002**, *Trends Parasitol.* 18, 360–366.
31 Gimenez F, de Lagerie SB, Fernandez C, Pino P, Mazier D **2003**, *Cell. Mol. Life Sci.* 60, 1623–1635.
32 Medana IM, Chan-Ling T, Hunt NH **2000**, *Am. J. Pathol.* 156, 1055–1065.
33 Medana IM, Hunt NH, Chan-Ling T **1997**, *Glia* 19, 91–103.
34 Kennan RP, Machado FS, Lee SC, Desruisseaux MS, Wittner M, Tsuji M, Tanowitz HB **2005**, *Parasitol. Res.* 96, 302–307.
35 Penet MF, Viola A, Confort-Gouny S, Le Fur Y, Duhamel G, Kober F, Ibarrola D, Izquierdo M, Coltel N, Gharib B, Grau GE, Cozzone PJ **2005**, *J. Neurosci* 25, 7352–7358.
36 Machado FS, Kennan RP, Desruisseaux MS, Lee SC, Wittner M, Nagajyothi F, Tsuji M, Tanowitz HB **2005**, *Int. Conf. Endothelin* 9, abstract.
37 Hermsen CC, Mommers E, van de Wiel T, Sauerwein RW, Eling WMC **1998**, *J. Infect. Dis.* 178, 1225–1227.
38 WHO Expert Committee **1998**, *Control and Surveillance of African Trypanosomiasis*, WHO, Geneva.
39 Pentreath VW **1995**, *Trans. R. Soc. Trop. Med. Hyg.* 89, 9.
40 Enanga B, Burchmore RJS, Stewart ML, Barrett MP **2002**, *Cell. Mol. Life Sci.* 59, 845.
41 Kennedy PGE **2004**, *J. Clin. Invest.* 113, 496.
42 Silva GC, Nagib PRA, Chiari E, Van Rooijen N, Machado CRS, Camargos ERS **2004**, *J. Neuroimmunol.* 149, 50.
43 Da Mata JR, Camargos ERS, Chiari E, Machado CRS **2000**, *Brain Res. Bull.* 53, 153.
44 Lury KM, Castillo M **2005**, *Am. J. Roent.* 185, 550.
45 Greenwood BM, Whittle HC **1980**, *Trans. R. Soc. Trop. Med. Hyg.* 74, 716.
46 Dumas M, Bouteille B **1996**, *C. R. Seances Soc. Biol. Fil.* 190, 395.
47 Philip KA, Dascombe MJ, Fraser PA, Pentreath VW **1994**, *Ann. Trop. Med. Parasitol.* 88, 607.

48 Masocha W, Robertson B, Rottenberg ME, Mhlanga J, Sorokin L, Kristensson K **2004**, *J. Clin. Invest.* 114, 689.
49 Mulenga C, Mhlanga JDM, Kristensson K, Robertson B **2001**, *Neuropathol. Appl. Neurobiol.* 27, 77.
50 Lejon V, Buscher P **2005**, *Trop. Med. Int. Health* 10, 395.
51 Buguet A, Burlet S, Auzelle F, Montmayeur A, Jouvet M, Cespuglio R **1996**, *C. R. Acad. Sci. Paris III* 319, 201.
52 Girard M, Giraud S, Courtioux B, Jauberteau-Marchan M, Bouteille B **2005**, *Mol. Biochem. Parasitol.* 139, 41.
53 Sternberg JM, Rodgers J, Bradley B, MacLean L, Murray M, Kennedy PGE **2005**, *J. Neuroimmunol.* in press.
54 Ching S, He L, Lai W, Quan N **2005**, *Brain Behav. Immun.* 19, 127.
55 Kalaria RN **1999**, *Am. J. Pathol.* 154, 1311.
56 Rock RB, Gekker G, Hu S, Sheng WS, Cheeran M, Lokensgard JR, Peterson KP **2004**, *Clin. Microbiol. Rev.* 17, 942.
57 Chianella S, Semprevivo M, Peng ZC, Zaccheo D, Bentivoglio M, Grassi-Zucconi G **1999**, *Brain Res.* 832, 54.
58 Viswambharan H, Seebeck T, Yang Z **2003**, *Int. J. Parasitol.* 33, 1099.
59 Girard M, Ayed Z, Preux P, Bouteille B, Preud'Homme J, Dumas M, Jaubertau M **2000**, *Parasite Immunol.* 22, 7.
60 Keita M, Vincendeau P, Buguet A, Cespuglio R, Vallat J, Dumas M, Bouteille B **2000**, *Exp. Parasitol.* 95, 19.
61 Vaidian AK, Weiss LM, Tanowitz HB **2004**, *Kinetoplastid Biol. Dis.* 3, 2.
62 Tanowitz HB, Gumprecht JP, Spurr D, Calderon TM, Ventura MC, Raventos-Suarez C, Factor SM, Hatcher V, Wittner M, Berman JW **1992**, *J. Infect. Dis.* 166, 598–603.
63 Petkova SB, Huang H, Factor SM, Pestell RG, Bouzahzha B, Jelicks LA, Weiss LM, Douglas SA, Wittner M, Tanowitz HB **2001**, *Int. J. Parasitol.* 31, 499–511.
64 Spray DC, Moreno AP, Roy C, Saez JC, Burt JM, Hertzberg E, Campos de Carvalho AC, Dermietzel R, Hatcher V, Wittner M, Tanowitz HB **1989**, *Physiologist* 32, 212.
65 Michailowsky V, Silva NM, Rocha CD, Vieira LQ, Lannes-Vieira J, Gazzinelli RT **2001**, *Am. J. Pathol.* 1159, 1723–1733.
66 Zenner L, Darcy F, Capron A, Cesbron-Delauw MF **1998**, *Exp. Parasitol.* 90, 86–94.
67 Courret N, Darche S, Sonigo P, Milon G, Buzoni-Gatel D, Tardieux I **2005**, *Blood* in press.
68 Barragan A, Brossier F, Sibley LD **2005**, *Cell Microbiol.* 7, 561–568.
69 Ferguson DJP, Hutchison WM **1987**, *Virchows Arch. A* 411, 39–43.
70 Ferguson DJP, Hutchison WM, Peterson E **1989**, *Parasitol. Res.* 75, 599–603.
71 Frenkel JK **1988**, *Parasitol. Today* 4, 273–278.
72 Luft BJ, Remington JS **1992**, *Clin. Infect. Dis.* 15, 211–222.
73 Suzuki Y, Orellana MA, Schreiber RD, Remington JS **1988**, *Science* 240, 516–518.
74 Suzuki Y, Remington JS **1990**, *J. Immunol.* 144, 1954–1956.

75 Scharton-Kerston TM, Wynn TA, Denkers EY, Bara S, Shoue L, Grunvald E, Hieny E, Gazzinelli RT, Sher A **1996**, *J. Immunol.* 157, 4045–4054.
76 Hunter CA, Roberts CW, Murray M, Alexander J **1992**, *Parasite Immunol.* 14, 405–413.
77 Hunter CA, Litton MJ, Remington JS, Abrams JS **1994**, *Infect. Immun.* 65, 2339–2345.
78 Deckert-Schluter M, Albrecht S, Hof H, Wiestler OD, Schluter D **1995**, *Immunology* 85, 408–418.
79 Suzuki Y, Conley FK, Remington JS **1989**, *J. Immunol.* 143, 2045–2050.
80 Gazzinelli R, Eltoum I, Wynn TA, Sher A **1993**, *J. Immunol.* 151, 3672–3681.
81 Yap GS, Sher A **1999**, *J. Exp. Med.* 189, 1083–1091.
82 Kang H, Suzuki Y **2001**, *Infect. Immun.* 69, 2920–2927.
83 Suzuki Y **2002**, *J. Infect. Dis.* 185, S58–S65.
84 De Simone R, Levi G, Aloisi F **1998**, *Cytokine* 10, 418–422.
85 Neumann H, Schmidt H, Wilharm E, Behrens L, Wekerle H **1997**, *J. Exp. Med.* 186, 2023–2031.
86 Deckert-Schluter M, Bluethmann H, Kaefer N, Rang A, Schluter D **1999**, *Am. J. Pathol.* 154, 1549–1561.
87 Channon JY, Sequin RM, Kasper LH **2000**, *Infect. Immun.* 68, 4822–4826.
88 Bliss SK, Gavrilescu LC, Alcaraz A, Denkers EY **2001**, *Infect. Immun.* 69, 4898–4905.
89 Del Rio L, Bennouna S, Salinas J, Denkers EY **2001**, *J. Immunol.* 167, 6503–6509.
90 Subauste CS, Wessendarp M **2000**, *J. Immunol.* 165, 1499–1505.
91 Reis e Sousa C, Hieny S, Scharton-Kerston T, Jankovic D, Charest H, Germain RN, Sher A **1997**, *J. Exp. Med.* 186, 1819–1829.
92 Denkers EY, Gazzinelli RT **1998**, *Clin. Microbiol. Rev.* 11, 569–588.
93 Hunter CA, Bermudez L, Beernink H, Waegell W, Remington JS **1995**, *Eur. J. Immunol.* 25, 994–1000.
94 Hunter CA, Chizzonite R, Remington JS **1995**, *J. Immunol.* 155, 4347–4354.
95 Alexander A, **1997**, *J. Immunol.* 62, 5449–5454.
96 Gazzinelli R, Hieny S, Wynn TA, Wolf S, Sher A **1993**, *Proc. Natl Acad. Sci. USA* 90, 6115–6119.
97 Ely KH, Kasper LH, Khan IA **1999**, *J. Immunol.* 169, 268–272.
98 Gazzinelli R, Hakim FT, Hieny S, Schearer GM, Sher A **1991**, *J. Immunol.* 146, 286–292.
99 Gazzinelli, R, Xu Y, Hieny S, Cheever A, Sher A **1992**, *J. Immunol.* 149, 175–180.
100 Hakim FT, Gazzinelli RT, Denkers E, Hieny S, Shearer GM, Sher A **1991**, *J. Immunol.* 147, 2310–2316.
101 Khan IA, Smith KA, Kasper LH **1990**, *J. Clin. Invest.* 85, 1879–1886.
102 Subauste CS, Koniaris AH, Remington JS **1991**, *J. Immunol.* 147, 3955–3959.
103 Montoya JG, Lowe KE, Clayberger C, et al. **1996**, *Infect. Immun.* 64, 176–181.

104 Chao C, Gekker G, Hu S, Peterson PH **1994**, *J. Immunol.* 152, 1246–1252.
105 Schluter D, Meyer T, Strack A, Reiter S, Kretschmar M, Wiestler CD, Hof H, Deckert M **2001**, *Brain Pathol.* 11, 44–55.
106 Halonen SK, Chiu FC, Weiss LM **1998**, *Infect. Immun.* 66, 4989–4993.
107 Fischer HG, Nitzgen B, Reichmann G, Hadding U **1997**, *Eur. J. Immunol.* 27, 1539–1548.
108 Daubener WK, Spors B, Hucke C, Adam R, Stins M, Kim KS, Schroten H **2001**, *Infect. Immun.* 69, 6527–6531.
109 Taylor TE, Fu WJ, Carr RA, Whitten RO, Mueller JS, Fosiko NG, Lewallen S, Liomba NG, Molyneux ME **2004**, *Nat. Med.* 10, 143–145.
110 Pino P, Taoufiq Z, Nitcheu J, Vouldoukis I, Mazier D **2005**, *Thromb. Haemost.* 94, 336–340.
111 Corti M **1990**, *AIDS Chagas Dis.* 14, 581–588.
112 Hoff R, et al. **1978**, *N. Engl. J. Med.* 298, 604–606.
113 Antunes ACM, et al. **2002**, *Arq. Neuro-Psiq.* 60, 730–733.
114 Pedreira de Freitas J, Lion M, Tartari J **1953**, *Rev. Hosp. Clin. Sao Paulo* 8, 81–92.
115 Del Castillo M, et al. **1990**, *Am. J. Med.* 88, 693–694.
116 Ferreira M, et al. **1997**, *Clin. Infect. Dis.* 25, 1397–1400.
117 Cohen J, et al. **1998**, *Surg. Neurol.* 49, 324–327.
118 Rocha A, de Meneses A, da Silva A **1994**, *Am. J. Trop. Med. Hyg.* 50, 261–268.
119 Yoo T, et al. **2004**, *Clin. Infect. Dis.* 39, 30–34.
120 Hatabu T, Kawazu S, Aikawa M, Kano S **2003**, *Proc. Natl Acad. Sci. USA* 100, 15942–15961.
121 Wassmer SC, Lepolard C, Traore B, Pouvelle B, Gysin J, Grau GE **2004**, *J. Infect. Dis.* 189, 180–191.
122 Kirchgatter K, Del Portillo HA **2005**, *An. Acad. Bras. Cienc.* 77, 455–751.
123 Wickham ME, Rug M, Ralph SA, Klonis N, McFadden GI, Tilley L, Cowman AF **2001**, *EMBO J.* 20, 5636–5649.
124 Biggs BA, Gooze L, Wycherley K, Wilkinson D, Boyd AW, Forsyth KP, Edelman L, Brown GV, Leech JH **1990**, *J. Exp. Med.* 171, 1883–1892.
125 Pongponratn E, Viriyavejakul P, Wilairatana P, Ferguson D, Chaisri U, Turner G, Looareesuwan S **2000**, *Southeast Asian J. Trop. Med. Public Health* 31, 829–835.
126 Ruangjirachuporn W, Afzelius BA, Paulie S, Wahlgren M, Berzins K, Perlmann P **1991**, *Parasitology* 102, 325–334.
127 N'Goran EK, Utzinger J, Gnaka HN, Yapi A, N'Guessan NA, Kigbafori SD, Lengeler C, Chollet J, Shuhua X, Tanner M **2003**, *Am. J. Trop. Med. Hyg.* 68, 24–32.
128 Utzinger J, N'Goran EK, N'Dri A, Lengeler C, Xiao S, Tanner M **2000**, *Lancet* 355, 1320–1325.
129 Xiao S, Tanner M, N'Goran EK, Utzinger J, Chollet J, Bergquist R, Chen M, Zheng J **2002**, *Acta Trop.* 82, 175–181.

28
The Blood Retinal Interface: Similarities and Contrasts with the Blood-Brain Interface

Tailoi Chan-Ling

28.1
Introduction

The mammalian retina forms as an outgrowth of the diencephalon during early embryonic development. Embryologically, the retina is thus part of the central nervous system (CNS) and its blood vessels share many characteristics with those of the brain, including barrier properties (Bill et al. 1980), the ability to autoregulate (Bill and Nilsson 1982, 1985; Bill and Sperber 1990), close contact with astrocytes (Ling and Stone 1988; Ling et al. 1989; Chu et al. 2001; Chan-Ling et al. 2004a,b), association with microglia (Hollander et al. 1991), pericyte ensheathment (Chan-Ling et al. 2004b; Hughes and Chan-Ling 2004), and involvement in immune responses (Chan-Ling et al. 1992; Hu et al. 1998a,b). However, whereas the brain is encased in the cranium with pial coverage and is bathed in cerebrospinal fluid, the retina is located inside the eyecup and its microvasculature is adjacent to and adjoins nonCNS vascular beds. These junctions between vasculatures with CNS and peripheral characteristics in the eye are unique in their cell biology, are inherently susceptible to breakdown of the blood-retinal barrier (BRB), and play a prominent role in immune activation.

The transparency of the ocular media makes it possible to view the retinal blood vessels in situ, in turn enabling the detection of cardiovascular pathology such as nipping of arterioles caused by high blood pressure and changes in arteriole and venular walls due to atherosclerosis or diabetic retinopathy. In contrast to the brain, the anterior location of the globe also renders it readily accessible to experimental manipulation, such as intraocular injection. For these reasons, the BRB has often been studied as a model to gain insight into the structure of the blood-brain barrier (BBB) as well as into various CNS disease processes. However, it is necessary to gain a full understanding of the unique features of the various blood-retinal interfaces before attempting to use the retina as an experimental model of the CNS. Changes in the BRB are of great importance in their own right, given that breakdown of this barrier is an inherent component of the pathogenesis of most sight-threatening retinopathies. This chapter reviews our current understanding of blood-retinal interfaces, highlighting both unique features and those shared with other CNS vessels.

28.2
The Inner and Outer BRB

The BRB is essential for control of the microenvironment of the retinal parenchyma, given that it physically impedes the diffusion of substances between the blood and nervous tissue. In humans and several other mammalian species, the BRB is composed of two parts (Fig. 28.1 A) because the retina has a dual vascular supply, the inner two-thirds of the retina being nourished by intraretinal vessels and the outer one-third being supplied by choroidal vessels. The barrier properties of the inner retina depend on the tight junctions between adjacent endothelial cells of the intraretinal capillaries and other vessels (Fig. 28.1 B, H), whereas those of the outer retina depend on the tight junctions between adjacent cuboidal cells of the retinal pigment epithelium (RPE; Fig. 28.1 C, D). These two barriers have been described as the inner and outer BRB, respectively (Cunha-Vaz 1976). The inner BRB separates the retina from circulating blood, whereas the outer BRB, which is formed by the single layer of cuboidal cells that constitute the RPE, separates the neural retina from the fenestrated capillaries of the choroids (Tyler and Burns 1991). The capillaries of the choroid are highly permeable and manifest only a low level of expression of the tight junction protein occludin between adjacent vascular endothelial cells (Flage 1977; Morcos et al. 1999) (Fig. 28.1 E).

The vessels of the retinal circulation derive from mesenchymal tissue that enters the eyecup through the optic fissure and spreads through the superficial layers of the retina (Ashton 1970; Chan-Ling et al. 1990). The vessels therefore form initially at the inner surface of the retina, with branches subsequently extending from these superficial vessels into the deeper layers (Michaelson 1954). The retinal vasculature resembles the cerebral circulation in terms of vessel structure and permeability (Bill et al. 1980), flow rate, oxygen clearance (with the venous partial pressure of oxygen typically being 60–65% of the arterial value; Bill and Nilsson 1982), and the ability to autoregulate flow in response to changes in arterial pressure and oxygen tension (Alm and Bill 1973; Bill and Nilsson 1982).

28.3
The Choroidal Vasculature

The choroidal vasculature lies outside the CNS, supplying the retina by diffusion through Bruch's membrane and the RPE. The outer region of the retina is the only part of the CNS of an adult mammal that is supplied via diffusion from a distance by extrinsic vessels, rather than by intrinsic vessels. The choroidal vascular plexus thus differs from the cerebral and retinal circulations and its vessels are distinct from CNS vessels in terms of their location, structure (lacking glial ensheathment), flow rate, permeability, and oxygen clearance (Bill et al. 1983). Direct studies of blood flow have shown that the flow rate in the choroidal circulation does not decrease during hyperoxia or increase during hypoxia, but rather varies linearly with vascular resistance (Bill and Nilsson 1982). These observations suggest that the choroidal circulation, unlike the CNS circulation, does not autoregulate its flow. However, more recent studies based on fluorescein angiography (Flower 1990; Flower and Klein 1990) suggest that choroidal blood flow does indeed decrease during hyperoxia. Furthermore, the pH of the subretinal space was found to change in response to an increase in intraocular pressure (Yamamoto and Steinberg 1992), suggesting that the choroidal circulation may be able to regulate its flow, albeit to a small extent and only in the vicinity of the optic disk. Not only do choroidal capillaries lack CNS-like barrier properties, as indicated by the extravasation of intravascularly introduced horseradish peroxidase (HRP) into the parenchyma surrounding these vessels (Morcos et al. 1999), but they are fenestrated and highly permeable. Their permeability to plasma proteins is five times that of the capillaries of the kidney, which are also fenestrated, and 10–30 times that of the capillaries of striated muscle, which are not fenestrated (Bill et al. 1980). The permeability of choroidal capillaries to smaller substances, such as sodium ions, is about 30 times and 50 times that of the capillaries in heart and skeletal muscle, respectively. The tight junctions between adjacent cells of the RPE thus separate the retina from the leaky blood vessels of the choroid.

28.4
Characteristics of Intraretinal Blood Vessels

The intraretinal vasculature consists largely of two layers, the inner or superficial retinal plexus, and the outer or deep retinal plexus. The superficial plexus is located in the ganglion cell and nerve fiber layers; and the deep plexus is located at the junction between the inner nuclear and outer plexiform layers (Fig. 28.1A). The superficial retinal plexus contains vessels of different calibers, including arterioles, venules, capillaries, and postcapillary venules, whereas the deep plexus consists mostly of capillary-sized vessels (Fig. 28.1A). The deep retinal plexus has been described as being more venous in nature, with postcapillary venules also being evident in this plexus.

28.5
Ensheathment and Induction of the Inner BRB by Astrocytes and Müller Glia

Although the precise nature of the interaction between blood vessels and surrounding astrocytes in the CNS is not completely understood, *in vitro* evidence suggests that astrocytes or substances released by them contribute directly to the formation and maintenance of functional barrier properties (Tao-Cheng et al. 1987, 1988; Laterra et al. 1990; Tontsch and Bauer 1991; Raub et al. 1992;

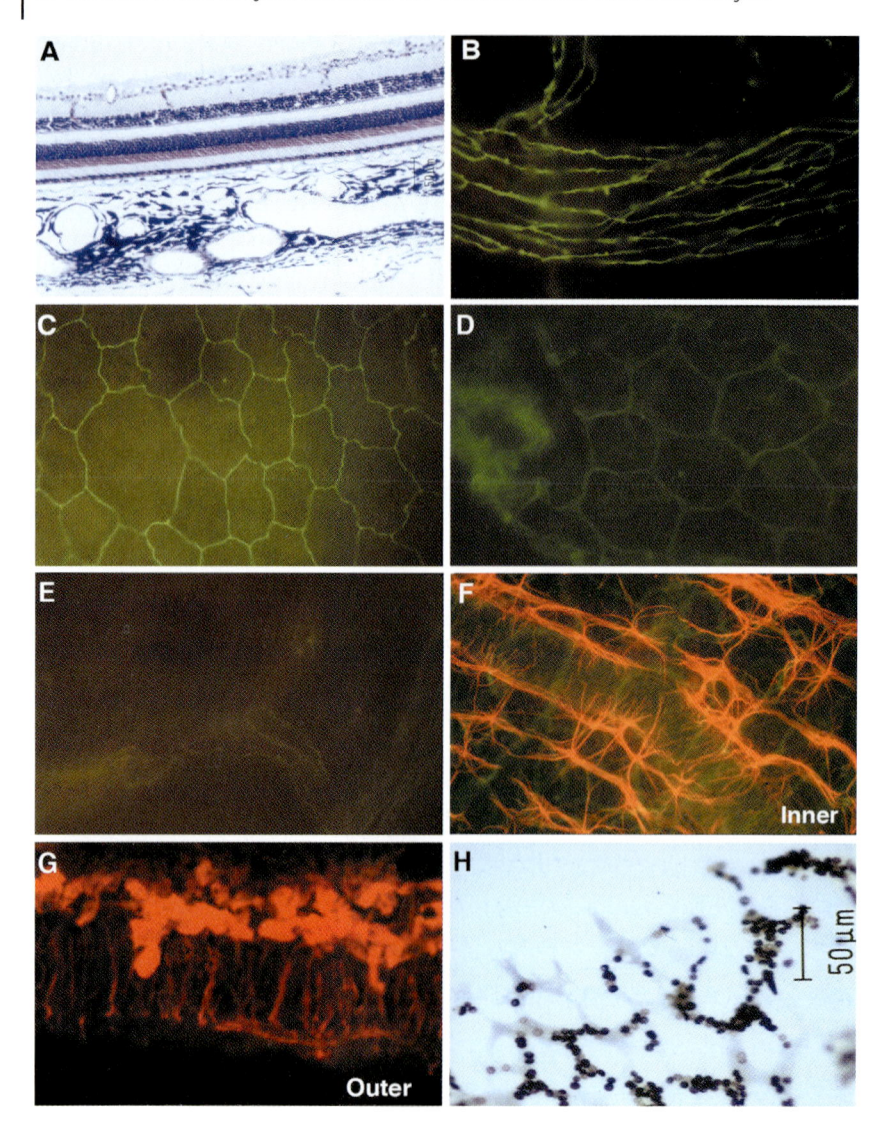

Fig. 28.1 (A) The intraretinal plexus is found predominantly in two layers, the inner or more commonly known as the superficial plexus, and the outer or deep retinal plexus. The superficial plexus is located in the ganglion cell and nerve fibre layers and the deep retinal plexus is located at the junction between the inner nuclear and outer plexiform layers. The outer one-third of the retina is nourished via diffusion from the choroidal vessels. The outer blood retinal barrier is formed by the simple layer of cuboidal cells that constitute the retinal pigment epithelium (RPE). The RPE separates the neural retina from the fenestrated capillaries of the choroid.
(B–E) Light microscopic analysis of the expression of the tight junction proteins, occludin and claudin-1, in the rabbit eye (Morcos et al. 2001).
(B) Expression of occludin in the retinal blood vessels within the myelinated region of the adult rabbit retina. The continuous expression of the protein was evident at the

Wolburg et al. 1994; Gardner et al. 1997). Studies with an animal model of retinopathy of prematurity have revealed a relation between breakdown of the BRB and degeneration of astrocytes (Chan-Ling and Stone 1992; Chan-Ling et al. 1992). Furthermore, astrocytes express occludin in vitro (Bauer et al. 1999). The level of occludin expression in retinal endothelial cells decreases in animals with experimental diabetes at the same time as retinal permeability increases (Antonetti et al. 1998). Moreover, occludin expression decreases together with that of zonula occludens-1 (ZO-1), a peripheral membrane protein associated with tight junctions, in the cerebral vascular endothelium during neutrophil-induced breakdown of the BBB (Bolton et al. 1998).

Since the first demonstration that the barrier properties of blood vessels are influenced by the vessel environment (Stewart and Wiley 1981), evidence has accumulated to suggest that astrocytes induce endothelial cells to develop barrier properties, specifically to form complex tight junctions with their neighbors and to express various barrier-specific enzymes and molecular transporters (Beck et al. 1984, 1986; Stewart and Coomber 1986; Maxwell et al. 1987; Stewart and Hayakawa 1987; Tao-Cheng et al. 1987; Brightman 1989; Dehouck et al. 1990; Hollander et al. 1991; Tontsch and Bauer 1991). Iris vessels have been shown to become impermeable on association with implanted astrocytes (Olson et al. 1983; Janzer and Raff 1987; Tout et al. 1993). The arteries of the retina, iris, and ciliary processes are all branches of the ophthalmic artery. However, ocular ves-

Fig. 28.1 (continued)
interfaces between adjacent endothelial cells of vessels in the myelinated streak supportive of their tight barrier properties, demonstrated using intravascular injection of the tracer horse radish peroxidase.
(C) Expression of the tight junction protein, occludin, in the RPE of the rabbit eye. Occludin was detected around the entire circumference of RPE cells.
(D) Expression of the tight junction protein, claudin-1, in the RPE of the rabbit eye. Claudin-1 was detected around the entire circumference of RPE cells.
(E) Expression of occludin in vessels of the choroid of the rabbit eye. The choroidal vessels exhibited only weak and irregularly distributed occludin immunoreactivity, supportive of the leakiness of choroidal blood vessels. Occludin was not evident at most endothelial cell junctions in the choroidal vascular plexus, in marked contrast to the patterns of expression apparent in vessels with barrier properties.
(F) Relation of astrocytes to blood vessels in the superficial retinal plexus shown with an antibody to GFAP conjugated to Texas red and *Griffonia simplicifolia* isolectin B4 conjugated to FITC. A vein runs obliquely across this field and is wrapped by many astrocyte processes.
(G) Muller cells forming the glia limitans of the deep retinal vessels in adult cat retina, shown by immunohistochemistry. This field shows the broken edge of a wholemount preparation, labelled with 4D6, a Müller cell-specific antibody. The inner endfeet of Müller cells are brightly fluorescent, at top. Their radial processes can be traced across the inner plexiform and nuclear layers. At the outer margin of the inner nuclear layer, they outline a capillary of the outer layer of retinal vasculature.
(H) Vessels at the edge of the region vascularized in a normal P12 retina. Blood cells and the HRP are trapped within the vessel lumen; and the spaces between vessels appear free of peroxidase.

sels vary from being highly permeable in the choroid, ciliary processes, and meningeal aggregates, to less permeable in the iris, to impermeable in the retina.

Müller cells are the radial glia of the retina, with the inner and outer endfeet of these cells expanding to form the external and internal limiting membranes of this tissue. Glial ensheathment of the superficial retinal vascular plexus is mediated by astrocytes, Müller cells, and, to a much lesser extent, perivascular microglia (Bussow 1980; Stone and Dreher 1987; Schnitzer 1988; Ling et al. 1989; Hollander et al. 1991) (Fig. 28.1 F), whereas vessels of the outer plexus are ensheathed by Müller cells only (Kondo et al. 1984; Hollander et al. 1991) (Fig. 28.1 G). Despite the difference in glial ensheathment between these two vascular beds, occludin expression is evident in both the inner and outer plexuses of the adult rat retina (Chan-Ling et al. 2005). These observations led us (Tout et al. 1993) to suggest that Müller cells share the ability of astrocytes (Tao-Cheng et al. 1987; Bouchard et al. 1989; Brightman 1989; Dehouck et al. 1990) to induce the development of barrier properties in endothelial cells. Implantation of cultured Müller cells into the anterior chamber of the rat eye revealed that the cells formed aggregates and became associated with iris vessels (Olson et al. 1983; Janzer and Raff 1987; Tout et al. 1993). Assessment of the barrier properties of the vessels associated with the cell aggregates by intravascular perfusion either with Evans blue to visualize serum albumin by light microscopy or with HRP to detect vessel leakage by electron microscopy showed that Müller cells indeed share with astrocytes the ability to induce the development of barrier properties in endothelial cells (Tout et al. 1993). Müller cells thus appear to induce the formation of tight junctions.

28.6
BRB Properties of Newly Formed Vessels

Astrocytes are intimately associated with the vasculature during development of the mammalian retina (Chan-Ling 1994). Both endothelial cells and astrocytes proliferate during retinal vascular development (Chan-Ling et al. 1995; Sandercoe et al. 1999). Astrocyte migration precedes the formation of patent vessels during normal development in both the rat and cat retinas (Ling and Stone 1988; Ling et al. 1989). A close relation between astrocytes and the forming vasculature has also been demonstrated at the ultrastructural level, with astrocytic processes being shown to associate with both well differentiated and immature vessels during normal human retinal vasculogenesis (Penfold et al. 1990).

Astrocytes play a role both in endothelial cell differentiation (Jiang et al. 1995) and in BRB function (Chan-Ling and Stone 1992). They contribute to the developing vascular complex at an early stage, contacting the collagenous matrix surrounding capillaries at a time coincident with the opening of the capillary lumen. The proximity of astrocytes to the differentiating vascular endothelium at the time of lumen formation suggests that these cells, as well as Müller cells, are ideally located for induction of the BRB during development (Bellhorn 1980;

Hughes et al. 2000; Chan-Ling et al. 2004a). Conversely, the endothelium may also influence astrocyte differentiation. Vascular endothelial cells have been shown to induce the expression of glial fibrillary acidic protein (GFAP) in astrocyte precursor cells from the rat optic nerve *in vitro*, an effect that is likely mediated by the production of leukemia inhibitory factor (Mi et al. 2001).

Vascular segments are produced in excess relative to the metabolic requirements of neurons and glia during CNS development. The vascular tree subsequently undergoes substantial remodeling, during which excess capillary segments are pruned and major vessels are selected (Hughes and Chan-Ling 2000). It is not known whether vessel regression is associated with a transient, focal loss of the BBB. Segments of vessels at various stages of formation and regression in the retina lack occludin immunoreactivity (Morcos et al. 2001) (Fig. 28.2A–D). Discontinuous occludin expression is most evident in remodeling vascular segments in regions of the retina central to the leading edge of vessel formation, where vessel retraction occurs. A small proportion of vessels at the leading edge of vessel formation, where endothelial cell migration takes place in association with vessel formation, also exhibit discontinuous occludin expression. These observations suggest that loss of continuous occludin expression is associated with the migration of endothelial cells during developmental remodeling of the vascular plexus, but that most of the endothelial cell migration associated with vessel formation is completed prior to the expression of occludin.

These changes in occludin expression can be interpreted in at least three ways. First, the loss of occludin expression reflects a focal, transient breakdown of the BRB. Second, such a loss occurs only in vessel segments that lack a lumen. Third, the absence of occludin in forming or regressing vascular segments is an artifact attributable to the inability of the monoclonal antibody used for these studies to detect all forms of occludin.

28.7
Pericytes and the BRB

The capillary wall consists of an endothelial tube ensheathed by pericytes, which lie within the basal lamina of the vessel; and mural cells have been shown to be associated with newly formed retinal vessels (Chan-Ling et al. 2004a, b). Postulated functions of pericytes include vessel stabilization, control of capillary blood flow or macrophage activity, and modulation of vascular permeability in the CNS (see Chapter 5). In the CNS, pericytes express BBB-specific enzymes, such as a specific isoform of aminopeptidase N (pAPN; Kunz et al. 1994) and glutamyltranspeptidase (Frey et al. 1991). In vitro studies showed that culture medium conditioned by a brain pericyte cell line increased the level of occludin expression in brain capillary endothelial cell lines (Hori et al. 2004). Furthermore, coculture of a brain endothelial cell line with pericytes triggered upregulation of BBB function (Dohgu et al. 2005). Mice that lack either the B chain of platelet-

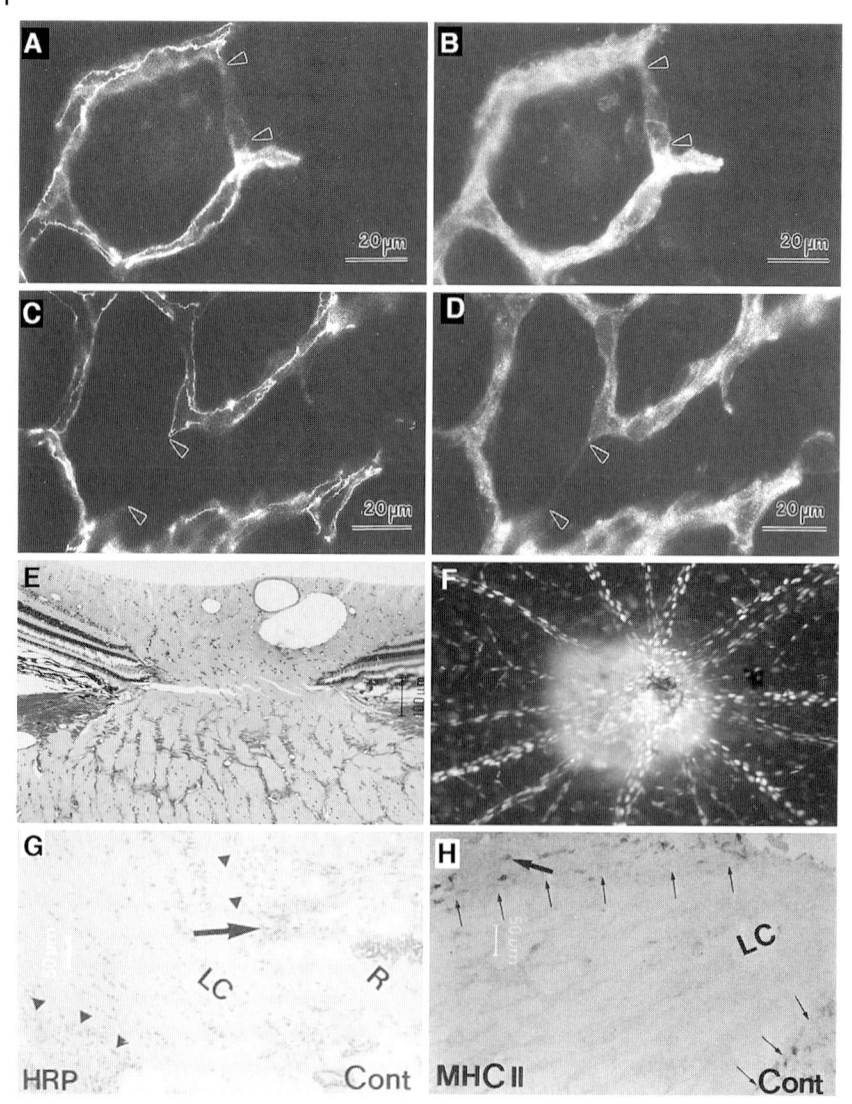

Fig. 28.2 (A–D) Light microscopic analysis of occludin expression in vessels of the rat retina during development. (A) and (C) show occludin expression, while (B) and (D) show the GS lectin appearance of the vascular segment in the respective fields of view. (A) and (B) show vascular segments at the leading edge of vessel formation. The majority of the vascular endothelial cells showed continuous occludin expression between adjacent vascular endothelial cells. The arrowheads demarcate the segment of the blood vessel, which lacks occludin expression. The occludin-free segment in (A) appears to lack a vessel lumen and also shows a weak expression for GS lectin (B). The avascular retina is on the right of the field of view.

The occludin-free segment shown in (C) has the characteristic appearance of a basal lamina when visualized with GS lectin (D). (E) H&E-stained transverse section of the region of the optic nerve head in a monkey. The retina and optic nerve head is in a unique position where CNS microvasculature is adjacent to leaky peripheral vascular beds, derived growth factor (PDGF) or the β chain of the PDGF receptor manifest a reduced number of pericytes as well as edema and evidence of the immaturity of tight and adherens junctions in CNS vessels (as revealed by the patterns of occludin and VE-cadherin immunostaining, respectively) and increased transcellular transport. Immunostaining for other tight junction proteins was unaffected in these mice, however, and there was no morphological evidence of altered tight junction permeability. The junctional phenotype of these knockout mice was thus suggested to be a result of the increased expression of the vascular endothelial growth factor-A (VEGF-A), which was also apparent in these animals, rather than of the reduced number of pericytes (Hellstrom et al. 2001).

The extent of pericyte coverage of retinal vessels is markedly greater than that of choriocapillaries (Tilton et al. 1985) or cerebral capillaries (Frank et al. 1987, 1990). The increased ratio of pericytes to endothelial cells in the retina has been proposed to account for the observation that the BRB is less vulnerable to acute hypertension than is the BBB (Laties et al. 1979). It has also been suggested that retinal vessels require an increased level of pericyte coverage in order to maintain a robust BRB, given that they exhibit a higher density of inter-endothelial cell junctions than do cerebral vessels (Stewart and Tuor 1994). The location of pericytes on the abluminal surface of the endothelium has also led others to postulate that pericytes may play a role in the vascular response to BRB breakdown (Sakagami et al. 1999). Given that these cells express tissue factor, they may also contribute to blood coagulation (Bouchard et al. 1997; Rucker et al. 2000).

Fig. 28.2 (continued)
including the choroid. This continuity between structures with CNS and peripheral characteristics in the eye results in regions with an inherent breakdown of the blood-retinal barrier and immune activation. The region of the optic nerve head is one such region, making it prone to inflammatory attack in autoimmune diseases, such as multiple sclerosis (MS) and experimental allergic encephalomyelitis (EAE). Optic neuritis is the clinical manifestation of the inherent weakness of the BRB in the region of the optic nerve head.
(F) The peripapillary region of the control CBA mice is characterised by a marked increase in permeability due to the discontinuity of the glia limitations. The increased permeability is evidenced by the higher fluorescence from the bisbenzimide around the optic disc.
(G) HRP leakage in the optic nerve of control rats. Small arrows indicate meningeal surface. Retina (Ergun et al. 1997) shows HRP leakage from meninges in the lamina cribrosa (LC) region (large arrow) of control rats. (H) MHC class II$^+$ cells with extensive processes (large arrow) in the meninges of control rats. The expression of MHC class II on the surface of a cell is indicative of its potential for antigen presentation. The constitutive expression of MHC class II in the meninges of the retrobulbar portion of the optic nerve, observed in control rats, likely contributes to the development of inflammatory lesions in this region in rats with EAE.

28.8
Membrane Proteins of Tight Junctions

Tight junctions are important in the paracellular transport pathway and maintain cellular polarity (Brightman and Tao-Cheng 1993; Vorbrodt 1993). Occludin was the first integral membrane protein shown to be localized exclusively to tight junctions in both epithelial and endothelial cells (Furuse et al. 1993; Hirase et al. 1997). Occludin contributes to both the structure (Fujimoto 1995; Furuse et al. 1996) and function (Balda et al. 1996; McCarthy et al. 1996; Chen et al. 1997; Wong and Gumbiner 1997) of tight junctions and is expressed in a wide variety of ocular tissues, including the rabbit ciliary epithelium and iris vascular endothelium (Wu et al. 2000).

Two additional integral membrane proteins, claudin-1 and claudin-2, were subsequently shown to be associated with tight junctions, with the former protein being more widely distributed than the latter (Furuse et al. 1998). Claudin-1 and claudin-2 have been shown to belong to a large family of proteins, with more than 20 members identified to date (Furuse et al. 1998; Morita et al. 1999a, b; Simon et al. 1999; Tsukita and Furuse 1999; Mitic et al. 2000). Overall, the claudins are widely distributed in the body, but only claudin-1 and claudin-5 have been shown to be expressed by endothelial cells (Morita et al. 1999a, b) Evidence from cultured cells suggests that claudin-1 and claudin-2 are primarily responsible for the formation of tight junction strands and that occludin is an accessory protein that plays only a minor role in strand formation (Furuse et al. 1998). Whereas ZO-1 is expressed by vascular endothelial cells of the CNS, its expression is poorly correlated with tight junctional complexity or with permeability of endothelial cell monolayers (see Chapters 1 and 4 for further information).

28.9
Localization of Occludin and Claudin-1 to Tight Junctions of Retinal Vascular Endothelial Cells

Information on the expression of occludin and claudins in the retinal vasculature is limited. Increased occludin expression appears to be associated with the development of CNS barrier properties. The extent of occludin expression in brain endothelial cells is thus markedly greater than that in endothelial cells of nonneural origin (Hirase et al. 1997); and downregulation of occludin expression correlates both with breakdown of the BRB in experimentally induced diabetes (Antonetti et al. 1998) and with neutrophil-induced BBB breakdown (Bolton et al. 1998). In human gliomas, expression of claudin-1, but not that of claudin-5, is downregulated in association with an increase in microvascular permeability (Liebner et al. 2000). In contrast, expression of claudin-5 and occludin was found to be downregulated in hyperplastic vessels (Liebner et al. 2000).

The application of occludin and claudin-1 immunohistochemistry and immunoelectron microscopy to vessels of the rabbit eye has provided information on the molecular structure of tight junctions during vessel formation and retraction of vascular segments as well as on the influence of various glial cells on junctional protein expression in different vascular beds (Morcos et al. 2001). Barrier properties and glial ensheathment of vessels were examined by intravascular injection of HRP and by immunohistochemistry, respectively. The distribution of occludin was also examined both in the vessels of the choroid, which lack barrier properties (Flage 1977; Morcos et al. 1999), and in the RPE.

Both occludin (Fig. 28.1 B) and claudin-1 were detected at the junctions of adjacent vascular endothelial cells in the myelinated region of the rabbit retina. Both proteins were specific to vascular endothelial cells, not being expressed by any other cell type within the myelinated streak. In the adult rat retina, consistent with the known barrier properties of intraretinal blood vessels, occludin is expressed by endothelial cells of both the superficial and deep retinal plexuses (Chan-Ling et al. 2005). Consistent with the association of occludin expression with vessels that exhibit functional tight junctions, this protein was found to be expressed at a reduced level in vessels of the aged rat retina, which were also found to exhibit breakdown of the BRB, as revealed by intravascular perfusion with HRP (Chan-Ling et al. 2005).

Immunoelectron microscopy showed that claudin-1 immunoreactivity occupied a slightly greater percentage of the area of tight junctions than did occludin immunoreactivity and that it also extended slightly into the cytoplasm of endothelial cells in the myelinated streak of the rabbit retina (Morcos et al. 2001). These observations suggested that the structural role of claudin-1 may differ from that of occludin. The sites of occludin and claudin-1 immunoreactivity were separated by similar regular intervals (79 nm and 82 nm, respectively), suggesting that the two proteins are localized to the same tight junction strands. These observations also suggested that the paracellular pathway, by which material is transported across tight junctions (Madara 1998; Tsukita and Furuse 2000), consists of the occludin- and claudin-1-positive tight junction strands and the regular intervals between these strands.

28.10
Expression of Occludin by RPE Cells and Lack of Occludin Expression by Choroidal Vessels

RPE cells of the rabbit were found to express occludin around their entire circumference, consistent with the function of these cells as a barrier separating the retina from the leaky vessels of the choroid (Morcos et al. 2001) (Fig. 28.1 C). The rabbit RPE also showed continuous claudin-1 expression, consistent with the formation of tight junctions between adjacent RPE cells (Fig. 28.1 D). Expression of these two tight junction proteins thus likely contributes to the functional boundary between the retina and the leaky blood vessels of the choroid. In contrast, occludin immunoreactivity was detected at only a low level and showed an irregular distribution along the vessels of the choroid (Fig. 28.1 E), a vascular bed that lacks BBB properties.

28.11
Inherent Weakness of the BRB and Existence of Resident MHC Class II⁺ Cells Predisposes the Optic Nerve Head to Inflammatory Attack

Blood vessels in the retrobulbar optic nerve are permeable to intravascular tracers, such as Evans blue and HRP (Tso et al. 1975; Chan-Ling et al. 1992a, b). This inherent weakness of the BBB has been suggested to be responsible for the susceptibility of this region to lesion formation in both experimental allergic encephalomyelitis (EAE) and multiple sclerosis (MS; Guy and Rao 1984; Hu et al. 1998a, b). Indeed, regions of the CNS, such as spinal nerve root sheaths (Pettersson 1993), that lack an inherent BBB often show early accumulation of inflammatory cells in animals with EAE (Shin et al. 1995; Hu et al. 1998a, b). In addition, the periventricular white matter is susceptible to the formation of plaques in individuals with MS (Adams 1977). These observations are consistent with the hypothesis that sites of inherent breakdown of the BBB are susceptible to lesion formation in EAE and MS.

HRP and bizbenzimide leakage from the meninges are apparent only in the region of the lamina cribrosa in control animals (Fig. 28.2 F, G). However, intravascular injection of Monastral blue revealed no evidence of hemorrhage in the lamina cribrosa region of control animals. These observations show that this region has a mild weakness of the BRB, as indicated by leakage of HRP and bizbenzimide, but is not of a nature that results in franked hemorrhaging as would be evident by leakage of Monastral blue.

This site of inherent weakness of the BRB is also characterized by the presence of a resident population of MHC class II⁺ cells. This resident population of MHC class II⁺ cells is present in the meninges surrounding the optic nerve and the lamina cribrosa region of the optic nerve in naive rats (Fig. 28.2 H). These cells possess extensive processes and are possibly the equivalent of the MHC class II⁺ dendritic cells previously detected in the choroid plexus of the eye (Forrester et al. 1994). The expression of MHC class II on the surface of a cell is indicative of its potential for antigen presentation (Lassmann et al. 1994). Expression of MHC class II is associated with the formation of inflammatory lesions in EAE (Molleston et al. 1993) and autoimmune diseases of the peripheral nervous system (Pollard et al. 1986). The choroid plexus of the eye is not a part of the CNS, nor does its vasculature show embryological or functional characteristics of CNS microvessels (Chan-Ling and Stone 1993). However, because of its proximity to, and continuity with, the meninges of the optic nerve (Fig. 28.2 E), the characteristics of the choroid are important in the development of optic neuritis. Therefore, inherent breakdown of the BRB and constitutive expression of MHC class II in the meninges of the retrobulbar portion of the optic nerve, observed in naïve subjects, predisposes this region to the development of inflammatory lesions in this region in rats with EAE and patients with MS.

28.11.1
Compromised BBB Where CNS Meets Peripheral Vascular Bed

Other regions of inherent weakness of the BBB include the choroid plexus within the ventricles (Brightman and Reese 1969). Vessels of the ventricle are highly permeable due to their function in the formation of cerebral spinal fluid; and we also have demonstrated the existence of a substantial population of constitutive MHC class II$^+$ cells in this region in naïve control adult JC Lewis rats (Ping Hu and Tailoi Chan-Ling, unpublished data). This region of the CNS is also prone to inflammatory attack during acute EAE. A third region of special interest is the region of the retina adjacent to the ciliary margin where the retina is adjacent to the leaky vascular bed of the ciliary processes, a bed responsible for secretion of the aqueous humor.

28.12
Clinical and Experimental Determination of the Blood-Retinal Barrier

The eyes have been colloquially identified as the 'window' to the brain for many decades. This is because of the transparency of the ocular media, making it possible to view the retina in a living organism. In clinical studies, the integrity of the BRB can be evaluated by fluorescein angiography and vitreous fluorophotometry. These methods are able to provide qualitative and quantitative information of the BRB and in particular the exact location of the breakdown of the BRB. Further, the arterial/venous pairs that characterise mammalian retinal vascular tree makes it possible to determine whether it is an arterial or venous vessel that is compromised or whether it is a capillary-sized vessel that is affected.

While these techniques are useful in the clinic, far greater details of the functional alteration of the BRB can be gained utilizing invasive procedures on experimental animals. The majority of methods employed involve intravenous injection of intravascular tracers prior to the sacrifice of the experimental animals. The eye is then enucleated and the retina dissected and examined as a retinal wholemount preparation (Chan-Ling 1997). The retina is unique in that it is only around 200–250 µm in total thickness. As such, it can be examined as a retinal wholemount where the entire vascular tree can be examined with the relationship to neurons and glia intact. This approach has made possible the detection of a very mild and transient breakdown in the BRB induced by activated T cells of nonneural specificity which would not be possible if studying the brain (Hu et al. 2000) as well as the early detection of a breakdown in the BRB in an experimental model of cerebral malaria (Chan-Ling et al. 1992a). Experimental determination of the BRB involves the injection of intravascular tracers prior to sacrifice of the animal. Tracers frequently utilised include horse radish peroxidase with a molecular mass of 44 000 kDa, which is similar to that of plasma albumin (Fig. 28.3 A), lanthanum hydroxide, and FITC-conjugated dextran of various known molecular weights.

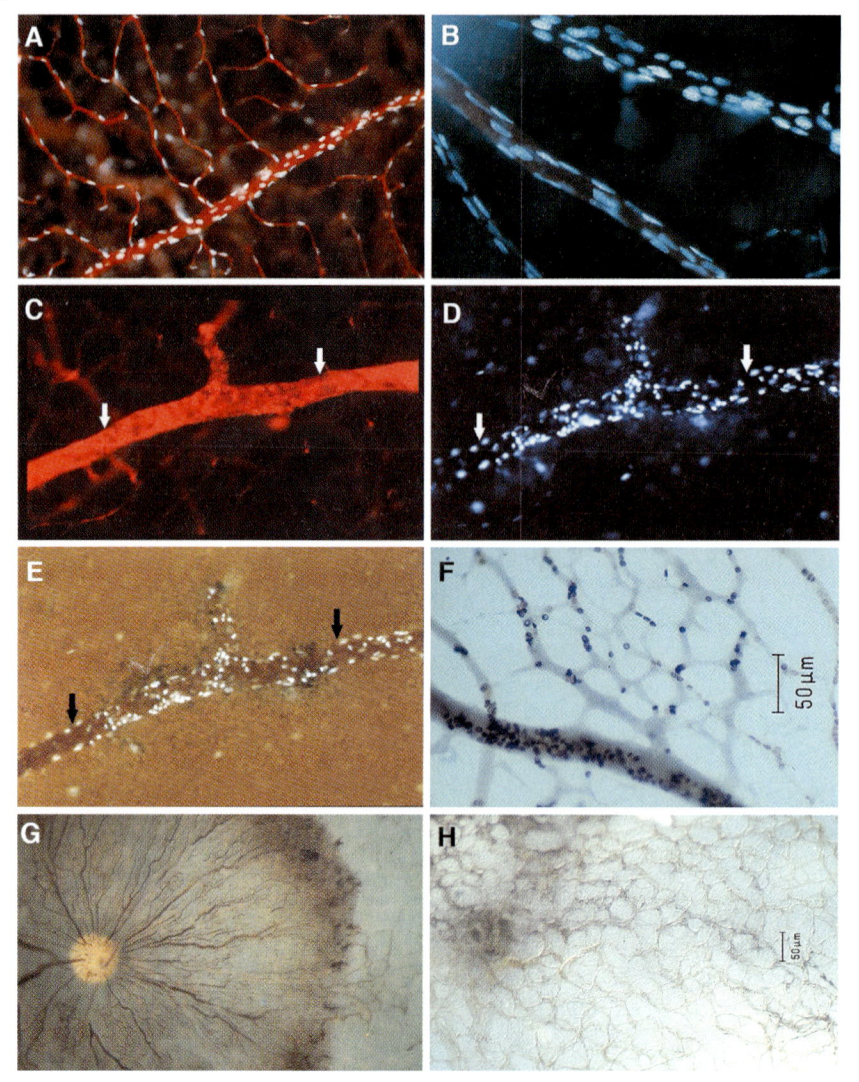

Fig. 28.3 (A) The superficial plexus of the kitten retina in a postnatal day 12 kitten retina visualized by intravenous injection of HRP prior to sacrifice. The tightness of the BRB in this region close to the optic disc is evidenced by inherent peroxidase activity of the red blood cells and HRP within the vessel lumens. The absence of HRP in the retinal parenchyma results in a tissue background that is almost transparent when the BRB is intact. This is in marked contrast to that seen in (B) and (C), where the BRB is breached.

(B) Lack of barrier properties in proliferative neovasculature in the kitten model of retinopathy of prematurity. The panel shows the extent of the neovascular rosette at 5 days after return of the kitten to room air, demonstrated via intravenous injection of HRP prior to sacrifice. The vessels at the edge of the rosette and throughout the retinal parenchyma are leaky.

(C) At 10 days after return of the kitten to room air, the loss of HRP from within the vessel lumen is still evident, shown here in a mid-peripheral region of the retina.

The permeability of retinal vessels can be assessed, following perfusion with HRP, a long-established tracer used in studies of the BBB (for a review, see Broadwell 1989) either at the light or electron microscopic level. Briefly, HRP (200 mg kg^{-1} body mass in 1 ml of physiological saline) is injected into the tail vein 20 min prior to sacrifice. The retinas are dissected and then incubated in buffer containing diaminobenzidine, to visualize the reaction product. It can be mounted and examined with a transmitted light microscope at this stage or further processed for electron microscopy (Chan-Ling et al. 1992b). Examination at the light microscopic level offers significant information of the exact sites of breakdown of the BRB, including the types of vessels affected as well as the region of the retina affected (Fig. 28.3B, C), whereas studies at the ultrastructural level permits the localization of sites of HRP leakage with the morphology of the junctional complexes in the same vessel.

The modified trichrome technique developed in our laboratory (Chan-Ling et al. 1992a) has proven a useful tool for determination of the BRB. This technique involves the intravenous administration of Monastral blue, a colloidal dye, 2 h prior to sacrifice, followed by the intravenous injection of bisbenzimide and Evan's blue 30 min prior to sacrifice. Only small quantities of dyes are injected, taking care not to increase the total blood volume, as small changes in blood volume can be fatal in fragile experimental animals. Monastral blue is a colloidal dye, the particles of which coat, or are ingested by, activated monocytes (Neill and Hunt 1992). It is

Fig. 28.3 (continued)
HRP is seen both within the vessel lumen and as a brown reaction product in the parenchymal tissue.
(D–E) The barrier properties of various retinal vascular beds demonstrated utilising the modified Trichrome technique.
(D) The inner vascular plexus in control CBA mice visualized with intravascular perfusion of Evan's blue and Bisbenzimide (double exposure). The nuclei of retinal vascular endothelial cells are visible as pale blue nuclei. The intravascularly injected Evan's blue is bound to plasma albumin. All Evan's blue is retained within the vessel lumens, indicative of the tight barrier properties of intra-retinal vessels in control CBA mice.
(E) Where the BRB is intact, intravascular perfusion of bisbenzimide (a nuclear stain) results in staining of only vascular endothelial cells and intravascular leukocytes. Endothelial cells in arteries have a characteristic elongated morphology, whereas venous endothelial cells have a rounded morphology. Where the BRB is breached, resulting in leakage of the bisbenzimide, retinal somas adjacent to the vessels are also lightly stained, as seen in (G).
(F–H) The same field of view of a retinal vein, utilising the modified trichrome technique, at day 3 post-induction with ovalbumin (OVA)-activated T cells combined with intraocular injection of OVA. The intensity of the breakdown of the BRB, accumulation of monocytes, and franked haemorrhaging evident in these panels is typical of that observed in antigen-specific inflammatory responses. Arrows point to the same segment of the vein in each field of view and mark the location of adherent monocytes.
(F) Evan's blue show sites of leakage of plasma albumin.
(G) Hoechst stain show endothelial nuclear morphology and leukocyte adhesion in the affected region of the retinal vein.
(H) Monastral blue, Hoechst stain and transmitted light microscopy show the site of franked haemorrhaging as blue particulate matter on the outside of the vessel wall.

also used to detect sites of marked breakdown of the BRB, as in frank hemorrhage. Briefly, rats are injected intravenously with 1 ml of 3% (w/v) solid Monastral blue solution in 0.9% (w/v) sodium chloride, 2 h prior to sacrifice, taking care to sonicate the dye immediately before injection, to disrupt aggregates. Evan's blue binds strongly to serum albumin (68 500 kDa) (Wolman et al. 1981); and its distribution after intravascular injection reflects the distribution of serum albumin (Radius and Anderson 1980). Bisbenzimide, also known as Hoechst stain is a nonspecific DNA marker. Where the BRB is intact, intravascular injection of the dye restricts its access to contents within the vessel lumen, resulting in the nuclear staining of vascular endothelial cells as well as intravascular leukocytes. Where the BRB is breached, the bisbenzimide leaks out of the vessel lumen, thus exposing all retinal somas, resulting in the staining of neurons, glia, and microglia alike. The retinas are dissected (Chan-Ling 1997) and placed in fixative for 20 min prior to mounting and examination under a fluorescent microscope. Evan's blue is observed under green fluorescence microscopy (excitation filter 545 nm, barrier filter 590 nm) and appears red (Fig. 28.3 D, F). The Hoechst stain is examined under ultraviolet fluorescence microscopy (excitation filter 360 nm, barrier filter 455 nm) and appears light blue (Fig. 28.3 D, E, G, H). Transmitted light microscopic examination is used to examine the rheologic aspects of the blood column; and, in particular, erythrocytes can be visualized easily through their brown hemoglobin and monocytes by their phagocytic granules of Monastral blue and their characteristic nuclear morphology (Fig. 28.3 H).

The modified trichrome technique has been successfully applied for the study of the BRB in a number of experimental models (Chan-Ling et al. 1992a; Hu et al. 1998a, b, 2000). Compared with histological sections of the brain or retina, the retinal wholemount technique allows earlier detection and localisation of marginating monocytic cells, congestion, and small hemorrhages and is an extremely sensitive tool for the detection of even the mildest breakdown of the BRB. Other changes to the microvasculature, such as occlusion of vessel segments to plasma perfusion are only detectable using the modified trichrome technique in retinal wholemounts.

28.13
Conclusions

In humans and several other mammalian species, the BRB is composed of two parts because the retina has a dual vascular supply, the inner two-thirds of the retina being nourished by intraretinal vessels and the outer one-third being supplied by choroidal vessels. The barrier properties of the inner retina depend on the tight junctions between adjacent endothelial cells of the intraretinal capillaries, whereas those of the outer retina depend on the tight junctions between adjacent cells of the RPE. These two barriers have been described as the inner and outer BRB, respectively.

The retina is an extension of the CNS and, as a result, the characteristics of the blood-retinal interface is similar to that of the blood-brain interface, including its barrier properties, its ability to autoregulate its blood flow, ensheathment by astrocytes, presence of microglia, pericyte association, and immune responses. However, by virtue of the fact that the retina is found within the eyecup, differences do exist. Müller cells, the radial glia of the retina, share in the ensheathment and induction of the BRB with astrocytes. Further, the proximity and continuity of retinal vascular beds with leaky peripheral vascular beds results in sites with inherent breakdown of the BRB and priming of the immune cells in the eye in the region of the optic nerve head, predisposing this region to inflammatory attack in autoimmune diseases, such as MS. The retina is unique in that it is only around 200–250 µm in total thickness. As such, it can be examined as a retinal wholemount where the entire vascular tree can be examined with the relationship to neurons and glia intact. Experimental determination of the BRB involves the injection of intravascular tracers, including horse radish peroxidase, lanthanum hydroxide, FITC-conjugated dextran, and the modified trichrome technique. Studies of the BRB continue to contribute to the understanding of various CNS and retinal pathologies.

References

Adams, C. W. **1977**, Pathology of multiple sclerosis: progression of the lesion (review), *Br. Med. Bull.* 33, 15–20.

Alm, A., Bill, A. **1973**, Ocular and optic nerve blood flow at normal and increased intraocular pressures in monkeys (*Macaca irus*): a study with radioactively labelled microspheres including flow determinations in brain and some other tissues, *Exp. Eye Res.* 15, 15–29.

Antonetti, D. A., Barber, A. J., et al. **1998**, Vascular permeability in experimental diabetes is associated with reduced endothelial occludin content: vascular endothelial growth factor decreases occludin in retinal endothelial cells, *Diabetes* 47, 1953–1959.

Ashton, N. **1970**, Retinal angiogenesis in the human embryo, *Brit. Med. Bull.* 26, 103–106.

Balda, M. S., Whitney, J. A., et al. **1996**, Functional dissociation of paracellular permeability and transepithelial electrical resistance and disruption of the apical-basolateral intramembrane diffusion barrier by expression of a mutant tight junction membrane protein, *J. Cell Biol.* 134, 1031–1049.

Bauer, H., Stelzhammer, W., et al. **1999**, Astrocytes and neurons express the tight junction-specific protein occludin in vitro, *Exp. Cell Res.* 250, 434–438.

Beck, D. W., Vinters, H. V., et al. **1984**, Glial cells influence polarity of the blood-brain barrier. *J. Neuropathol. Exp. Neurol.* 43, 219–224.

Beck, D. W., Roberts, R. L., et al. **1986**, Glial cells influence membrane-associated enzyme activity at the blood-brain barrier, *Brain Res.* 381, 131–137.

Bellhorn, R. **1980**, Permeability of the blood-ocular barriers of neonatal and adult cats to sodium fluorescein. *Invest. Ophth. Vis. Sci.* 1980, 870–877.

Bill, A., Nilsson S. **1982**, The blood supply of the eye and its regulation, in *Basic Aspects of Glaucoma Research*, ed. E. Lutjen-Drecoll, Schattaller, Stuttgart, p. 39–48.

Bill, A., Nilsson S. **1985**, Control of ocular blood flow, *J. Cardiovasc. Pharmacol.* 7[Suppl 3], S96–S102.

Bill, A., Sperber, G. **1990**, Control of retinal and choroidal blood flow, *Eye* 4, 319–325.

Bill, A., Tornquist, P. et al. **1980**, Permeability of the intraocular blood vessels, *Trans. Ophth. Soc. UK* 100, 332–336.

Bill, A., Sperber, G., et al. **1983**, Physiology of the choroidal vascular bed, *Int. Ophth.* 6, 101–107.

Bolton, S. J., Anthony, D. C., et al. **1998**, Loss of the tight junction proteins occludin and zonula occludens-1 from cerebral vascular endothelium during neutrophil-induced blood-brain barrier breakdown in vivo, *Neuroscience* 86, 1245–1257.

Bouchard, B. A., Shatos, M. A., et al. **1997**, Human brain pericytes differentially regulate expression of procoagulant enzyme complexes comprising the extrinsic pathway of blood coagulation, *Neuroscience* 85, 1–9.

Bouchard, C., Lebert, M., et al. **1989**, Are close contacts between astrocytes and endothelial cells a prerequisite condition of a blood brain barrier – the rat subfornical organ as an example, *Biol. Cell* 67, 159–165.

Brightman, M. W. **1989**, The anatomic basis of the blood-brain barrier, in *Implications of the Blood-Brain Barrier and its Manipulation*, vol. 1, ed. E. A. Neuwelt, Plenum Medical, New York, p. 53–78.

Brightman, M. W., Reese, T. S. **1969**, Junctions between intimately opposed cell membranes in the vertebrate brain, *J. Cell Biol.* 40, 648–677.

Brightman, M. W., Tao-Cheng, J. H. **1993**, Tight junctions of brain endothelium and epithelium, in *The Blood-Brain Barrier, Cellular and Molecular Biology*, ed. W. M. Pardridge, Raven Press, New York, p. 107-125.

Broadwell, R. D. **1989**, Transcytosis of macromolecules through the blood-brain barrier: a cell biological perspective and critical appraisal, *Acta Neuropathol.* 79, 117–128.

Bussow, H. **1980**, The astrocytes in the retina and optic nerve head of mammals: a special glia for the ganglion cell axons, *Cell Tissue Res.* 206, 367–378.

Chan-Ling, T. **1994**, Glial, neuronal and vascular interactions in the mammalian retina, in *Progress in Retinal Research*, vol. 13, eds. N. Osborne, G. Chader, Pergamon Press, Oxford, p. 357–389.

Chan-Ling, T. **1997**, Glial, vascular, and neuronal cytogenesis in whole-mounted cat retina, *Microsc. Res. Tech.* 36, 1–16.

Chan-Ling, T., Stone, J. **1992**, Degeneration of astrocytes in feline retinopathy of prematurity causes failure of the blood-retinal barrier, *Invest. Ophth. Vis. Sci.* 33, 2148–2159.

Chan-Ling, T., Stone, J. **1993**, Retinopathy of prematurity: origins in the architecture of the retina, *Prog. Retinal Res.* 12, 155–177.

Chan-Ling, T., Halasz, P., et al. **1990**, Development of retinal vasculature in the cat: processes and mechanisms, *Curr. Eye Res.* 9, 459–478.

Chan-Ling, T., Neill, A. L., et al. **1992**, Early microvascular changes in murine cerebral malaria detected in retinal wholemounts, *Am. J. Pathol.* 140, 1121–1130.

Chan-Ling, T., Tout, S., et al. **1992**, Vascular changes and their mechanisms in the feline model of retinopathy of prematurity, *Invest. Ophth. Vis. Sci.* 33, 2128–2147.

Chan-Ling, T., Gock, B., et al. **1995**, The effect of oxygen on vasoformative cell division: evidence that physiological hypoxia is the stimulus for normal retinal vasculogenesis, *Invest. Ophth. Vis. Sci.* 36, 1201–1214.

Chan-Ling, T., McLeod, D. S., et al. **2004 a**, Astrocyte-endothelial cell relationships during human retinal vascular development, *Invest. Ophth. Vis. Sci.* 46, 2020–2032.

Chan-Ling, T., Page, M. P., et al. **2004 b**, Desmin ensheathment ratio as an indicator of vessel stability: evidence in normal development and in retinopathy of prematurity, *Am. J. Pathol.* 165, 1301–1313.

Chan-Ling, T., Hughes, S., et al. **2005**, Inflammation and breakdown of the blood-retinal barrier during physiological aging in the rat retina, *Neurobiol. Aging*, in press.

Chen, Y.-H., Merzdorf, C., et al. **1997**, COOH terminus of occludin is required for tight junction barrier function in early *Xenopus* embryos, *J. Cell Biol.* 138, 891–899.

Chu, Y., Hughes, S., et al. **2001**, Differentiation and migration of astrocyte precursor cells and astrocytes in human fetal retina: relevance to optic nerve coloboma, *FASEB J.* 15, 2013–2015.

Cunha-Vaz, J. G. **1976**, The blood-retinal barriers, *Doc. Ophth.* 41, 287–327.

Dehouck, M. P., Meresse, S., et al. **1990**, An easier, reproducible, and mass-production method to study the blood-brain barrier in vitro, *J. Neurochem.* 54, 1798–1801.

Dohgu, S., Takata, F., et al. **2005**, Brain pericytes contribute to the induction and up-regulation of blood-brain barrier functions through transforming growth factor-beta production, *Am. J. Pathol.* 166, 208–215.

Ergun, S., Bruns, T., et al. **1997**, Angioarchitecture of the human spermatic cord, *Adv. Exp. Med. Biol.* 424, 183–184.

Flage, T. **1977**, Permeability properties of the tissues in the optic nerve head region in the rabbit and the monkey: an ultrastructural study, *Acta Ophth.* 55, 652–664.

Flower, R. W. **1990**, Perinatal ocular physiology and ROP in the experimental animal model, *Doc. Ophth.* 74, 153–162.

Flower, R. W., Klein, G. J. **1990**, Pulsatile flow in the choroidal circulation: a preliminary investigation, *Eye* 4, 310–318.

Forrester, J. V., McMenamin, P. G., et al. **1994**, Localization and characterization of major histocompatibility complex class II – positive cells in the posterior segment of the eye: implications for induction of autoimmune uveoretinitis, *Invest. Ophth. Vis. Sci.* 35, 64–77.

Frank, R. N., Dutta, S., et al. **1987**, Pericyte coverage is greater in the retinal than in the cerebral capillaries of the rat, *J. Cell Sci.* 1987, 1086–1091.

Frank, R. N., Turczyn, T. J., et al. **1990**, Pericyte coverage of retinal and cerebral capillaries, *Invest. Ophth. Vis. Sci.* 31, 999–1007.

Frey, A., Meckelein, B., et al. **1991**, Pericytes of the brain microvasculature express gamma-glutamyl transpeptidase, *J. Cell Sci.* 1991, 421–429.

Fujimoto, K. **1995**, Freeze-fracture replica electron microscopy combined with SDS digestion for cytochemical labeling of integral membrane proteins: application to the immunogold labeling of intercellular junctional complexes, *J. Cell Sci.* 108, 3443–3449.

Furuse, M., Fujimoto, K., et al. **1996**, Overexpression of occludin, a tight junction-associated integral membrane protein, induces the formation of intracellular multilamellar bodies bearing tight junction-like structures, *J. Cell Sci.* 109, 429–435.

Furuse, M., Fujita, K., et al. **1998**, Claudin-1 and -2: novel integral membrane proteins localizing at tight junctions with no sequence similarity to occludin, *J. Cell Biol.* 141, 1539–1550.

Furuse, M., Hirase, T., et al. **1993**, Occludin: a novel integral membrane protein localizing at tight junctions [see comments], *J. Cell Biol.* 123, 1777–1788.

Furuse, M., Sasaki, H., et al. **1998**, A single gene product, claudin-1 or -2, reconstitutes tight junction strands and recruits occludin in fibroblasts, *J. Cell Biol.* 143, 391–401.

Gardner, T. W., Lieth, E., et al. **1997**, Astrocytes increase barrier properties and ZO-1 expression in retinal vascular endothelial cells, *Invest. Ophth. Vis. Sci.* 38, 2423–2427.

Guy, J., Rao, N. A. **1984**, Acute and chronic experimental optic neuritis: alteration in the blood-nerve barrier, *Arch. Ophth.* 102, 450–454.

Hellstrom, M., Gerhardt, H., et al. **2001**, Lack of pericytes leads to endothelial hyperplasia and abnormal vascular morphogenesis, *J. Cell Biol.* 153, 543–553.

Hirase, T., Staddon, J. M., et al. **1997**, Occludin as a possible determinant of tight junction permeability in endothelial cells, *J. Cell Sci.* 110, 1603–1613.

Hollander, H., Makarov, F., et al. **1991**, Structure of the macroglia of the retina: sharing and division of labour between astrocytes and Muller cells, *J. Comp. Neurol.* 313, 587–603.

Hori, S., Ohtsuki, S., et al. **2004**, A pericyte-derived angiopoietin-1 multimeric complex induces occludin gene expression in brain capillary endothelial cells through Tie-2 activation in vitro, *J. Neurochem.* 89, 503–513.

Hu, P., Pollard, J., et al. **1998a**, Microvascular and cellular responses in the retina of rats with acute experimental allergic encephalomyelitis (EAE), *Brain Pathol.* 8, 487–498.

Hu, P., Pollard, J., et al. **1998b**, Microvascular and cellular responses in the optic nerve of rats with acute experimental allergic encephalomyelitis (EAE), *Brain Pathol.* 8, 475–486.

Hu, P., Pollard, J. D., et al. **2000**, Breakdown of the blood-retinal barrier induced by activated T cells of nonneural specificity, *Am. J. Pathol.* 156, 1139–1149.

Hughes, S., Chan-Ling, T. **2000**, Roles of endothelial cell migration and apoptosis in vascular remodeling during development of the central nervous system, *Microcirculation* 7, 317–333.

Hughes, S., Chan-Ling, T. **2004**, Characterization of smooth muscle cell and pericyte differentiation in the rat retina in vivo, *Invest. Ophth. Vis. Sci.* 45, 2795–2806.

Hughes, S., Yang, H., et al. **2000**, Vascularization of the human fetal retina: roles of vasculogenesis and angiogenesis, *Invest. Ophth. Vis. Sci.* 41, 1217–1228.

Janzer, R. C., Raff, M. C. **1987**, Astrocytes induce blood-brain barrier properties in endothelial cells, *Nature* 325, 253–257.

Jiang, B., Bezhadian, M. A., et al. **1995**, Astrocytes modulate retinal vasculogenesis: effects on endothelial cell differentiation, *Glia* 15, 1–10.

Kondo, H., Takahashi, H., et al. **1984**, Immunohistochemical study of S-100 protein in the postnatal development of Muller cells and astrocytes in the rat retina, *Cell Tissue Res.* 238, 503–508.

Kunz, J., Krause, D., et al. **1994**, The 140-kDa protein of blood-brain barrier-associated pericytes is identical to aminopeptidase N, *Cell Tissue Res.* 248, 2375–2386.

Lassmann, H., Rinner, W., et al. **1994**, Differential role of hematogenous macrophages, resident microglia and astrocytes in antigen presentation and tissue damage during autoimmune encephalomyelitis, *Neuropath. Appl. Neurobiol.* 20, 195–196.

Laterra, J., Guerin, C., et al. **1990**, Astrocytes induce neural microvascular endothelial cells to form capillary-like structures in vitro, *J. Cell. Physiol.* 144, 204–215.

Laties, A. M., Rapoport, S. I., et al. **1979**, Hypertensive breakdown of cerebral but not of retinal blood vessels in rhesus monkey, *Cell Tissue Res.* 233, 1511–1514.

Liebner, S., Fischmann, A., et al. **2000**, Claudin-1 and claudin-5 expression and tight junction morphology are altered in blood vessels of human glioblastoma multiforme, *Acta Neuropathol.* 100, 323–331.

Ling, T., Mitrofanis, J., et al. **1989**, Origin of retinal astrocytes in the rat: evidence of migration from the optic nerve, *J. Comp. Neurol.* 286, 345–352.

Ling, T. L., Stone, J. **1988**, The development of astrocytes in the cat retina: evidence of migration from the optic nerve, *Brain Res. Dev. Brain Res.* 44, 73–85.

Madara, J. L. **1998**, Regulation of the movement of solutes across tight junctions, *Annu. Rev. Physiol.* 60, 143–159.

Maxwell, K., Berliner, J. A., et al. **1987**, Induction of gamma-glutamyl transpeptidase in cultured cerebral endothelial cells by a product released by astrocytes, *Brain Res.* 410, 309–314.

McCarthy, K. M., Skare, I. B., et al. **1996**, Occludin is a functional component of the tight junction, *J. Cell Sci.* 109, 2287–2298.

Mi, H., Haeberle, H., et al. **2001**, Induction of astrocyte differentiation by endothelial cells, *J. Neurosci.* 21, 1538–1547.

Michaelson, I. C. **1954**, *Retinal Circulation in Man and Animals*, Charles C. Thomas, New York.

Mitic, L. L., Van Itallie, C. M., et al. **2000**, Molecular physiology and pathophysiology of tight junctions I, tight junction structure and function: lessons from mutant animals and proteins, *Am. J. Physiol. Gastroint. Liver Physiol.* 279, G250–G254.

Molleston, M. C., Thomas, M. L., et al. **1993**, Novel major histocompatibility complex expression by microglia and site-specific experimental allergic encephalomyelitis lesions in the rat central nervous system after optic nerve transection, *Adv. Neurol.* 59, 337–348.

Morcos, Y., Shorey, C. D., et al. **1999**, Contribution of O^{4+} oligodendrocyte precursors and astrocytes to the glial ensheathment of vessels in the rabbit myelinated streak, *Glia* 27, 1–14.

Morcos, Y., Hosie, M. J., et al. **2001**, Immunolocalization of occludin and claudin-1 to tight junctions in intact CNS vessels of mammalian retina, *J. Neurocytol.* 30, 107–123.

Morita, K., Sasaki, H., et al. **1999a**, Claudin-11/OSP-based tight junctions of myelin sheaths in brain and Sertoli cells in testis, *J. Cell Biol.* 145, 579–588.

Morita, K., Sasaki, H., et al. **1999b**, Endothelial claudin: claudin-5/TMVCF constitutes tight junction strands in endothelial cells, *J. Cell Biol.* 147, 185–194.

Neill, A. L., Hunt, N. H. **1992**, Pathology of fatal and resolving *Plasmodium berghei* cerebral malaria in mice, *Parasitology* 105, 165–175.

Olson, L., Seiger, A., et al. **1983**, Intraocular transplantation in rodents: a detailed account of the procedure and examples of its use in neurobiology with special reference to brain tissue grafting, *Adv. Cell. Neurobiol.* 4, 407–442.

Penfold, P. L., Provis, J. M., et al. **1990**, Angiogenesis in normal human retinal development: the involvement of astrocytes and macrophages, *Arch. Clin. Exp. Ophth.* 228, 255–263.

Pettersson, C. A. **1993**, Sheaths of the spinal nerve roots: permeability and structural characteristics of dorsal and ventral spinal nerve roots of the rat, *Acta Neuropathol.* 85, 129–137.

Pollard, J. D., McCombe, P. A., et al. **1986**, Class II antigen expression and T lymphocyte subsets in chronic inflammatory demyelinating polyneuropathy, *J. Neuroimmunol.* 13, 123–134.

Radius, R. L., Anderson, D. R., **1980**, Distribution of albumin in the normal monkey eye as revealed by Evans Blue fluorescence microscopy, *Invest. Ophth. Vis. Sci.* 19, 238–243.

Raub, T. J., Kuentzel, S. L., et al. **1992**, Permeability of bovine brain microvessel endothelial cells in vitro: barrier tightening by a factor released from astroglioma cells, *Exp. Cell Res.* 199, 330–340.

Rucker, H. K., Wynder, H. J., et al. **2000**, Cellular mechanisms of CNS pericytes. *Exp. Cell Res.* 207, 363–369.

Sakagami, K., Wu, D. M., et al. **1999**, Physiology of rat retinal pericytes: modulation of ion channel activity by serum-derived molecules, *J. Physiol.* 521, 637–650.

Sandercoe, T. M., Madigan, M. C., et al. **1999**, Astrocyte proliferation during development of the human retinal vasculature, *Exp. Eye Res.* 69, 511–523.

Schnitzer, J. **1988**, Astrocytes in mammalian retina, in *Progress in Retinal Research*, ed. C. Osbourne, Pergamon Press, New York, p. 209–231.

Shin, T., Kojima, T., et al. **1995**, The subarachnoid space as a site for precursor T cell proliferation and effector T cell selection in experimental autoimmune encephalomyelitis, *J. Neuroimmunol.* 56, 171–178.

Simon, D. B., Lu, Y., et al. **1999**, Paracellin-1, a renal tight junction protein required for paracellular Mg^{2+} resorption [see comment], *J. Neuroimmunol.* 60, 103–106.

Stewart, P. A., Coomber, B. L. **1986**, Astrocytes and the blood-brain barrier, *Astrocytes* 1, 311–327.

Stewart, P. A., Hayakawa, E. M. **1987**, Interendothelial junctional changes underlie the developmental 'tightening' of the blood-brain barrier, *Brain Res.* 429, 271–281.

Stewart, P. A., Tuor, U. I. **1994**, Blood-eye barriers in the rat: correlation of ultrastructure with function, *J. Comp. Neurol.* 340, 566–576.

Stewart, P. A., Wiley, M. J. **1981**, Developing nervous tissue induces formation of blood-brain barrier characteristics in invading endothelial cells: a study using quail-chick transplantation chimeras, *Dev. Biol.* 84, 183–192.

Stone, J., Dreher, Z. **1987**, Relationship between astrocytes, ganglion cells, and vasculature of the retina, *J. Comp. Neurol.* 255, 35–49.

Tao-Cheng, J. H., Brightman, M. W. **1988**, Development of membrane interactions between brain endothelial cells and astrocytes *in vitro*, *Int. J. Dev. Neurosci.* 6, 25–37.

Tao-Cheng, J. H., Nagy, Z., et al. **1987**, Tight junctions of brain endothelium in vitro are enhanced by astroglia, *J. Neurosci.* 7, 3293–3299.

Tilton, R. G., Miller, E. J., et al. **1985**, Pericyte form and distribution in rat retinal and uveal capillaries, *J. Neurosci.* 9, 68–73.

Tontsch, U., Bauer, H. C. **1991**, Glial cells and neurons induce blood-brain barrier related enzymes in cultured cerebral endothelial cells, *Brain Res.* 539, 247–253.

Tout, S., Chan-Ling, T., et al. **1993**, The role of Muller cells in the formation of the blood-retinal barrier, *Neuroscience* 55, 291–301.

Tso, M. O., Shih, C. Y., et al. **1975**, Is there a blood-brain barrier at the optic nerve head? *Arch. Ophth.* 93, 815–825.

Tsukita, S., Furuse, M. **1999**, Occludin and claudins in tight-junction strands: leading or supporting players? *Trends Cell Biol.* 9, 268–273.

Tsukita, S., Furuse, M. **2000**, Pores in the wall: claudins constitute tight junction strands containing aqueous pores, *J. Cell Biol.* 149, 13–16.

Tyler, N. K., Burns, M. S. **1991**, Comparison of lectin reactivity in the vessel beds of the rat eye, *Curr. Eye Res.* 10, 801–810.

Vorbrodt, A. W. **1993**, Morphological evidence of the functional polarization of brain microvascular endothelium, in *The Blood Brain Barrier, Cellular and Molecular biology*, ed. W. M. Pardridge, Raven Press, New York, p. 137–164.

Wolburg, H., Neuhaus, J., et al. **1994**, Modulation of tight junction structure in blood-brain barrier endothelial cells: effects of tissue culture, second messengers and cocultured astrocytes, *J. Cell Sci.* 107, 1347–1357.

Wolman, M., Klatzo, I., et al. **1981**, Evaluation of the dye-protein tracers in pathophysiology of the blood-brain barrier, *Acta Neuropathol.* 54, 55–61.

Wong, V., Gumbiner, B. M. **1997**, A synthetic peptide corresponding to the extracellular domain of occludin perturbs the tight junction permeability barrier, *J. Cell Biol.* 136, 399–409.

Wu, P., Gong, H., et al. **2000**, Localization of occludin, ZO-1, and pan-cadherin in rabbit ciliary epithelium and iris vascular endothelium, *Histochem. Cell Biol.* 114, 303–310.

Yamamoto, F., Steinberg, R. H. **1992**, Effects of systemic hypoxia on pH outside rod photoreceptors in the cat retina, *Exp. Eye Res.* 54, 699–709.

Subject Index

a

Aβ amyloidose 279
A cell 151
ABC
– superfamily 432
– transporter 425, 442
absorption, distribution, metabolism and excretion, see ADME
Acanthamoeba 680
Acanthamoeba castellanii 683
acetate 573
acetylcholine 274, 613, 654
acidic fibroblast growth factor (aFGF) 18
acidosis 653
acoustic stimulation 574
actin 443
actin cytoskeleton 622
– mediator of cerebral edema 626
– paracellular transport 626
– pinocytotic vesicle formation 626
– regulated permeability 626
– tight junction 626
– transcellular flux 626
– zona occludens 626
active
– influx/efflux 418
– transport 423
acute
– lymphoblastic leukaemia (ALL) 445
– trauma 612
ADAM (a disintegrin and metalloproteinase) 317
ADAMT (a disintegrin and metalloproteinase with a thrombospondin motif) 317
addiction 580
adherens junction 15
adhesion molecule 117, 181, 290
ADME 404, 424, 436
– property 404, 421

adrenomedullin (AM) 53
advanced glycation end-product 90
aerobic glycolysis 578
African trypanosomiasis 672 ff., 689 ff.
agent, HIV-associated 440
agrin 23, 96
AIDS 275 ff., 325, 693 f.
AKT signal pathway 46
albumin 90, 654
– extravasation 242
ALK-1 212
alkaline phosphatase (ALKP) 211, 214, 241, 321
Alzheimer's disease (AD) 277 ff., 440, 448, 480, 486, 612 ff., 659
– inflammatory response 613
American trypanosomiasis 689 ff.
amino acid 654 f.
– transporter 313
aminopeptidase A (APA) 321
aminopeptidase A (EC 3.4.11.7) 314
aminopeptidase M 314
aminopeptidase N 113
aminopeptidase N (EC 3.4.11.2) 314
aminopeptidase N pericytic pAPN 321 f.
aminopeptidase N(M) (APN) 321
amoebae 678
amphiphilicity 436
β-amyloid (Aβ) 448
β-amyloid aggregation 278
amyloid A precursor protein 278
amyloid peptide precursor (A4P) 613
β-amyloid protein 613
amyotrophic lateral sclerosis (ALS) 256
analgesia 485, 488
analgesic 487
androgen 481
Ang-1 26, 122
Ang-2 122

Subject Index

angioblast 17
angiogenesis 17, 21, 28, 46, 49, 52, 61 ff., 120, 151, 321 f.
– androgen-induced 63
– estrogen-induced 63
– persistent 62
angiogenesis-neurogenesis interaction 65
angiogenic factor 47 f.
– neuroglia-derived 47
angiogenic vessel 42, 440
angiopoietin-1 (Ang-1) 19, 119, 211 f.
angiopoietin-1 (Ang-1/Tie-2) 51
angiopoietin-2 (Ang-2) 19
Angiostrongylus cantonensis 681
angiotensin I 315
angiotensin II 315, 321 f.
angiotensin III 321
angiotensin-converting enzyme (ACE, EC 3.4.15.1) 314 f.
Anisakis 678
annexin-1 293
anorexia 274 ff.
antibody 179
anticancer agent 435
antigen presentation 169, 172
antigen-presenting cell (APC) 138, 169
antioxidant 656
AP-1 299
A4 peptide 613
apolipoprotein E (apoE) 612 ff.
apoptosis 216
Aβ protein 614
– expression during ischemia 625
aquaporin (AQP) 621
– AQP1 216
– AQP4 94, 96, 98, 209 ff., 212 ff., 218, 224 f.
– AQP9 218
– expression 216, 224
– expression during ischemia (AQP4) 625
– function 216
– water channel 197
arachnoid 140
ARD1 44
aromatic l-amino acid decarboxilase (AADC) 592
aromaticity 436
arteriole 112 f., 655
– vasoconstrictive response 655
artificial neural network (ANN) 419 f.
ascariasis 678
ascorbate 653

astrocytoma 224
astrocyte 26, 78, 83, 90, 94, 110, 118, 120, 131, 209 f., 225 ff., 239, 320, 322, 357, 385, 431, 466, 486, 686, 691 ff.
– activation 215
– aquaporine 624
– bipolar 189, 195
– boundary layers 195
– co-culture 225
– differentiation 49, 195
– endfeet 209 ff.
– functional polarity 197
– inductive molecule 199
– morphology 193
– multipolar 195
– perivascular end feet 198
– plasticity 195
– polarity 194
– precursor cell 191
– protective of endothelium 624
– protective of neuron 624
– reporter molecule 573
– secondary barrier 624
– secreted molecule 200
– secretory function 199
– signaling molecule 199
– spacing 195
– territory 195
astrocyte-endothelial co-culture 225
astrocyte-endothelial signaling 202
astrocyte-to-neuron lactate shuttle 578
astrocytic
– endfeet 23
– foot process 13
astrocytoma 216
astrocytosis 691
ATP 66 f., 566
ATPase 654
ATP-binding cassette (ABC) 405, 432
ATP-binding site 433
autoimmune 173, 255
– neuroinflammation 181
autoimmunity 179
AZT 475

b

Babesia 678
bacteria 606
Balamuthia mandrillaris 683
barbiturate 487
barrier permeability 21
barriergenesis 28, 41 ff.
basal lamina 117, 129

baseline 565
– activity 578
basement membrane 110, 171, 655
basic fibroblast growth factor (bFGF) 18, 42, 144
BBB *see* blood-brain barrier
BDNF 62 ff.
behavior circuit 580
bEnd3 227
Benton's visual retention test 659
benzylpenicillin 478
beta-adrenergic receptor 657
17beta-estradiol (E2) 298
bi-dimensional 384
bioavailability 403
blood-brain barrier (BBB, blood-brain interface) 11, 41, 49, 51 f., 77, 109, 143, 189, 209 ff., 247, 265 ff., 287, 313, 321, 357, 375, 403 ff., 431 ff., 444, 446, 463 f., 501 f., 552, 603 ff., 611, 614, 619, 621, 649 ff., 671 ff.
– aquaporin 4 621
– astrocyte 621
– barrier function 654
– BBB+ 422
– BBB– 420, 422
– biochemical change 656
– biophysical change 656
– damage 215
– development 45, 209
– diabetes-related changes in transport 651 ff.
– differentiation 212
– dysfunction 27, 612
– enzyme 466
– function 650
– functional change 649 ff.
– functioning 209
– hemodynamic change 655
– induction 199, 201
– inflammation 351
– in silico prediction model 403
– insulin crossing 269
– integrity 658, 693
– in vitro blood-brain barrier 337
– in vivo study 337
– maintenance 27, 210 f., 216
– maturation 21
– pathophysiology 272
– permeability 269, 318, 431, 441 f., 612, 654, 684, 693
– permeation 403, 414, 418, 422
– phenotype 201

– property 211 ff., 224 f., 389
– transport 272, 274 f.
– transport function 654
– transport system 265
– water transport 211
blood-brain barrier model 384, 421
– astrocyte conditioned medium 343
– brain vessel 338
– capillary 338
– coculture 347
– confluent monolayer 347
– species 346
– transport 349
blood-brain interface *see* blood-brain barrier
blood-brain transfer 523 ff., 535
blood-cerebrospinal fluid 552
– barrier (BCSFB) 265, 432
blood flow 115, 536 ff., 540 ff.
blood-liquor barrier 324
blood-oxygen level-dependent (BOLD) 557
blood-testis barrier 438
blood-tissue barrier 438
Bluthirnschranke 175
bone marrow transplantation 168, 174
bradykinin 315, 323
brain 49, 61 ff., 431, 438, 686
– activation 565
– active influx 405
– adult 61 ff.
– angiogenesis 25, 41 ff., 212
– capillary 465
– capillary endothelial cell (BCEC) 431
– development 42 ff.
– disease 325, 440 f.
– drug 448
– endothelium 199
– endothelium function 202
– endothelium receptor 202
– glucose utilization 274
– gonadal steroid 62
– imaging methodology 554
– injury 612
– insulin-resistant state 279
– ischemic 215
– malignant tumor 440
– maturation 274
– metastasis 441
– microenvironment 49
– microvasculature 313 ff., 321
– nutrient transport 210
– oedema 612
– oxygen 523
– oxygenation 49

728 | Subject Index

- pathology 445
- perivasculature 130
- resolution 625
- response 49
- substrate 448
- targeted redox analog (BTRA) 489
- targeting 463
- tissue 529, 542
- trauma 255
- tumor 325ff., 440
- vascular permeability 687
- work 565

brain edema 216ff., 220, 619f.
- AQP1 625
- AQP4 624f.
- cytotoxic 620, 622
- during stroke 620
- ischemia 623
- Na-K-ATPase 622
- vasogenic 620

brain parenchyma cell 338
- astrocyte 347
- glial cell population influence 338
- pericyte 346
- regulation of blood-brain barrier properties 347

brain-targeting 473
brain vessel, maturation 212
brain water homeostasis 223
Brugia 678
bryostatin 1 614

c

$^{45}Ca^{2+}$ 584
cadherin 88
- vascular endothelial 213
calcification 114
calcitonin gene-related peptide (CGRP) 53, 612
calnexin 443
canalicular multispecific organic anion transporter (cMOAT/MRP2) 442
cancer 576
capacitance 367
capillary 112f., 542, 650ff.
- adventitial layer 130
- cerebral 650 ff
- cortical 650
- lumen protein 210
- model 525
- profile 464
capillary-like structure 228
CAR 88

carboxypeptidase N 314f.
carrier 467
caspase-1 inhibitor 611
CATALYST pharmacophore model 424
caveolae 89, 442, 449
caveolar
- protein PV-1 215
- trafficking 449
caveolin 251, 442f.
caveolin-1 90, 449
CCL 177f.
CCR 177f.
CD36 684
cell
- culture 379
- differentiation 604
- immortalized 380
- line 380
- substrate contact 364
- surface antigen CD44 149
cell polarity, origin 196
cell-cell contact 357, 364
cellular
- adhesion molecule 288
- basis of glucose utilization 572
- differentiation 388
- process 589
central nervous system (CNS) 49, 62, 265ff., 278, 313, 325, 404, 419f., 431, 438, 446, 448, 603ff., 649ff., 658, 672ff., 682ff., 689ff., 692f.
- acid-base balance 657
- active drug 419
- BBB$^+$ 419f.
- CNS$^+$ 419f.
- CNS$^-$ 419f.
- disease 658, 215
- homeostasis 49, 210
- nonactive drug 419
- physiology 211
ceramide 449
cerebral
- blood flow (CBF) 535, 566, 653
- cortex 239
- edema formation 622
- endothelial cell 357
- glucose metabolism rate (CMR$_{glc}$) 653
- ischemia 326
- metabolic rate for glucose (CMR$_{glc}$) 566, 568
- metabolic rate for oxygen (CMR$_{O2}$) 566, 568
- metabolism 652

- microvasculature   650
- microvessel   649 ff.
cerebro-spinal fluid (CSF)   77, 211, 266 ff., 324, 603 ff., 613, 655
- diagnostic feature   603
- glucose   605
- protein concentration   605
cerebrovascular accident   659
ceruloplasmin   90
Chaga's disease   689 ff.
channel   209
characteristics for in vitro BBB model   347
- claudin   349
- JAM-A   349
- occluding   349
- TEER   349
- tight junction and parcellular permeability   349
- transcellular transport   350
- ZO-1   349
chemical delivery system   463, 468, 472
chemoattractant protein-1 (MCP-1)   252
chemokine   176, 181, 250, 608, 685 ff., 689
chemotaxis   607
chemotherapeutic agent   432
chemotherapy   440
cholesterol   443, 487, 656
choline   654
chondroitin sulfate   684
choroid plexus (CP)   12, 77, 129, 136, 194, 657
- epithelium   216
ciliary neurotrophic factor   249
cingulin   84, 89
circadian   254
circumventricular organ (CVO)   12, 77, 130, 196
clathrin   251
claudin   14, 84 f.
claudin-1   26, 86, 242
claudin-5   319
CMR_{O2}/CMR_{glc} 568
- mismatch   578
CNS see central nervous system
coculture   359, 385
cognitive
- activity   576
- impairment   659
coimmunoprecipitation   443
collagen   140, 369
compartment model   533 f., 541
- insufficient oxygen delivery   541
- mitochondrial oxygen   533 f.

- oxygen transfer   528
- tissue oxygen   533 f.
complement factor   179
computational model   410
computer tomography   556
connexin (Cx)43   141
contractility   115
cortical
- freezing injury   574
- spreading depression (CSD)   328
Crane, R. K.   569
CSF see cerebro-spinal fluid
CXCL   178
CXCR   177 f.
CXCR4   216
cyclodextrin   484
cyclosporin A (CsA)   436
cysticercosis   672 ff.
cytochrome   422
- oxidation   529
- oxydase   531 ff., 542
- P450 enzyme complex (CYP)   404
cytochrome c   651
cytokine   181, 247, 323, 436, 607 f., 683, 689 ff.
- angiogenic   70, 212
- anti-inflammatory   611, 689
- molecule   290
- neurotrophic   62, 70
- proinflammatory   275 f., 318, 608 ff., 686, 689
cytokine-induced neutrophil chemo-attractant-1 (CINC1)   250
cytoplasmic accessory protein   210
cytoskeleton   115

d
DA transporter   582
DADLE   485
dehydroascorbate   653 ff.
dementia   325, 475, 480, 580
dendritic cell   129, 138, 168, 173
2-deoxy-d-glucose (DG)   569
depression   659
descriptor for hydrophobic interaction   422
descriptor of polarity   421
destructive infection   694
dexamethasone   289
diabetes   274, 649 ff.,
- insulin-resistant   277
- noninsulin-dependent   279
- type 2 diabetes mellitus   277
diabetic neuropathy   660

diapedesis 180
differentiation 115, 118
– of the BBB 25
diffusibility 541
diffusible tracer 568
diffusion
– passive 431
– relation 534
digit span 659
diphtheria toxin receptor 501, 511
Dirofilaria immitis 678
dissociated GC 302
distributed model 531
– insufficient oxygen delivery 538
– mitochondrial 531 f.
– tissue 531 f.
DIV-BBB 384, 387 f.
DOPA 592
DOPA-decarboxylase 93
dopamine 274, 582
dopaminergic system 580
DP71 223
Dracunculus medinensis 678
drug 433, 438, 445, 449, 654
– delivery system 470
– delivery to the brain 501
– of abuse 580
– resistance 671
– targeting 470
– transport 443
Duchenne muscular dystrophy (DMD) 220 ff.
Duffy 176
dynamic
– function 551
– movement 589
dyslipidemia 275 f.
α-dystroglycan 96, 224
dystrophin 97, 198
– protein 220 f.
dystrophin-associated protein (DAPs) 223
dystrophin-dystroglycan complex 96

e

E3 ubiquitin ligase 44
EC *see* endothelial cell
Echinococcus 678
ECIS 367
ECM 150
ECV304 292
edema 657 f., 685
efflux 650 f.
– pump 432, 467
– transporter 441 f.
EGF 323, 443
EGF/TGFα 66
Ehrlich, Paul 11, 175
electric cell-substrate impedance sensing (ECIS) 359
electrical resistance 13, 358
electrolyte transport 584
electrostatic interaction energy 417
encephalitis 325
endocytosis 253
– receptor-mediated 431
endogenous
– fluorofluor 564
– substrate 436
– tumor necrosis factor (TNF-á) 683
endogenously derived ECM 368
endoglin 212
endothelial
– secretion 448
– tight junction 385
endothelial barrier antigen (EBA) 656
endothelial cell (EC) 41, 77, 172, 179, 226 ff., 239, 250, 272, 287, 313, 321, 357, 431, 464, 609 ff., 650 ff.
– activation 690
– capillary 440
– cerebral 318
– microviscosity 656
endothelial cell isolation 338
– aortic 338
– BBCEC 340
– brain capillary 338
– cerebral microvessel 338
– contamination with nonendothelial cells 344
– enzymatic 338
– filtration step 338
– glial cell 347
– immortalization 342
– MBCEC 340
– mechanical 338
– PECAM-1 346
– primary endothelial cell 341
– subculture 341
– von Willebrand factor 346
endothelial cell-selective adhesion molecule (ESAM) 15
endothelin A-like receptor 658
endothelin-1 (EN-1) 657, 688
endotoxin 275
energy consumption 652
energy currency 566

enkephalin 484
eNOS 443
Entamoeba histolytica 678 ff.
ependyma 192 f.
– bipolar 193
ependymocyte 147
ependymoglia
– derivative 196
– polarity 194
ependymoglial cell 191 f.
– Müller cell 192
– tanycyte 192
Eph/ephrin family 48, 213
EphrinB-EphB families 19
epidermal growth factor 249
17-Epiestrol 299
epilepsy 440
EPO *see* erythropoietin
erbB 66
erythropoietin (EPO) 42
ESAM 84, 88
Escherichia coli 606
E-selectin 690
estradiol 479
estrogen 287, 479
– receptor (ER) 297
ethylenediaminetetraacetic acid (EDTA) 654
exocytoplasmic face (the E-face) 13
experimental allergic encephalomyelitis (EAE) 177, 295
externally applied molecule 563
extracellular matrix 82, 129, 313, 325, 357
– protein 119
extrinsic 563
ezrin 443

f

factor inhibiting HIF-1 (FIH-1) 44
fatty acid 656
FDG-PET 576
feeding 255
FGF *see* fibroblast growth factor
fibroblast 138
fibroblast growth factor (FGF) 47
– FGF/FGFR system 47
– growth factor receptor (FGFR) 239
– growth factor-2 (FGF-2) 211, 239
– growth factor-2 (FGF-2) 212
– growth factor-5 (FGF-5) 242
fibronectin 607
Fick principle 568
field-based method 421

filarial infection 678
filopodial extension 589
Fischer, Emil 569
flow-metabolism 535, 543 f.
[^{18}F]fluorodeoxyglucose (FDG) 569
[^{18}F]fluoro-l-*m*-tyrosine (FMT) 592
flux capacity 540
[^{18}F]FMISO 576
focal brain ischemia 242
fractalkine 176, 684
fractone 148
free energy of solvation 411
freeze-fracture technique 83, 211
FSH 481
functional
– activity 565
– brain imaging 551
– magnetic resonance imaging 554
– MRI 557
funnel-shape 433

g

GAG 176, 684
ganciclovir 476 f., 592
gap junction 112
gap junctional communication 141
GC therapy 296
GDNF 241
genetic
– algorithm (GA) 414
– engineering 592
– function approximation (GFA) 413
geohelminth 678
GFAP *see* glial fibrillary acidic protein
glia 385
glia limitans 172, 181, 191, 198, 209
– perivascularis 170
glial
– activity 66
– cell 23, 313
– cell-derived neurotrophic factor 249
– endfeet protein 213
– limitans 138
– limiting membrane 218
– perivascular endfeet differentiation 213
glial fibrillary acidic protein (GFAP) 212 ff., 241, 438 f.
glioblastoma 440
gliogenesis 61
glioma 220, 440
glioma C6 228
gliovascular unit 66
global method 567

glucocorticoid hormone   287
glucose   209, 214f., 277f., 567f., 605
– analog   569
– transport   652ff.
– transporter (GLUT-1)   16, 92, 313, 614, 660
– transporter insulin-sensitive   278
– transporter-1 protein   298
glucuronoxylomannan (GXM)   608
GLUT1   22, 241, 653ff.
glutamate
– pool   573
– transporter   198
γ-glutamyl transpeptidase (γ-GT)   211, 241, 321
glutathione   93
Goldman, Edwin E.   12
Gombar-Polli rule   424
GR-1   684
granulocyte colony-stimulating factor (G-CSF)   301
granulomatous amoebic encephalitis   681
green tea   438
GRID   421
growth factor   46, 318, 323
Guillain Barré syndrome   605

h

H^+-ATPase   657
Haemophilus influenzae   606
HBEC   292
H-bond
– accepting   436
– acceptor/donor feature   404
– property   411
HDL   90
heat-shock protein   288
hemodynamic change   655
hemodynamic force   634
– flow cessation   634
– myogenic tone   634
– pinocytotic vesicle formation   635
– reperfusion-induced hyperemia   634
– shear stress alteration   634
hemoglobin
– saturability   541
– saturation   525
hemorrhage   685ff.
heparan sulfate proteoglycans (HSPG)   140
heparin-binding
– epidermal growth factor-like growth factor (HB-EGF)   501

– growth factor   148
hexose transporter   653ff.
HIF *see* hypoxia-inducible factor
HIF-responsive element (HRE)   44
highly charged polar surface area (HCPSA)   415
Hill
– coefficient   529ff.
– equation   525
hippocampal formation   239
histamine   318
HIV/AIDS   693
HIV-1   180
HIV-associated dementia   440
homeostasis   118, 209f., 220, 274, 278, 321, 686
hormone   654
– replacement therapy   479
horseradish peroxidase (HRP)   213, 651
housekeeping gene   566
HSP70   323, 443
HSP90   443
HSVEC   300
HT7-antigen   215
human
– brain EC   381
– brain endothelial cell   292
– brain microvascular endothelial cell (HBMEC)   51
– brain tumor   28
– endothelial-like cell   292
– ether-a-go-go related gene (hERG) potassium channel   404, 422
– parasitic disease   671ff.
– saphenous vein endothelial cell   300
– umbilical vein endothelial cell   292
human-applicable carrier protein (CRM197)   511
HUVEC   292
hydrocortisone   241, 358
hydrogen ion concentration   584
hyperglycemia   274, 276ff., 652ff., 658
hyperglycemic stroke, brain edema   633
hyperinsulinemia   277ff.
hyperosmolarity   655, 658
hypoglycaemia   605, 652, 660
hypoinsulinemia   274
hypoxia   42, 46, 627
– ATP   627
– depletion during ischemia   627
– permeability   627
– production of autocoids (NO, O^{-2}) during reperfusion   627

– state of actin 627
hypoxia-inducible factor (HIF) 42
hypoxic
– ischemia 69
– tissue 576

i

ICAM-1 *see* intercellular cellular adhesion molecule
IFN-beta 296
IFN-γ 690 ff.
IGF1 65 ff.
I*k* B 611
IL1-β *see* interleukin-1β
IL-6 *see* interleukin-6
image pH 586
immobilized artificial membrane (IAM) 379
immune
– clearance 684
– response 215, 611, 685
– system 681
immune-privileged 175
immunoconjugate 302
immunoglobulin G 654
immunoglobulin superfamily 685
immunosuppression 678
in silico predicition 403
in vitro blood-brain barrier model 28, 358, 378
increased glycation 649
infection 605
inflammation 93, 289, 609, 692
inflammatory
– cell 685
– condition 605
– process 613
– response 603 ff.
– state 275 f.
inhibitory domain 44
innate immunity 170
insulin 90, 265 ff., 274 ff., 431, 653
– CNS 274, 279
– dysregulation 278
– pathophysiology 272
– receptor 443, 504, 508
– receptor desensitization 279
– resistance 275 ff.
– resistance syndrome (IRS) 277
– sensivity 276
– signal transduction cascade 279
– TNF-induced resistance 276
– transport 272, 275 ff.

insulin-degrading enzyme (DIE) 278 f.
insulin-like growth factor 1 249
insulin-resistant brain state 279
integral membrane protein 210
integrin 290, 685
– receptor 369, 607
β-integrin signaling 48
intercellular
– adhesion molecule (ICAM) 290, 608
– cellular adhesion molecule 93, 613, 684 ff., 690, 694 ff.
interferon 296
– gamma (IFN-γ) 292, 318, 611, 684 ff.
interleukin 248, 691 ff.
interleukin-1 90, 607 ff.
interleukin-1β (IL-1β) 292, 318, 323
interleukin-6 (IL-6) 211, 241, 276, 683 ff.
interleukin-10 250
intermediate filament protein 113, 241
internalizing receptor 501
intracranial hemorrhage 605
intravital microscopy 651
intrinsic 563
– signal 587
inulin 654
invasive brain drug delivery strategy 504
ischemia 584
islet amyliod polypeptide (IAPP) 278

j

JACOP 84, 89
JAM 84, 87 f.
– protein 210
junctional adhesion molecule (JAM) 15, 291, 608
juxtavascular microglia 169, 172 f.

k

kainic acid 584
Karnowsky, Morris 12
K$^+$ channel 224
K$^+$ concentration 209
ketoacidosis 658
ketone body 653
ketosis 653, 658
Kety, Seymour 568
Kir 4.1 219
knockout mouse 448
Kolmer cell 130
Kolmer macrophage 136
Krogh cylinder 523
Krogh's diffusion coeffizient 531 f.
K$^+$ siphoning 211

K⁺ spatial buffering 220
kyotorphin 484, 487f.

l

lamina affixa 138
laminin 140, 369
LDL see low-density lipoprotein
leakiness 27
Leishmania 681
leptin 660
leptomeninges 170
leu-enkephalin 485
leukemia inhibitory factor (LIF) 26, 49, 249
leukocyte 288, 685 ff.
– migration 607
leukocyte-endothelial cell interaction 293
Lewandowsky, M. 12
LH 480 f.
LHRH 480
lipid
– peroxidation 656
– solubility 431
– transport 449
lipoaffinity 410
lipophilicity 404, 468, 485
lipopolysaccharide (LPS) 213, 224, 275, 291, 607
Listeria monoxytogenes 606
log$P$ 468
lordosis 481 f.
low-density lipoprotein (LDL) 90
– receptor-related protein (LRP) 320, 326
low-density lipoprotein (LDL) receptor 211
– family 320
– related protein 1 (LRP1) 501
Lowry, Oliver 568
LPS see lipopolysaccharide
LRP1 receptor 509
LRP2 501
LRP2 receptor 509
luciferase 592
luminal membrane 438
luteinizing hormone 480
lysosome 116

m

MAC-1 169
α-macroglobulin 315
macrophage 116, 129, 589, 607 ff., 694
– inflammatory protein 608
macrovascular disease 649
magnesium ATPase 657

magnetic resonance imaging 554, 556
MAGUK see membrane associated guanylate kinase
major histocompatibility complex (MHC) 138
– class II 173
malaria 671, 689, 683
MAO see monoamine oxidase
MAP kinase 658
– pathway 47
marker enzyme 359
marrow transplant 173
matrix metalloproteinase (MMP) 314 ff., 358
maturation 122, 580
mature BBB 25
MCAO 480
MCP-1 see chemotactic protein-1 or monocyte-chemoattractant protein-1
MDR 433
MDR1 gene 432
MDR1/ABCB1 444
– single nucleotide polymorphismus (SNP) 444 f.
mdr1a gene 438
MDR-associated protein 442
mdx mouse 212
mdx^{3cv} 220
$mdx^{\beta geo}$ 220
MECA 32 antigen 215
mediator of EC permeability 631
– arachidonic acid 632
– bradykinin 632
– during ischemia 631
– during reperfusion 631
– endothelin-1 632
– histamine 632
– property 631
– vascular endothelial growth factor 632
membrane associated guanylate kinase (MAGUK) 15, 88
memory 278, 487, 659
meninges 129
meningitis 325, 604 ff., 681, 689
– acute bacterial 605
– aseptic (viral) 611
– integrity 611
meningoencephalitis 693
menopause 479
metabolic
– brain image 567
– map 567
– X syndrome 277

metabolism 523 ff.
– cerebral 652
– oxygen 523 ff.
metalloproteinase 145, 313 ff., 322 ff.
– function 314
– types 314
metastasis 441
methylphenidate (MP) 582
methylprednisolone 291
MHC see major histocompatibility complex
Michaelis-Menten equation 529 ff.
microglia 78, 117, 181, 320, 322, 589, 611, 686, 691 ff.
– activation 168, 172, 687
– resting 168
microglial cell 129, 167, 685
microinvironment 147
microneme protein 2 (MIC2) 694
microvascular
– cell 381
– endothelial cell 239
– permeability 612
– wall 357
microvasculature 649 f.
– brain 313 ff.
– cerebral 649 ff.
microvessel 525, 528 ff., 531, 542, 614
– cerebral 649 ff.
– permeability 651
microviscosity 656
middle cerebral artery occlusion 242
migraine 328
mitochondria 529 ff., 614, 688
mitochondrial oxygen tension 525, 531 ff., 542
mitochondrion 541
MMP-9 145
Mn^{2+} 559
moesin 443
molar refractivity (MR) 414
molecular
– imaging 551 f.
– packaging 484
molecular weight (MW) 436
monoamine oxidase (MAO) 581
– MAO A 583
– MAO B 583
monocarboxylic acid 654
– transporter (MCT) 567
monoculture 382
monocyte 142, 167
– chemotactic protein-1 (MCP-1) 177, 608, 613

monocyte-chemoattractant protein-1 (MCP-1) 177, 294, 613
mono-dimensional 384
Monte Carlo simulation 412
mouse
– brain capillary endothelial cell 361
– model 689
– model malaria 683 ff.
– N-terminal acetyltransferase 1 (mNAT-1) 44
movement disorder 590
MPTP 592
MS 177, 180
multidrug
– resistance transporter P-glycoprotein 431 ff.
– resistance-associated protein Mrp1 93
– transport pump 405
multimodal assay 574
multiple linear regression (MLR) analysis 413
multiple sclerosis (MS) 176, 294, 325 ff., 605
myelination 688
myocardial infarction 660
myoinositol 649

n

Na,K-ATPase 197, 211, 654, 657
NADPH 587
Naegleria fowleri 683
NAHD 474, 587
naloxone 486
natural killer (NK) 694
N-cadherin 26, 79
NDIV-BBB 390
near-infrared spectroscopy (NIRS) 562
negative BOLD response 557
Neisseria meningitidis 606
nematode 678
nephropathy 649
netrin-1 48
neural
– progenitor cell 148
– stem cell 146
– tube, formation of 189 f.
neuroactive substance 448
neurobiochemical 660
neuroblast 136
neuroblastoma cell (N2A) 230
neuroepithelial cell 190
neurofibrillary tangle formation 278
neurogenesis 61 ff., 147, 190 f., 278
– mitotic 67

– persistent 65
neurogenesis (birth of neurons) 191
neurogenic niche 147
neuroglia 44, 46
neuroglial progenitor cell 41
neuroimmune function 210
neuroinflammatory response 612
neuron 320, 686
neuronal
– differentiation 67
– GT1 228
– marker NFH (neurofilament H) 219
– percursor cell 239
– protection 210
neuronal polarization 196
neuron-specific enolase (NSE) 658
neuropathy 649
neuropeptide 274, 468, 484
neurophilin (NRP) 46, 48
neuroprotection 480, 483
neurotransmitter 220, 315, 613, 651 ff.
– system 580
neurotrophin 216
neurovascular unit 94
neurulation 189 f.
neutral aminoacid (NAAs) 214
neutral endopeptidase 24.11 (enkephalinase, NEP, EC 3.4.24.11) 314
NF-*k*B 299, 611
– pathway 611
NG2 chondroitin sulfate proteoglycan 68
nitric oxide (NO) 67, 657, 691 ff.
nitric oxide synthase (NOS) 90, 92, 298, 487, 692
– synthase inhibition 67
nodule 114
noggin 67
nonCNS drug 419
nongenomic 289
non-metabolizable analog 572
non-receptor tyrosine kinase Lyn 215
norepinephrine 274
normoxic condition 44
NOS *see* nitric oxide synthase
Notch receptor 48
nucleotide triphosphate diphosphohydrolase (NTDPase) 66 f.
nucleotide-binding fold (NBF) 432
nutrient 650 ff.

o

OAP *see* orthogonal arrays of intramembranous particles
obesity 274 ff.
obligatory fuel 567
occludin 14, 84 ff., 242, 295, 319, 650 f.
oligodendrocyte 320
– precursor cell 191
3-*O*-methyl-d-glucose 572
Onchocerca volvulus 681
onchocerciasis 681
opiate 485
optical imaging 562
optico-acoustic imaging 564
organic anion transport protein (OATP) 423
orthogonal arrays of intramembranous particles (OAPs) 94, 96 ff., 211 ff.
osmosensitive organ 219
oxidative
– load 649
– stress 278
oxidized flavoprotein 587
oxygen 523 ff., 531 ff.
– arterial 535
– arterial delivery 542
– capillary 539 f.
– compartment model 528 ff., 541
– consumption 529 ff., 539, 542, 545
– diffusibility 532, 527 f., 544 ff.
– distributed model 531 ff.
– flow-metabolism 535
– flux 540 ff.
– ischemic limit 546
– metabolism 523 ff., 535, 539
– mitochondrial 531 ff., 546
– regulatory gene 50
– saturation 537
– supply 538
– tension 41 ff., 49, 525, 528 ff., 539, 542, 684
– transfer 525, 528
oxygenation response 49
oxygen-dependent degradation (ODD) domain 44

p

P100 659
p300 44, 659
p42/p44 mitogen-activated protein kinase (MAPK) 323
pain 660
– tolerance 659

panendothelial cell antigen MECA-32 22
pAPN   113f., 121, 327
paracellular pathway   404
paramagnetic agent   559
parasitic disease   671ff.
Parkinson's disease   440, 445, 590
partition coefficient   468f.
passive diffusion   404, 410, 418, 431
pathophysiology   272ff.
PDGF see platelet-derived growth factor
PDZ domain   84
PECAM-1   16, 88, 290
"peg and socket" contact   112
peptidase   485
β-peptide   613
pericyte   23, 26, 68f., 78, 82, 90, 109, 129, 170f., 209f., 313, 321f., 357
– marker   113
– microvascular   68
– positioning   110
– structure   110
perineural vascular plexus   17
peripheral neuropathy   660
perivascular
– astrocyte   51
– astrocytic endfeet   209
– cell   321, 611
– macrophage   169ff.
– microglia   170
– pericyte   51
– plexus   41f.
– space   170
perlican   148
permeability   118, 180, 358, 389, 432
– capillary   318
– coefficient   469
– microvascular   215
– of the blood-brain barrier   651
peroxisome proliferator-activated receptor-gamma (PPAR-gamma) agonist   280
PET   560
P-glycoprotein (P-gp)   16, 22, 79, 93, 211, 215, 313, 350, 404f., 420, 422, 431ff., 442ff., 467, 469f.
– AHNAK   350
– endothelial adhesion molecule   351
– isoform   432
– localization   438
– MECA-32   350
– modulator   436
– polymorphismus   444
– structure   432

– structure-activity relationship (SAR)   434ff.
– subcellular localization   442
– substrate   435
– transport of LDL   347
pH response element   653
phagocytosis   117, 140, 170, 687
phagocytotic role   116
pharmacophore model   424
phenylarsine oxyd (PAO)   319
phospholipid   656
photon emission tomography   554
pial
– macrophage   131
– transendothelial electrical resistance   624
– vessel BBB property   624
pig microvascular endothelial cell   358
pineal gland   239
pinocytosis   117, 140
– endothelial   215
pituitary gland   239
placenta   438
plasminogen activator (PA)   325
plasminogen/plasmin system enzyme   326
Plasmodium berghei   683ff.
Plasmodium falciparum   678ff.
Plasmodium malariae   678ff.
Plasmodium ovale   678ff.
Plasmodium vivax   678ff.
platelet-activating factor (PAF) receptor   606
platelet-derived growth factor (PDGF)   66
– growth factor B (PDGF-B)   119f.
– growth factor BB (PDGF-BB)   19
– growth factor-β (PDGF-β)   42, 52
pneumococcal surface protein (PspA)   606
pneumolysin   606f.
polar surface area   404, 410
polarizability   410
polymorphism   444
polymorphonuclear leukocyte (PMN)   604ff.
polyphenol   438
positive BOLD response   557
positron emission tomography   554
postcapillary venule   180
post-capillary venule   175
post-myocardial infarction   276f.
post-translational modification   656
potassium   574
– siphoning   211

presenilin 279
primary culture 358
principal component analysis (PCA) 412
prodrug 468, 478
progenitor cell 61 ff.
– microvascular influence 61
– mobilization 61, 66 f.
– neural 66
– nitric oxide 67
– parenchymal 68
– post-ischemic mobilization 69
– purinergic signaling 66
proinflammatory mediator 607
protein glycation 649, 656
protein kinase C (PKC) 443, 614, 657
– hyperglycemia-induced PKC activation 634
– pathway 649
– permeability during hyperglycemia 633
– tight junction permeability 633
– vascular permeability 633
proteoglycan NG2 151
proton basicity 436
protoplasmic face (the P-face) 13
PSD95-disc large ZO-1 (PDZ) 223
pulsatile flow 388
purinergic signaling 66
pVHL-degradation pathway 44

q

QSAR analysis, membrane interaction (MI-QSAR) 413 ff.
QSAR model 403 ff., 424
quantitative
– autoradiographic method 568
– autoradiography 560
quiescence 694

r

[^{11}C]raclopride 581
radial
– glia (RG) 190 f.
– glial cell 147
radioactive tracer 560
radionuclide imaging 560
radixin 443
Raichle, M. E. 578
Ras/Raf/MEK/MAPK pathway 66
rat brain endothelial cell (RBE4) 294
receptor-mediated
– drug delivery 506
– transcytosis 504

redox analog 489
Reese, Thomas 12
regression model 410
remodeling 122
remyelination 243
reperfusion injury 628
– cerebral edema 628
– nitric oxide 628
– peroxynitrite 628
– reactive oxygen species 628
– vascular permeability 629
reporter molecule 563
resistance 367
resting state 565
retinal
– Müller cell 97
– pigment epithelial cell 592
– pigment epithelium (RPE) 194, 197
retinopathy 649
reversal agent 436
RG2 228
Risau, Werner 17
rosiglitazone 279
rostral migratory stream (RMS) 191
rotable bond 404

s

S100B 658
scaffolding domain 443
Schistosoma 678
Schistosoma haematobium 680
schistosomiasis 681
seizure 485, 487
– threshold 220
SEL 147
selectin 290, 684
sensory threshold 660
serine peptidases dipeptidyl peptidase II and IV (DPP) 314 f.
serum
– effect 362
– withdrawal 360
sexual
– behavior 482
– dysfunction 481
shear stress 386 f.
– laminar 386
– pulsatile 386
signaling pathway 212
single nucleotide polymorphismus 444
site-targeting index (STI) 470 f.
SMAD 212
SMAD6 212

small
- interfering RNA 437
- ubiquitin-related modifier (SUMO) 44f.
smoker 583
smooth muscle
- actin (SMA) 113
- cell 112, 115
- α-smooth muscle actin (α-SMA) 114
Sokoloff, Louis 568
Sols, A. 569
soluble E-selectin (s-ELAM-1) 613
solvent-accessible surface area (SASA) 412
sonic hedgehog (Shh) 48
sorbitol 649
sparganosis 678
spatial resolution 555
SPECT 560
sphingomyelin (SM) 450
sphingomyelinase 450
spinal cord injury 255
sprouting 121
src-suppressed C kinase substrate (SSeCKS) 26, 50, 52, 211
SSeCKS see src-suppressed C kinase substrate
stabilization 122
stem cell 62
steroid 480
Streptococcus pneumoniae 605 f.
streptozotocin (STZ)-induced diabetic 650
stroke 255, 479, 619,
- hyperglycemic 633
stromal-derived factor-1 (SDF-1) 216
Strongyloides stercoralis 678
structure-activity relationship (SAR) 434ff.
subarachnoid hemorrhage 605
subependymal basal laminae 148
subgranular zone (SGZ) 65
substrate 435
SUBSTRUCT 420
subventricular zone (SVZ) 191
sucrose 651, 654
superparamagnetic iron oxide (SPIO) 559
support vector machine (SVM) 421
supraoptic nucleus 146
surface area 436
α-syn$^{-/-}$ mouse 220
α-syntrophin 223
α1-syntrophin 96f.

t
Taenia solium 678ff.
tanycyte 12, 194
targeting 463
targeting enhancement factor (TEF) 470f.
targetor (T) 472
tau phosphorylation 278
T-cell 215, 293, 318, 686
TEER see transendothelial electrical resistance
temperature gradient 564
temporal discrimination 555
TFG see transforming growth factor
T-helper
- cell response 684f.
- lymphocyte 688
therapeutic index (TI) 472
thermal imaging 564
thrombospondin 684
thrombospondin-1 (TSP-1) 53
thymidine kinase 592
thyreotropin-releasing hormone (TRH) 484, 486
Tie-1 212
Tie-2 19, 122
tight junction (TJ) 11, 49, 77, 82ff., 85ff., 88, 92f., 118, 172, 175, 179, 181, 196, 294, 318, 325, 357, 431, 609, 650ff.
- endothelial 201, 210, 213
- epithelial 201
- formation 211
- modulation 202
tissue inhibitor of metalloproteinase (TIMP) 315ff.
TIMP 322ff.
tissue-type plasminogen activator (t-PA) 326
T lymphocyte 143, 691
TNF see tumor necrosis factor
Toll-like receptor 4 170
Tower 569
toxin 650
Toxocara canis 678
Toxoplasma gondii 694
toxoplasmosis 672ff., 692ff.
- destructive infection 694
- quiescence 694
- transgression 694
trafficking 253
transactivation 288
- domain 44
transcellular transport 630

- actin depolymerization 631
- during ischemia 630
- during reperfusion 630
- mediator of cerebral edema 631
- pinocytosis in vascular permeability 630
transcription factor 611
- NFkB 323
transcytosis, receptor-mediated 405
trans-differentiation 129
transendothelial electrical resistance (TEER) 13, 318 f., 379, 387, 389, 465
transferrin 90, 431, 501
- receptor 215, 313, 504, 506
transforming growth factor (TGF) 48, 322
- alpha (TGFα) 251
- beta (TGF-β) 211 f., 611
- beta 1 (TGF-β1) 53, 119, 121, 144, 241
transgression 694
transport 248, 313, 572,
- activity 438
- β-amyloid (Aβ) 448
- carrier-mediated 405
- glucose 214 f.
- multidrug transport pump 405
- system 376
transporter 423, 438, 442, 466 f., 553
- ATP-binding cassette (ABC) 432
- endothelial 214 f.
- for low-density lipoprotein 504
- glucose 209, 660
- GLUT-1 653 ff.
- hexose 653
- insulin-sensitive glucose 278
- ketone 653
- multidrug resistance 431 ff.
transrepression 289
Transwell system 382, 407
trematode 678
TRH 486
Trichinella spiralis 682
tri-culture 120
[^{131}I]trifluoroiodomethane 568
trigonelline 474
Trojan horse 174
Trypanosoma brucei gambiense 689
Trypanosoma cruzi 689 ff.
trypanosome 692
trypanosomiasis 689
tumor necrosis factor (TNF) 276, 576
- receptor-2 687

- secretion 216
- TNF-α 292, 318, 323, 684 ff., 690 ff.
- tumor necrosis factor alpha (TNFα) 144, 248, 274, 607 ff.
tunica adventitia 135
two-dimensional gel electrophoresis 656
type 1 diabetes 652
type 2 diabetes 658

u

ultrasound-mediated imaging 564
urokinase-type plasminogen activator (uPA) 326

v

variable surface glycoprotein (VSG) 690
vascular
- basement membrane 655
- damage 688
- endothelial (VE)-cadherin 213
- route 504
- stability 119
vascular endothelial growth factor (VEGF) 18, 42, 51 f., 61 ff., 85, 90, 92, 120, 122, 144, 212, 242
- HIF1a-mediated production 65
- hypoxia-induced 46
- isoform 46
- receptor 46, 443
- receptor 1 (VEGFR1, flt-1) 18
- receptor 2 (VEGFR2, flk-1/KDR) 18
vascularization 42
vasculogenesis 28, 120
vasoactive agent 116
vasocinstrictive response 655
vasoconstrictor 116
vasogenic brain edema 220
vasorelaxant 116
VCAM-1 290, 690, 93
VE-cadherin 88
VEGF *see* vascular endothelial growth factor
ventricular zone (VZ) 190
venule 112 f.
vessel assembly 120
αV-integrin deficient mouse 27
Virchow-Robin space 133, 170, 172
viscosity 656
vitamin C 653, 661
vocal control nucleus HVC 63 ff.
VolSurf descriptor 421
voltage-sensitive dye 587

w

Walker A motif 432
Walker B motif 432
water
– channel 216
– transport 216
weakly rectifying potassium channel
 Kir4.1 97
white blood cell (WBC) 603 f.
Wnt signaling 19
World Drug Index (WDI) 420

x

xenobiotic 448
X-ray computed tomography 554

y

yolk sac 167

z

zidovudine 475
ZO-1 15, 26, 84, 87 f., 242, 650
ZO-2 15, 84
ZO-3 15, 84
zonula
– occludens 210
– occludens-1 (ZO-1) 212 f., 294, 319,
 650 f.
z-VAD-fmk (benzyloxycarbonyl-Val-Ala-Asp-
 fluoromethyl ketone) 611